SOIL IMPROVEMENT AND GROUND MODIFICATION METHODS

SOIL IMPROVEMENT AND GROUND MODIFICATION METHODS

PETER G. NICHOLSON

AMSTERDAM • BOSTON • HEIDELBERG • LONDON
NEW YORK • OXFORD • PARIS • SAN DIEGO
SAN FRANCISCO • SYDNEY • TOKYO
Butterworth-Heinemann is an imprint of Elsevier

ELSEVIER

Butterworth Heinemann:
Butterworth Heinemann is an imprint of Elsevier
The Boulevard, Langford Lane, Kidlington, Oxford OX5 1GB, UK225 Wyman Street,
Waltham, MA 02451, USA

Notices
Knowledge and best practice in this field are constantly changing. As new research and
experience broaden our understanding, changes in research methods, professional practices,
or medical treatment may become necessary.

Practitioners and researchers must always rely on their own experience and knowledge in
evaluating and using any information, methods, compounds, or experiments described
herein. In using such information or methods they should be mindful of their own safety and
the safety of others, including parties for whom they have a professional responsibility.

To the fullest extent of the law, neither the Publisher nor the authors, contributors, or editors,
assume any liability for any injury and/or damage to persons or property as a matter of
products liability, negligence or otherwise, or from any use or operation of any methods,
products, instructions, or ideas contained in the material herein.

Library of Congress Cataloging-in-Publication Data
Application Submitted

British Library Cataloguing-in-Publication Data
A catalogue record for this book is available from the British Library

ISBN: 978-0-12-408076-8

For information on all Butterworth-Heinemann publications
visit our website at store.elsevier.com

This book has been manufactured using Print On Demand technology.

Working together
to grow libraries in
developing countries

www.elsevier.com • www.bookaid.org

CONTENTS

PREFACE

With very few exceptions, everything we construct in our built environment lies in or on the ground. As a consequence, the earth materials involved must be evaluated to ensure that engineering properties will adequately provide for acceptable performance of a project. There are many different properties that must be assessed based on the requirements of the project, and these will vary greatly depending on the overall (and underlying) objectives. One must be careful not to overlook other components that may affect lives and property, including natural and man-made hazards. These may involve natural, constructed and cut slopes, potential flooding and storm surges, earthquakes, and so on.

For most structures, including buildings, bridges, roadways, and engineered earth structures, there are some fundamental "rules" pertaining to the ground that must be followed in order to ensure the "success" of the structure. For example, imposed loads must be supported without ground failure, and within maximum limits of acceptable settlement or deformations. To properly evaluate the capacity of earth materials to adequately support loads, one should have a basic understanding of soil mechanics and be able to perform relatively straightforward design exercises as required for the circumstances. However, this is *as long as the strength parameters are correctly evaluated and ground conditions are well defined.* While load applications can be defined with reasonably good accuracy, it is much more difficult to make an accurate evaluation of earth material properties and their response to imposed loadings. Engineers use a number of methods to estimate the response of earth materials to various loading conditions. While performing full-scale load tests at the field site can produce one of the best evaluations, these types of tests are not often performed due to feasibility and cost restraints. This forces the engineer to rely on interpretations based on experience and a combination of laboratory and in situ test results from field investigations and sampling. Unfortunately, this, in turn, creates a whole additional level of uncertainty arising from questions about everything from testing accuracy, sample disturbance, and sample representation, to natural spatial variability of conditions and material properties in the ground, for example. These are just the basic challenges that face all geotechnical engineers when dealing with designs in or on the ground.

In many cases, the ground conditions and the earth materials are not ideal for proposed or planned development. In these instances, the geotechnical engineer must thoughtfully consider how to address potential problems with plausible solutions. Modification of the earth materials or stabilization of soils can provide a means to achieve the desired goals of assuring adequate engineering properties and/or responses for a variety of applications and conditions. Depending on the initial ground conditions, soil properties, and desired outcomes, the engineer may select from a wide choice of ground improvement and soil stabilization techniques that will help solve challenges of poor site conditions, inadequate soil qualities, mitigation of potential problems, or remedial work.

As development continues throughout the world, many of the most ideal sites have already been built upon, leaving less desirable sites for future use. This is compounded with the desire and need to build larger and safer structures in urban areas, imposing greater loads and/or requiring greater reliability than previously considered. At many locations that may have previously been considered unsuitable for development due to poor soil conditions, ground improvement techniques provide suitable alternatives for new construction. In addition, precious resources of select earth materials may be preserved with better use of existing soils that can be treated to provide acceptable engineering properties and the reuse of waste material or industry by-products.

In many instances, the combination of ground conditions and objectives requires the use of more than one approach or methodology to achieve the desired goals. This may be particularly helpful for large projects where multiple problems exist, where different types of soil strata are encountered, or when surface treatments are needed after a stabilization of deeper materials is completed. For smaller projects, the use of multiple techniques may be cost prohibitive, and a single method, albeit comparatively more expensive in some respect than some others, may be the best solution.

This text presents an overview and discussion of a number of ground improvement methodologies that have been devised over the years. Many of these are fundamental and have existed in some form for a long, long time. Others have developed with advancements in technology, and still others are continually emerging with the ever changing engineering environment. Over the past few decades, there have been significant advancements made in the tools, technology, and materials available to the engineer faced with finding workable, economic solutions to the myriad geotechnical problems that arise. In addition, construction and development have reached new levels in size, loads, and complexity. Safety issues and environmental concerns have also played an important role in reshaping values and approaches that may have evolved since earlier references on the subjects of soil and ground improvement were published. The intent of this work is not necessarily to prepare one with all of the tools needed to plan and/or design a ground improvement program, but to provide insight to the general civil engineer, contractor, or construction manager as to what tools are now available and what approaches have been effective in solving a wide range of geotechnical challenges. With this background and knowledge, one can be better prepared to consider options when faced with difficult or less than optimal soil and site conditions.

Section I provides a short background on why and where ground improvement is needed; a brief description of the various categories of methods as they apply to controlling one or more types of engineering properties and/or performance; an overview of the types of applications available and tried; an outline of typically desired improvements (objectives); an outline of factors that control the choice of improvement method that may be most suitable; and information about technological advancements that allow new tools and materials to be implemented. Included in Chapter 3 is a brief overview of basic soil mechanics fundamentals, such as soil strength, compressibility (settlement), and fluid flow (permeability) topics. In addition, this chapter describes the performance of typical field investigations with commonly collected and reported data. It is from this data (usually contained in boring logs and geotechnical reports) and correlations with soil characteristics that many ground improvement designs are formulated.

Section II provides full coverage of the topic of soil densification. Beginning with Chapter 4 is a description of objectives and improvements attained by densification of soil, including fundamental soil engineering properties and an overview of liquefaction phenomenon, followed by a thorough explanation of the principles of shallow compaction theory, control of compacted soil engineering properties, and finally, a discussion of field applications, contractual specifications, and quality control described in Chapter 5.

Section III on hydraulic modification provides an overview of how the control of water within the ground can be used to improve soil and site conditions. Strength gains, stability in slopes, seepage, drainage, and consolidation are addressed. A wide variety of methods, including conventional drain and pumps for dewatering, are outlined, along with more innovative techniques utilizing geosynthetics (for drainage and filtration) and electrokinetics (for dewatering, forced consolidation, stabilization of wastes, contaminant control and/or removal, and as an aid to grouting).

Section IV is devoted to the broad subjects related to stabilization by the addition of admixtures, grouting, and thermal treatments. Chapter 11 outlines the various available materials commonly mixed with soil to improve engineering properties, including natural soils and waste products. The applicability of mixing these various materials with different soil types is also discussed. In addition, Chapter 11 provides a general overview of the mixing methods, engineering property improvement objectives, and common applications for admixture treatments. Chapter 12 is devoted to describing a variety of grouting techniques, objectives, and various applications. Without the addition of materials to be mixed with existing soil, a discussion of thermal treatments is included in Chapter 13. Both heating and freezing methods are described, although heat treatments are usually cost-prohibitive and are rarely used anymore except for use in some geoenvironmental applications. Freezing technologies, on the other hand, have now advanced to the point of being incorporated as viable solutions for a variety of applications and have been used in a number of notable high-profile projects, as described in Section 13.4.3.

Section V outlines techniques that incorporate the use of structural elements for stabilization of existing ground as well as for new construction. Chapter 14 describes how geosynthetic reinforcement materials are utilized for construction of stabilized earth walls, slopes, and as an aid for construction over weak or soft foundation materials. Chapter 15 provides an overview of in situ reinforcement with structural members in the form of soil nails, anchors, tiebacks, and bolts. These methods are primarily used for slope stabilization, tunneling support, and excavation/subsurface foundation support. Chapter 16 describes the relatively simplistic, but very functional, practical, and often aesthetic applications of soil confinement. Cribs, gabions and mattresses provide multiple functions and have been widely used for retaining walls, slope stabilization, earth structure foundations, and erosion control in channels or other high-energy flow regimes. Cellular confinement with geosynthetic materials and the use of fabricated modular blocks have also added some structural components to traditional geotechnical applications. Then Chapter 17 discusses the use of relatively newer, lightweight technologies and materials, including expanded polystyrene (EPS) foam, industrial wastes, and recycled materials.

Finally, Chapter 18 touches on the ongoing advancements, emerging trends, and new ideas that foster future advancements to soil and ground improvement. With continuing advancements in ground modification technology, application ideas, and materials, there will certainly be new innovations and untested applications to push new limits of our understanding and appreciation for what can be accomplished.

Note regarding ASTM Standards: Throughout this text, references are made to test standards published by the American Standards of Testing and Materials, current at the time of writing. Rather than referencing each test standard in the list of references for each chapter, the Book of Standards is referenced as a whole and a listing of topic related standards is provided at the end of each chapter in which the standards are mentioned.

This is my gift to the practice, students, and anyone interested in an array of amazing ways we can consciously work with our environment while advancing engineering and achieving new successes.

Peter G. Nicholson
March 2014

ACKNOWLEDGMENTS

I would like to acknowledge all in the industry who have contributed photos, images, and other information that have helped immensely in the preparation of this text. This includes: Colby Barnett and Haley Clanton—GeoStabilization International; Stuart Bowes—Broons; Horst Brandes—University of Hawaii; Mary Burk—ACH Foam Technologies; Christina Burns—Atlas Copco (Dynapac); Larry Cole, Carmeuse Lime and Stone; Jim Collin—The CollinGroup; Michele Curry—The Reinforced Earth Co.; David Dennison—Bomag; Lisa Edwards—American Wick Drain; Peter Faust—Malcom Drilling; Anne Fleming—Tensar International Corp.; Craig Foster—Maccaferri, Inc.; Blair Gohl—Explosive Compaction Inc.; Jeff Kalani—Yogi Kwong Engineers; Joshua Kolz and Logan Bessette—HB Wick Drains; Frederic Masse—Menard USA; Prof. Paul Mayne—Georgia Tech; Kevin McCann—Landpac Technologies; Kelly McGonagle—Layne Christensen; Clayton Mimura—Geolabs Hawaii; Rick Monroe—DGSI Slope Indicator; Matt Nolan—Hogentogler (SmartDrain); Daniel Mageau—SoilFreeze; Silas Nichols—FHWA; Lisa Render—Atlas EPS (Elevation Geofoam); Marisa Schleter—Hayward Baker; Paul Schmall—Moretrench; Julie Shuster—Wacker Neuson; Don Smith—Pump Hire, Ltd.; Patricia Stelter—Presto Geosystems; Mike Tobin—AFM Corporation; Dana Trierweiler—Infrastructure Alternatives; Billy Troxler—Troxler Laboratories; Robin Vodenlic—GSE Environmental; Kord Wissmann—Geopier Foundations; and Chris Woods—Densification, Inc. Please accept my apology if I omitted anyone.

I would also especially like to thank Shailesh Namekar of Yogi Kwong Engineers for assisting with some of the figures and tables, and Kristie Kehoe for helping me out with emergency detailed editing when I needed it.

Last, I would like to acknowledge the support of my wife Gigi and daughter Emily, for their patience and understanding while "Daddy" worked so many long hours.

Peter G. Nicholson
April 2014

ABBREVIATIONS AND ACRONYMS

AASHTO	American Association of State and Highway Transportation Officials
ADSC	International Association of Foundation Drilling
ASTM	American Standards of Testing and Materials
BPT	Becker penetration test
CBR	California bearing ratio
CDW	continuous diaphragm wall
CFA	continuous flight auger
CIP	cast-in-place
CIR	compaction impact response
CIS	continuous impact settlement
CKD	cement kiln dust
CPT	cone penetration test
CPTU	cone penetration test (with pore pressures)
CSV	soil stabilization with vertical columns
CVM	compaction meter value
DCP	dynamic cone penetrometer
DDC	deep dynamic compaction
DMT (flat plate)	dilatometer test
DOT	Department of Transportation
EERC	Earthquake Engineering Research Center
FHWA	Federal Highway Administration
FWD	falling weight deflectometer
GCL	geosynthetic clay liner
GCS	geosynthetically confined soil
GEER	Geotechnical Extreme Events Reconnaissance
GGBFS	ground granulated blast furnace slag
GPS	Global Positioning System
GRS	geosynthetically reinforced soil
HEIC	high energy impact compaction
IC	intelligent compaction
LKD	lime kiln dust
LRFD	load and resistant factor design
MIT	Massachusetts Institute of Technology
MSE	mechanically stabilized earth
MSW	municipal solid waste
NCHRP	National Cooperative Highway Research Program
NDM	National Deep Mixing Cooperative Research Program
OMC	optimum moisture content
PMT	pressuremeter test
PVD	prefabricated vertical drains
QA	quality assurance
QC	quality control
RAP	Rammed Aggregate Pier

RC	relative compaction
RDC	rolling dynamic compaction
RIC	rapid impact compaction
SASW	spectral analysis of surface waves
SCPTU	seismic cone penetration test (with pore pressures)
SPT	standard penetration test
TRB	Transportation Research Board
USCS	Unified Soil Classification System
USDA	U.S. Department of Agriculture
VST	vane shear test
ZAV	zero air voids

Introduction to Ground Improvement and Soil Stabilization

CHAPTER 1

What is "Ground Improvement?"

In this chapter, the subject of ground improvement is introduced along with a discussion of the engineering parameters that can be addressed and a brief history of ancient practices. An overview of the objectives of designing a ground improvement plan is provided with a description of how ground improvement methods may be implemented into a project. The general categories and objectives of ground improvement techniques are also described.

1.1 INTRODUCTION

While one of the most important criterion for establishing the value of a parcel of land has often been expressed as "location, location, location," the practical and economic feasibility of developing and building upon the land must be at least of equal (or greater) importance. When one considers developing a site either for construction, rehabilitation, preservation/protection, or other use, there needs to be consideration given to the effects of loads imposed and the behavior or response of the ground and soil to those loads. In some cases, the loads may be man-made, while in others forces of nature may be the driving mechanism. Either way, there are some fundamental engineering parameters that generally fall under the expertise of geotechnical engineers that can be evaluated and analyzed to predict what effects a variety of possible loading conditions may have on the ground. These engineers spend much of their careers devising solutions to prevent deleterious effects (or worse, failures) from occurring. Most commonly, these effects can be related to a limited number of soil behaviors or responses now reasonably well understood by geotechnical engineers. These include: *shear strength* of soils, responsible for sustaining loads (static and dynamic) without excessive deformation or failure; *compressibility* of soils, which manifests in settlement, slumping, and volume change of soil masses; *permeability* of soils, which is the rate at which a fluid may flow through the void (open) spaces in a soil mass; and *shrink/swell* potential in soils, which is a phenomenon whereby a soil mass may substantially change volume typically associated with intake or loss of moisture. Other properties, such as stiffness, durability, erodibility, and creep, are also of relative importance depending on the specifics of the application.

Soil Improvement and Ground Modification Methods

3

1.2 ALTERNATIVES TO "POOR" SOIL CONDITIONS

A soil or site may be considered "poor" if it fails to have minimum required engineering properties and/or has been evaluated to provide inadequate performance for the design requirements. A soil may be considered "marginal" if it possesses near the minimal requirements. When "poor" or inadequate soil and/or site conditions prevail, one must consider the available alternatives for the situation. These alternatives may include:

(1) Abandon the project. This might be considered a practical solution only when another suitable site can be found and no compelling commitments require the project to remain at the location in question, or when the cost estimates are considered to be impractical.
(2) Excavate and replace the existing "poor" soil. This method was common practice for many years, but has declined in use due to cost restraints for materials and hauling, availability and cost of select materials, and environmental issues.
(3) Redesign the project or design (often including structural members) to accommodate the soil and site conditions. A common example is the use of driven piles and drilled shafts to bypass soft, weak, and compressible soils by transferring substantial applied loads to a suitable bearing strata.
(4) Modify the soil (or rock) to improve its properties and/or behavior through the use of available ground improvement technologies.

Ground improvement methods have been used to address and solve many ground condition problems and improve desired engineering properties of existing or available soils. In addition, they have often provided economical and environmentally responsible alternatives to more traditional approaches.

There are a number of terms that have been used to describe making changes to the ground and/or soil to improve them for engineering purposes. These include: soil improvement, ground improvement, ground modification, soil stabilization, and so forth. Various authors have attempted to define these terms to differentiate between them, but, generally, there is such overlap between the applications that the terms are often used interchangeably. In general, ground/soil improvement is a process carried out to achieve improved geotechnical properties (and engineering response) of a soil (or earth material) at a site. The processes can be achieved by methods that can be considered to fall into one of three categories:

(1) modification without the addition of any other material,
(2) modification including adding certain materials to the soil/ground, or
(3) modification by providing reinforcement or "inclusions" into the soil/ground.

The purpose of soil and ground Improvement is essentially to alter the natural properties of soil (and/or rock) and/or control the behavior of a geotechnical feature or earthwork in order to improve the behavior and performance of a project. Among the properties that are usually targeted for improvement are:

- Reducing compressibility to avoid settlement
- Increasing strength to improve stability, bearing capacity, or durability
- Reducing permeability to restrict groundwater flow
- Increasing permeability to allow drainage
- Mitigating the potential for (earthquake-induced) liquefaction

Each of these fundamental improvements may be achieved by a variety of methods that will be described in this text. Improvements will be done during one of three phases of a project:

Preconstruction improvements are often the most desirable and cost-effective. These types of improvements would be done to prepare a site for construction and would generally be a part of the planning and design to ensure the success of a project. Examples of preconstruction improvements are ground densification, preconsolidation, drainage, dewatering and modification of hydraulic flows, planned underpinning, and various grouting techniques.

Part-of-construction improvements are those improvement techniques that are done during the construction of the project and could become permanent components of a project. Examples of part-of-construction improvements are compacted gravel columns, shallow soil treatment (including gradation control, shallow compaction, and treatment with admixtures), ground freezing, construction with geosynthetics, soil nails, tie-backs and anchors for cuts, excavation, lightweight fills (including geofoam), and so on. Earthwork construction may involve a number of different methodologies and improvement processes for achieving one or more improvement objectives. These would include engineered fills such as constructed slopes and embankments, retaining wall backfill, and roadways. These would also be encompassed under the category of part-of-construction improvements.

Postconstruction improvements are done after completion of the construction phase of a project and are often remedial processes. These applications can be very costly, but are used as last choice alternatives to rectify problems encountered after (or long after) the completion of a project or to stabilize natural features that have failed or become hazardous. Examples include methods to stabilize settlement problems, failed or near-failure slopes, seepage problems, and so forth. Processes used for postconstruction improvements include grouting, soil nails, drainage, dewatering and modification of hydraulic flows, and so on.

1.3 HISTORICAL SOIL AND GROUND IMPROVEMENT

The fundamental idea of improving the engineering properties of soils or modifying earth materials to perform a desired function is not new. Some of the basic principles of ground improvement, such as densification, dewatering, and use of admixtures, have existed for thousands of years. The use of wood and straw inclusions mixed with mud for "Adobe" construction has been reported for civil works in ancient times of Mesopotamia (the productive "fertile triangle" formed between the Tigris and Euphrates rivers, now Iraq) and ancient Egypt (BCE). Written works from Chinese civilizations (3000-2000 BCE) described use of stone and timber inclusions (ASCE, 1978). Lime mixed with soil was used in construction with Rome's famous Appian Way, built around 600 AD during the height of the Roman Empire. That roadway has endured the test of time and is still fully functional today. An early application soil improvement by addition of infilling material was reportedly used for seepage control in construction of gravelly/rockfill dams in Egypt around 1900, where fine-grained soil was sluiced into the coarse aggregate to lower permeability.

As many of the soil and ground improvement techniques fall in a relatively new area of geotechnical specialization with only a limited database of case histories, some would argue that some methods are the "interaction of engineering science and experience-based technologies" (Charles, 2002). Burland et al. (1976) described the implementation of ground treatment in a "rational context" with the basic stages:
(1) Define the required ground behavior for a particular use of the ground.
(2) Identify any deficiencies in the ground behavior.
(3) Design and implement appropriate ground treatment to remedy any deficiencies.
While these steps may seem very simple and obvious, they are the essential basics to follow when addressing a site for new construction. But in the current field, we must also consider treatment techniques that can be used to remediate existing construction and/or to rehabilitate sites for rebuilding or new types of construction not considered feasible previously.

REFERENCES

ASCE, 1978. Soil improvement: history, capabilities and outlook. Report by the Committee on Placement and Improvement of Soils, Geotechnical Engineering Division. ASCE, 182 pp.

Burland, J.B., McKenna, J.M., Thomlinson, M.J., 1976. Preface: ground treatment by deep compaction. Geotechnique 25 (1), 1–2.

Charles, J.A., 2002. Ground improvement: the interaction of engineering science and experience-based technology. Geotechnique 52 (7), 527–532.

http://www.astm.org/Standards (accessed 02.11.14.).

CHAPTER 2

Ground Improvement Techniques and Applications

This chapter introduces the general categories of ground improvement along with descriptions of the main application techniques for each. An overview is provided of the most common and typical objectives to using improvement methods and what types of results may be reasonably expected. A discussion of the various factors and variables that an engineer needs to consider when selecting and ultimately making the choice of possible improvement method(s) is also included. This is followed by descriptions of common applications used. This chapter concludes with a brief discussion of a number of emerging trends and promising technologies that continue to be developed. These include sustainable reuse of waste materials and other "green" approaches that can be integrated with improvement techniques.

2.1 CATEGORIES OF GROUND IMPROVEMENT

The approaches incorporating ground improvement processes can generally be divided into four categories grouped by the techniques or methods by which improvements are achieved (Hausmann, 1990).

Mechanical modification—Includes physical manipulation of earth materials, which most commonly refers to controlled densification either by placement and compaction of soils as designed "engineered fills," or "in situ" (in place) methods of improvement for deeper applications. Many engineering properties and behaviors can be improved by controlled densification of soils by *compaction* methods. Other in situ methods of improvement may involve adding material to the ground as is the case for strengthening and reinforcing the ground with nonstructural members.

Hydraulic modification—Where flow, seepage, and drainage characteristics in the ground are altered. This includes lowering of the water table by drainage or dewatering wells, increasing or decreasing permeability of soils, forcing consolidation and preconsolidation to minimize future settlements, reducing compressibility and increasing strength, filtering groundwater flow, controlling seepage gradients, and creating hydraulic

Soil Improvement and Ground Modification Methods

9

barriers. Control or alteration of hydraulic characteristics may be attained through a variety of techniques, which may well incorporate improvement methods associated with other ground improvement categories.

Physical and chemical modification—"Stabilization" of soils caused by a variety of physiochemical changes in the structure and/or chemical makeup of the soil materials or ground. Soil properties and/or behavior are modified with the addition of materials that alter basic soil properties through physical mixing processes or injection of materials (grouting), or by thermal treatments involving temperature extremes. The changes tend to be permanent (with the exception of ground freezing), resulting in a material that can have significantly improved characteristics. Recent work with *biostabilization*, which would include adding/introducing microbial methods, may also be placed in this category.

Modification by inclusions, confinement, and reinforcement—Includes use of structural members or other manufactured materials integrated with the ground. These may consist of reinforcement with tensile elements; soil anchors and "nails"; reinforcing geosynthetics; confinement of (usually granular) materials with cribs, gabions, and "webs"; and use of lightweight materials such as polystyrene foam or other lightweight fills. In general, this type of ground improvement is purely physical through the use of structural components. Reinforcing soil by vegetating the ground surface could also fall into this category.

In fact, the division of ground improvement techniques may not always be so easily categorized as to fall completely within one category or another. Oftentimes an improvement method may have attributes or benefits that can arguably fall into more than one category by achieving a number of different engineering goals. Because of this, there will necessarily be some overlap between categories of techniques and applications. In fact, in looking at defining improvement methodologies, it very quickly becomes apparent that there are a broad array of cross-applications of technologies, methods, and processes. As will be described, the best approach is often to first address a particular geotechnical problem and identify the specific engineering needs of the application. Then a variety of improvement approaches may be considered along with applicability and economics.

2.2 TYPICAL/COMMON GROUND IMPROVEMENT OBJECTIVES

The most common (historically) traditional objectives include improvement of the soil and ground for use as a foundation and/or construction material.

The typical engineering objectives have been (1) increasing shear strength, durability, stiffness, and stability; (2) mitigating undesirable properties (e.g., shrink/swell potential, compressibility, liquefiability); (3) modifying permeability, the rate of fluid to flow through a medium; and (4) improving efficiency and productivity by using methods that save time and expense. Each of these broad engineering objectives are integrally embedded in the basic, everyday designs within the realm of the geotechnical engineer. The engineer must make a determination on how best to achieve the desired goal(s) required by providing a workable solution for each project encountered. Ground improvement methods provide a diverse choice of approaches to solving these challenges.

In many cases, the use of soil improvement techniques has provided economical alternatives to more conventional engineering solutions or has made feasible some projects that would have previously been abandoned due to excessive costs or lack of any physically viable solutions.

Some newer challenges and solutions have added to the list of applications and objectives where ground improvement may be applicable. This is in part a result of technological advancements in equipment, understanding of processes, new or renewed materials, and so forth. Some newer issues include environmental impacts, contaminant control (and clean up), "dirty" runoff water, dust and erosion control, sustainability, reuse of waste materials, and so on.

2.3 FACTORS AFFECTING CHOICE OF IMPROVEMENT METHOD

When approaching a difficult or challenging geotechnical problem, the engineer must consider a number of variables in determining the type of solution(s) that will best achieve the desired results. Both physical attributes of the soil and site conditions, as well as social, political, and economic factors, are important in determining a proposed course of action. These include:

(1) Soil type—This is one of the most important parameters that will control what approach or materials will be applicable. As will be described throughout this text, certain ground improvement methods are applicable to only certain soil types and/or grain sizes. A classic figure was presented by Mitchell (1981) to graphically represent various ground improvement methods suitable for ranges of soil grain sizes. While somewhat outdated, this simple figure exemplified the fundamental dependence of soil improvement applicability to soil type and grain size. An updated version of that figure is provided in Figure 2.1.

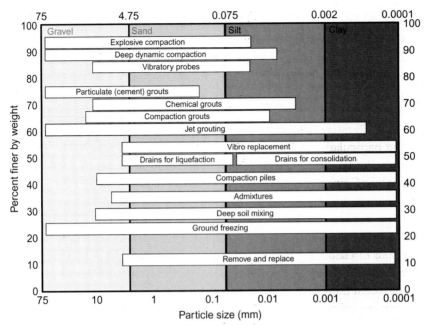

Figure 2.1 Soil improvement methods applicable to different ranges of soil sizes.

(2) Area, depth, and location of treatment required—Many ground improvement methods have depth limitations that render them unsuitable for application to deeper soil horizons. Depending on the areal extent of the project, economic and equipment capabilities may also play an important role in the decision as to what process is best suited for the project. Location may play a significant role in the choice of method, particularly if there are adjacent structures, concerns of noise and vibrations, or if temperature and/or availability of water is a factor.

(3) Desired/required soil properties—Obviously, different methods are used to achieve different engineering properties, and certain methods will provide various levels of improvement and uniformity to improved sites.

(4) Availability of materials—Depending on the location of the project and materials required for each feasible ground improvement approach, some materials may not be readily available or cost and logistics of transportation may rule out certain methods.

(5) Availability of skills, local experience, and local preferences—While the engineer may possess the knowledge and understanding of a preferred method, some localities and project owners may resist trying something that is unfamiliar and locally "unproven." This is primarily a social issue,

but should not be underestimated or dismissed, especially in more remote and less developed locations.

(6) Environmental concerns—With a better understanding and greater awareness of effects on the natural environment, more attention has been placed on methods that assure less environmental impact. This concern has greatly changed the way that construction projects are undertaken and has had a significant effect on methods, equipment, and particularly materials used for ground improvement.

(7) Economics—When all else has been considered, the final decision on choice of improvement method will often come down to the ultimate cost of a proposed method, or cost will be the deciding factor in choosing between two or more otherwise suitable methods. Included in this category may be time constraints, in that a more costly method may be chosen if it results in a faster completion allowing earlier use of the completed project.

All of these factors may play a role in determining the best choice(s) of improvement method(s) to be proposed. Each project needs to be addressed on a case-specific basis when making this decision.

2.4 COMMON APPLICATIONS

Within the categories outlined in Section 2.1, there are a range of commonplace soil and ground improvement techniques in daily use. Some need only readily available construction equipment, while others require specialized equipment. Due to the steady increase in acceptance, experience, and proven solutions utilizing these techniques, there are now many industry specialists from which to draw for improvement needs leading to healthy competition in the market.

Soil densification under various conditions is perhaps one of the oldest, and likely the most common, of all soil improvement methods. Consequently, a significant portion of this text is dedicated to describing the details of the theory, mechanics, and practice of soil densification techniques. Densification includes both shallow compaction methods and deep (in situ) techniques, which will be addressed individually. Densification provides for improving a number of fundamental properties that control characteristics of soil responses critical to the most fundamental geotechnical engineering analyses and designs. In many cases, densification will allow more efficient and cost-effective solutions for both the construction and remediation of civil engineering projects. Significant efforts have incorporated in situ densification techniques to alleviate or mitigate soil liquefaction, a dramatic and often devastating or catastrophic consequence of earthquake loading. This has been a driving force for remediation at coastal port facilities and high-hazard earth dams throughout the world.

Drainage and filtering of fluids (usually water) through or over the ground has also proven to be a rather conceptually simple solution to many ground engineering issues, including slope stability, ground strengthening, performance of water conveyance and other hydraulic structures (such as dams, levees, flood control, shorelines, etc.), environmental geotechnics (landfill construction, contaminated site remediation, and contaminant confinement), and construction dewatering, which often requires hydraulic barriers. Geotechnical engineering legend Ralph Peck used to say, "Water in the ground is the cause of most geotechnical engineering problems." Drainage applications may be "simply" draining water from a soil to reduce its weight and unwanted water pressure to increase strength while reducing load. Drainage may also relate to (1) dewatering for purposes of creating a (dry) workable construction site where there is either standing water or a relatively high water table that would otherwise be encountered during excavation, or (2) creating a situation that allows water to continually drain out and away from a structure such as a roadway or foundation. A third application of dewatering involves forcing water out of a saturated clayey soil in order to reduce compressibility, reduce settlement, and increase strength of the clayey strata. For each application there may be one or more different approaches to achieving desired objectives. While the fundamental concepts may at first appear straightforward, due to the high variability of soil permeability and the often difficult task of estimating intricate three-dimensional ground water flow by simplified idealized assumptions, solutions dependent on accurate flow estimates will often have the greatest uncertainty. A consequence of draining water or controlling water flow through the ground is the need to provide adequate *filtering* of the flow such that the soil structure is not negatively impacted by erosion. Proper drainage and filtering so as to ensure long-term stability is critical to water retention and conveyance structures, and may be achieved by a combination of improvement techniques, including soil grain size and gradation control and the use of geosynthetic materials.

In contrast to drainage, the objective of some hydraulic improvements is to retain or convey water by reducing the permeability of the ground. For these applications, a number of soil improvement and ground modification options are available. These options include soil densification techniques as well as treating the soil with additives and constructing soil "systems" with manufactured hydraulic barriers of both natural and manufactured (i.e., geosynthetic) materials.

Admixture stabilization has existed in some form for thousands of years, historically concentrated using lime, cement, fly ash, and asphalts. The area of soil additives and mixing continues to evolve with the advent of new materials and the desire to utilize and recycle waste materials. As will be discussed in some detail, soil additives can have profound effects on the engineering properties of earth materials. With the proper combination of soil type and admixture material, nearly any soil can be improved to make use of otherwise unsuitable materials, ground conditions, and/or save time and money. Much of the key to success with soil admixture improvement is the type and quality of the mixing process(s). Shallow surface mixing of admixture materials has been tremendously successful in improving the quality and reducing required maintenance of roadways and other transportation facilities which rely on strength, stability, and durability of near surface soils and/or placed engineered fill. Shallow surface mixing is typically limited to the top 0.6 m. Deep mixing is an in situ method that has been growing steadily in popularity and with improved technologies. Deep mixing techniques now attain depths of 30 m or more.

Within the realm of admixture improvement is the concept of *grouting*, which in the context of admixtures usually means a method whereby the grout material permeates and mixes with the natural soil materials, causing both physical and/or chemical improvements. *Jet grouting* is another type of process that involves the use of admixture materials. Grouting as a ground improvement process is addressed in its own chapter.

Geosynthetic reinforcement is commonly used to construct walls and slopes, eliminating the need for heavy structural retaining walls and allowing steeper stable slopes. Soil reinforcement is also being used for scour/erosion control and foundation support. Reinforcement provides load distribution and transfer between concentrated load points and a broader area, allowing construction of loads over weaker materials or to deep foundation support with reduced settlement problems and higher capacity.

Use of *structural inclusions* has become a common and practical solution for many ground improvement applications, especially for improving stability of slopes, cuts, and excavations. Structural inclusions can be incorporated as an integral part of constructed earthworks, such as embankments, slopes, and retaining walls, or placed into existing ground to improve stability with the use of "anchors," "nails," or columns/piles. Structural inclusions are also commonly used for temporary stabilization of excavations and for underpinning of existing structures.

Lightweight fill materials have become widely accepted for embankment construction and bridge approaches where conventional fill materials would impose too large a load to be accommodated by the underlying soil. Expanded polystyrene foam, or *geofoam*, has been effectively utilized for major transportation projects, such as the Boston Artery and Utah's I-15 reconstruction, as well as for many other smaller projects. Other lightweight fill materials have also been used to reduce applied loads, settlement, bearing capacity, and lateral earth pressure concerns.

Technological advancements in the use of *artificial ground freezing* techniques, once considered a novelty, have made it a competitive and viable option for temporary construction support, "undisturbed" sampling of difficult soils, and as an interim stabilization technique for active landslides and other ground failure situations.

2.5 EMERGING TRENDS AND PROMISING TECHNOLOGIES

A number of *"green" initiatives* have found their way into soil and ground improvement practice in recent years. Issues with environmental and potential health issues have resulted in a shift away from (and in some cases the discontinuation of) using additives that have been deemed to be potentially hazardous or toxic to people, livestock, groundwater supply, and agriculture. This also includes efforts to monitor, collect, and/or filter runoff from construction sites resulting from ground improvement activities. In addition, reduction of waste through reuse and recycling approaches has led to better utilization of resources as well as reduced volume of material in the often overtaxed waste stream. In fact, significant benefits have been realized by efforts striving for more environmental consciousness.

A wide array of new "environmentally correct" materials have become available for use as admixtures. Industry manufacturers are paying special attention to public concern by providing materials that are either inert, "natural," or in some cases, even biodegradable. Reuse of recycled pavements has decreased the demand on valuable pavement material resources and/or the need to import costly select materials.

Blast furnace slag is a by-product of the production of iron (Nidzam and Kinuthia, 2010), and is used as construction aggregate in concrete. Ground granulated blast furnace slag (GGBS) has been used as aggregate for use in lightweight fills, and as riprap and fill for gabion baskets. Steel slag fines (material passing the 9.5 mm sieve) are the by-product of commercial scale crushing and screening operations of steel mills. Recent research has shown that use of steel slag fines mixed with coastal dredged materials not only provides a source of good quality fill, but has the capability to bind heavy metals such that leached fluids are well below acceptable EPA levels (Ruiz et al., 2012).

New equipment design and technological advances in operations, monitoring, and quality control have all assisted in improving such soil and ground treatment techniques as *deep mixing* for bearing support, excavation support, hydraulic cutoffs, and in-place wall/foundations, providing new capabilities and levels of reliability. Advancements include the ability to mix at greater depths, more difficult locations, and with materials that had previously been beyond limitations.

The still relatively young practice of designing with geosynthetics for geotechnical applications is emerging with new materials and applications every year. It is expected that this area will continue to develop rapidly for many years to come.

The above is just a sampling of the activity in this still developing field of soil and ground improvement. While the fundamentals and basic theories of several improvement techniques are ancient, modern engineering design continues to advance the possibilities for problem solving using soil and ground improvement methodologies.

Another emerging technology that has attracted growing interest has been the field of "bioremediation." This topic includes a number of interesting approaches for stabilizing soils. One of these involves the use of organisms that would precipitate calcium-forming bonds to increase strength through a cementing process. Other bioremediation applications involve slope stabilization and erosion control through the use of vegetation to physically retain surface soils by their root systems. Vegetation can have both beneficial as well as adverse effects on slope stability. These technologies are described in Chapter 18.

REFERENCES

Hausmann, M.R., 1990. Engineering Principles of Ground Modification. McGraw-Hill, Inc, 632 pp.

Mitchell, J.K., 1981. State of the art – soil improvement. In: Proceedings of the 10th ICSMFE. Stockholm, vol. 4, pp. 509–565.

Nidzam, R.M., Kinuthia, J.M., 2010. Sustainable soil stabilisation with blastfurnace slag. Proc. ICE: Constr. Mater. 163 (3), 157–165.

Ruiz, C.E., Grubb, D.G., Acevedo-Acevedo, D., 2012. Recycling on the waterfront II. Geostrata. (July/August), ASCE Press.

http://www.nationalslag.org/blastfurnace.htm (accessed 06.08.13.).

CHAPTER 3

Soil Mechanics Basics, Field Investigations, and Preliminary Ground Modification Design

The first half of this chapter provides a brief overview of soil mechanics fundamentals such as soil strength, compressibility (settlement), and fluid flow (permeability) topics, as they pertain to some of the basic parameters and properties that are used to evaluate the engineering response of soils. Also included is a brief discussion of some field and laboratory methods typically used to obtain these values.

The second half of the chapter is principally dedicated to the information that should be obtained from typical site or field investigations and explorations in order to provide the engineer with the parameters necessary to perform analyses and initiate preliminary ground improvement selection. It is from this data (typically contained in boring logs, soil test results, and geotechnical reports) and correlations with soil characteristics that many ground improvement designs are formulated.

3.1 SOIL MECHANICS FUNDAMENTALS OVERVIEW

Presented here is a brief description of typical soil types and a review of soil mechanics basics that is necessary to understand the fundamentals used in soil improvement and ground modification design. This may be elementary for those with a strong background and/or education in geotechnical engineering, but will provide others with the background necessary for understanding the concepts and methods described throughout the remainder of this text.

3.1.1 Soil Type and Classification

Generally, most soil can be characterized as being made up of either or both of two distinctive types of grains. "Rounded" or "bulky" grains have a relatively small surface area with respect to their volume, similar to that of a sphere. These soil grains typically have little intragranular attraction (or bonds) and

Soil Improvement and Ground Modification Methods

are therefore termed "cohesionless," referring to lack of tendency to "stick" together. Soil with these grain characteristics may also be called "granular." This soil group includes sands and gravels. Clay particles are very different, and are made of very thin plate-like grains, which generally have a very high surface to volume ratio. Because of this, the surface charges play a critical role in their intragranular attractive behavior and are termed "cohesive." As will be discussed in much more depth in later chapters, this difference between grain types has a profound effect on behavior of a soil and the methodology by which improvement techniques can be effective.

3.1.1.1 Soil Classification Systems

There are a number of different soil classification systems that have been devised by various groups, which vary in definitions and categories of soil type. The Unified Soil Classification System (USCS; ASTM D2487) is dominant for most geotechnical engineers, as its soil type designations correlate well with many soils engineering properties. Thus, knowing a USCS designation may well be enough for a seasoned geotechnical engineer to be able to envision the types of properties such a soil may possess. The USCS will be used as the primary classification system throughout this text. Another common classification system, derived for use with roadway materials, is the American Association of State Highway and Transportation Officials (AASHTO) system (ASTM D3282, AASHTO M145). The AASHTO classification designations categorize soil types based on their usefulness in roadway construction applications. Another classification system is used by the US Department of Agriculture (USDA) for defining soil categories important for agricultural applications. The Massachusetts Institute of Technology also developed a soil classification system in which grain size definitions are nearly the same as the AASHTO. Table 3.1 and Figure 3.1 depict grain size definitions by various particle-size classification schemes. Soil classifications are typically limited to particle sizes less than about 76 mm (3 in).

Soil type and classification usually begins with analyzing the sizes of grains contained, followed by further defining the characteristics of the clayey portion (if any) and/or distribution of grain sizes for the coarser, granular portion (if any). The effect of clay content and characteristics of the clay portion play a very important role in affecting the engineering properties of a soil; therefore, soil types and soil classifications may include qualifiers of the finer-grained portion when as little as 5% of the soil consists of fine-grain sizes.

Table 3.1 Grain Size Definitions by Various Particle-Size Classification Schemes
Particle-Size Classifications

Name of Organization	Grain Size (mm)			
	Gravel	Sand	Silt	Clay
Massachusetts Institute of Technology (MIT)	>2	2–0.06	0.06–0.002	<0.002
US Department of Agriculture (USDA)	>2	2–0.05	0.05–0.002	<0.002
American Association of State Highway and Transportation Officials (AASHTO)	76.2–2	2–0.075	0.075–0.002	<0.002
Unified Soil Classification System (US Army Corps of Engineers, US Bureau of Reclamation, and American Society for Testing and Materials)	76.2–4.75	4.75–0.075	Fines (i.e., silts and clays) <0.075	

3.1.1.2 Grain Sizes and Grain Size Distributions

At this point, one needs to clearly define a standard size to differentiate between coarse- and fine-grain sizes. This has been done for a number of classification systems using a standard screen mesh with 200 openings per inch, referred to as a #200 sieve. The effective opening size of a #200 sieve is 0.075 mm. Material able to pass through the #200 sieve is termed "fine-grained" while that retained on the sieve is termed "coarse-grained." This standardized differentiation is not

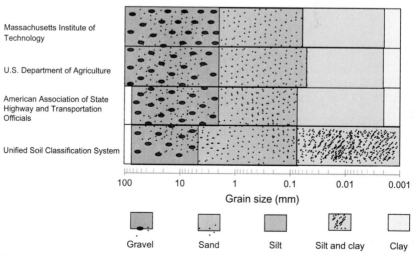

Figure 3.1 Grain size definitions by various particle-size classification schemes.

completely arbitrary or without merit as it is found that fine-grained soils tend to be more cohesive while coarse-grained soils are cohesionless. It is important to remember, however, that differentiation between clay and granular particles is not always represented by grain size and the #200 sieve!

Analyzing the amounts or percentages of various grain size categories can be used to further classify soil types. Much can be ascertained by knowing the distribution of grain sizes, as these differences are related to various engineering properties and characteristics of soil. Common practice for coarse-grained soils is to filter a known amount (weight) of dry soil through a set of mesh screens or *sieves* with progressively smaller openings of known size. This will separate the soil into portions that pass one sieve size and are retained on another. This approach is known as a "sieve analysis." Data of this type is collected such that the percentage passing each progressively smaller sieve opening size can be calculated. The results are presented as *gradation* plots or *grain size distribution* curves, plotted with percent passing versus nominal grain size. The grain size distribution is used for primary identification of coarse-grained soils and also can define gradation type.

Coarse-grained soils will generally fall into one of three different gradation types. Figure 3.2 depicts a representation of the general "shape" or trends of well-graded, poorly graded, and gap-graded soils. *Well-graded* soils span a wide range of grain sizes and include representation of percentages from intermediate sizes between the maximum and minimum sizes. Well-graded soils are often preferred as they are relatively easy to handle, can compact well, and often provide desirable engineering properties. *Poorly graded* (or well-sorted, or uniform) soils have a concentration of a limited range of grain sizes. This

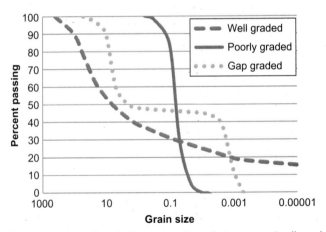

Figure 3.2 Representation of typical coarse soil gradation types (well-graded, poorly graded, gap-graded).

type of gradation can be found in nature due to natural phenomenon associated with depositional processes such as from *alluvial* and *fluvial* flows (rivers and deltas), waves (beach deposits), or wind (sand dunes). Poorly graded (or uniformly graded) soil gradations may be advantageous where seepage and ground water flow characteristics (drainage and filtering) are important. Uniformly graded soils can also be prepared manually by sieving techniques at small or large scale (such as for quarrying operations). A third category for gradation is known as *gap-graded*, which refers to a soil with various grain sizes but which lacks representation of a range of intermediate sizes. Usually, this type of gradation is never desirable as it can create problems with handling and construction due to its tendency to segregate and create nonuniform fills. For classification purposes, gap-graded soils are considered to be a subset of poorly graded soils, as they are not well-graded.

3.1.1.3 Plasticity and Soil Structure

Classification schemes based solely on grain sizes (i.e., USDA) are relatively simple, but do not take into account the importance of clay properties on the behavioral characteristics of a soil. Both the USCS and AASHTO classification systems utilize a combination soil grain size distribution along with clay properties identifiable by *plasticity* of the finer-grained fraction of a soil. Plasticity is the ability of a soil to act in a plastic manner and is identified by a range of moisture contents where the soil is between a semisolid and viscous liquid form. These limits are determined as the plastic limit (PL) and liquid limit (LL) from simple, standardized laboratory index tests. For a more detailed discussion of these and related tests, refer to an introductory soil mechanics text, laboratory manual, or ASTM specifications (ASTM D4318).

Plasticity is commonly referred to by the Plasticity Index (PI), where $PI = LL - PL$. A graphical representation of plasticity developed for the purposes of classifying fine-grained soils gives the PI plotted as a function of LL (Figure 3.3). The plot defines fine-grained soil classifications between clay and silt, and between high and low plasticity. There is a separating line called the A-line, defined by the equation $PI = 0.73 \, (LL - 20)$. Clay (C) is designated for soil with combinations of PI and LL above the "A-line" for soils with $PI > 7$. Soil below the A-line and $PI > 4$, and above the A-line with below $PI < 4$ are considered silt, designated "M." Another defining line is given for soils with LL above or below 50. Soils with $LL > 50$ are considered *high plasticity*, while those with $LL < 50$ are considered *low plasticity*. A special dual designation of CL-ML is given for soils above the A-line and $4 \le PI \le 7$.

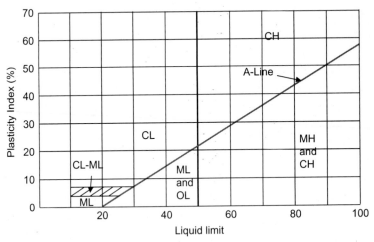

Figure 3.3 Plasticity chart for fine-grained soils.

3.1.1.4 Unified Soil Classification System

The USCS was originally developed by Casagrande in the 1940s to assist with airfield construction during World War II (Das, 2010) and has been modified a number of times since. In order to classify a soil according to the USCS, a number of relatively simple steps must be followed. Only one to three simple index tests need to be performed in order to fully classify a soil: a sieve analysis, and/or a LL test, and a PL test. In the USCS, soil is generally classified by a two-letter designation. (Note: Under special circumstances explained later, a soil may fall in between designations and will be given a dual classification.) The first letter denotes the primary designation and identifies the dominant grain size or soil type. The primary designations are G, gravel; S, sand; M, silt; C, clay, O, organic, and Pt, peat (a highly organic soil). The second letter denotes a qualifier that provides further information regarding more detailed information on the makeup and characteristics of the soil.

Coarse-grained soils are defined as those where more than 50% of the soil is retained on the No. 200 sieve. According to the USCS, coarse soil grains retained on the No. 4 sieve (nominal opening size of 4.75 mm) are defined as gravel while those grains passing the No. 4 and retained on the No. 200 sieve are defined as sand. A coarse-grained soil is defined as gravel or sand depending on the dominant grain size percentage of the coarse fraction of the soil (where the coarse fraction is the cumulative percentage coarser than the No. 200 sieve). For example, if more than 50% of the material coarser than the No. 200 sieve is retained on the No. 4 sieve, then the soil is classified as gravel (G). If 50% or more of the material coarser than the No. 200 sieve passes the No. 4 sieve, then the soil is classified as sand (S).

For coarse-grained soils (G or S), the second qualifier denotes the *type of gradation* (P, poorly graded; W, well-graded) or the type of fine-grained soil contained if significant (M or C), so that coarse-grained soils will generally be classified with designations of GP, GW, GM, GC, SP, SW, SM, or SC. As mentioned earlier, fine-grained soils ("fines") become significant to the engineering properties and soil characteristics when as little as 5% by weight is contained. According to USCS, when less than 5% fine-grained material is present in a soil, fines are insignificant, and the second qualifier should pertain to the gradation characteristics according to the definitions provided below. The definition of well-graded versus poorly graded is a function of various grain sizes as determined by the grain size distributions. The definition of well-graded is based on two coefficients determined by grain sizes taken from the gradation curves. These are the *coefficient of uniformity* (C_u) and *coefficient of curvature* (C_c). If one looks at a gradation curve for a specific soil, there is a grain diameter (size) where a certain percentage of the material grains are smaller. This is grain size for a given "percent finer." For example, if 30% of the grains of a material are smaller than 1 mm, then the grain size for 30% finer is equal to 1 mm. This is designated D_{30}. C_u and C_c are defined as:

$$C_u = \frac{D_{60}}{D_{10}} \tag{3.1}$$

$$C_c = \frac{(D_{30})^2}{D_{60} \times D_{10}} \tag{3.2}$$

For a soil to be designated as *well-graded*, the following must hold true:

$$1 < C_c < 3 \text{ and } C_u \geq 6 \text{ (for sand)}, \ C_u \geq 4 \text{ (for gravel)}$$

If either of these criteria fails, then the soil is designated as *poorly graded*.

If more than 12% of the soil is determined to be fine grained by sieve analysis, then the second qualifier refers to the type of fines present (C or M), as the soil characteristics and behavior will likely be more affected by the characteristics of the fine-grained material contained than the type of gradation. The "type" of fines is determined by classifying the fine-grained portion of the soil, and using the primary designation of those results from the *plasticity chart* (Figure 3.3), which provides information on the characteristics of the fine-grained fraction. For soils that contain between 5% and 12% fines, both the gradation type and properties of the fines may have important contributions to the engineering characteristics of the soil. Therefore, a dual classification is used whereby secondary qualifiers for both gradation and type of fines are used in addition to the primary designation for the soil. For instance, a soil that is primarily a well-graded sand but contains fines that plot above the A-line (clay) will be given a dual classification of SW-SC. Possible combinations for dual soil classifications would be: GW-GC, GW-GM, GP-GC, GP-GM, SW-SC, SW-SM, SP-SC, and SP-SM.

Fine-grained soils (those where more than 50% of the soil passes the #200 sieve) are defined according to the plasticity chart shown in Figure 3.3. Most fine-grained soils will have a primary designation based on the LL versus PI values and their relationship to the "A-line" on the chart, with secondary designation as high (H) or low (L) plasticity, determined by whether the LL is above or below 50, respectively. Special cases for fine-grained soils are *organic* (O) designations OL and OH. Soils are determined to be organic based on changes in the LL as determined before and after oven drying. Other special cases of classification for fine-grained soils occur with low PI and LL values as seen on the plasticity chart (and described previously). AASHTO soil classification of fine-grained soils also uses a variation of a plasticity chart (see ASTM D3282). Table 3.2 provides criteria for assigning USCS group symbols to soils.

Currently, ASTM D2487 utilizes the group symbol (two-letter designation) along with a group name, which can be determined using the same information gathered for classification designation, but adds a more detailed description that further elaborates on gradation. So for a complete classification and description including group name, one must know the percentages of gravel, sand and fines, and type of gradation (all based on sieve analyses), as well as LL and PI for fine-grained portions of the soil. Flowcharts for the complete USCS classifications for coarse-grained and fine-grained soils are given in Figures 3.4 and 3.5 respectively.

3.1.2 Principal Design Parameters

In order to develop a plan of approach for designing a practical and economical solution, a geotechnical engineer must first initiate a stepwise process of identifying fundamental project parameters. These include: (1) establishing the scope of the problem, (2) investigating the conditions at the proposed site, (3) establishing a model for the subsurface to be analyzed, (4) determining required soil properties needed for analyses to evaluate engineering response characteristics, and (5) formulating a design to solve the problem. A number of engineering parameters that play critical roles in how the ground responds to various applications and loads typically need to be determined for each situation. Values of each parameter may be evaluated by field or laboratory tests of soils, or may be prescribed by design guidelines. Fundamental to applicable analyses and designs are input of reasonably accurate parameters that provide an estimate of response of the ground to expected loading conditions. Some of the parameters forming the basis of design applications are reviewed here.

Table 3.2 Criteria for Assigning USCS Group Symbol (after ASTM D2487) USCS Group Symbol Criteria

Major Category	Major Classification	Soil Description	Specific Criteria	Group Symbol
Coarse-grained soils More than 50% retained on No. 200 sieve (coarse fraction)	*Gravels* More than 50% of coarse fraction retained on No. 4 sieve	Clean gravels Less than 5% fines[b]	$C_u \geq 4$ and $1 \leq C_c \leq 3$[a]	GW
			$C_u < 4$ and/or $1 > C_c > 3$[a]	GP
		Gravels with fines More than 12% fines[b]	$PI < 4$ or plots below "A" line	GM
			$PI > 7$ and plots on or above "A" line	GC
	Sands 50% or more of coarse fraction passes No. 4 sieve	Clean sands Less than 5% fines[c]	$C_u \geq 6$ and $1 \leq C_c \leq 3$[a]	SW
			$C_u < 6$ and/or $1 > C_c > 3$[a]	SP
		Sands with fines More than 12% fines[c]	$PI < 4$ or plots below "A" line	SM
			$PI > 7$ and plots on or above "A" line	SC
Fine-grained soils 50% or more passes No. 200 sieve	*Silts and clays* Liquid limit less than 50	Inorganic	$PI > 7$ and plots on or above "A" line	CL
			$PI < 4$ or plots below "A" line	ML
			$4 < PI < 7$ and plots on or above "A" line	CL-ML
		Significant organics	$\dfrac{\text{Liquid limit (oven dried)}}{\text{Liquid limit (not dried)}} < 0.75$	OL
	Silts and clays Liquid limit 50 or more	Inorganic	PI plots on or above "A" line	CH
			PI plots below "A" line	MH
		Significant organics	$\dfrac{\text{Liquid limit (oven dried)}}{\text{Liquid limit (not dried)}} < 0.75$	OH
Highly organic soils		Primarily organic matter, dark in color, and organic odor		Pt

[a] $C_u = \dfrac{D_{60}}{D_{10}}; C_c = \dfrac{(D_{30})^2}{D_{60} \times D_{10}}$.

[b] Gravels with 5–12% fine require dual symbols: GW-GM, GW-GC, GP-GM, GP-GC.

[c] Sands with 5–12% fine require dual symbols: SW-SM, SW-SC, SP-SM, SP-SC.

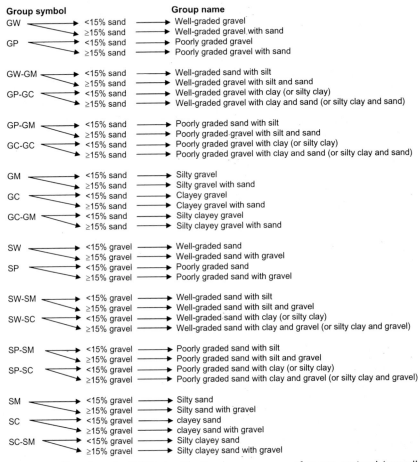

Figure 3.4 Flowchart for USCS classification group names of coarse-grained (gravelly and sandy) soil. *After ASTM D2487.*

Shear strength: Soil differs from most other engineering materials in that soil tends to fail in shear rather than a form of tension or compression. In fact, as soil exhibits very little tensile strength, convention is to take compression as positive and tension as negative, as opposed to standard mechanics of materials sign convention. Soil shear strength is then a function of the limiting shear stresses that may be induced without causing "failure." For the general case, shear strength is a function of frictional and cohesive parameters of a soil under given conditions of initial stresses and intergranular water pressures. Proper evaluation of shear strength is critical for many types of geotechnical designs and applications as it is fundamental to such

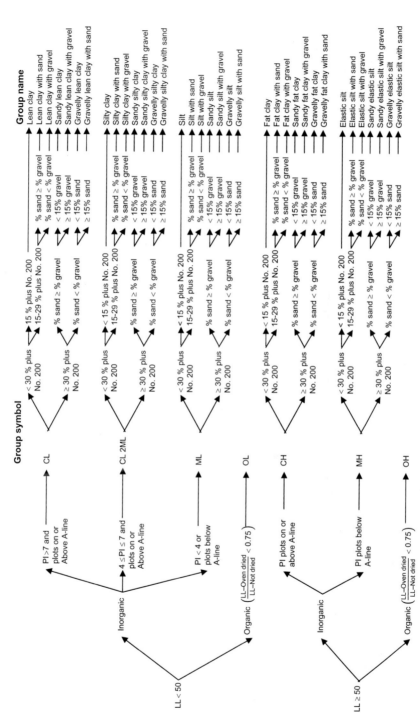

Figure 3.5 Flowchart for USCS classification group names of fine-grained (silty and clayey) soils. *After ASTM D2487.*

(Continued)

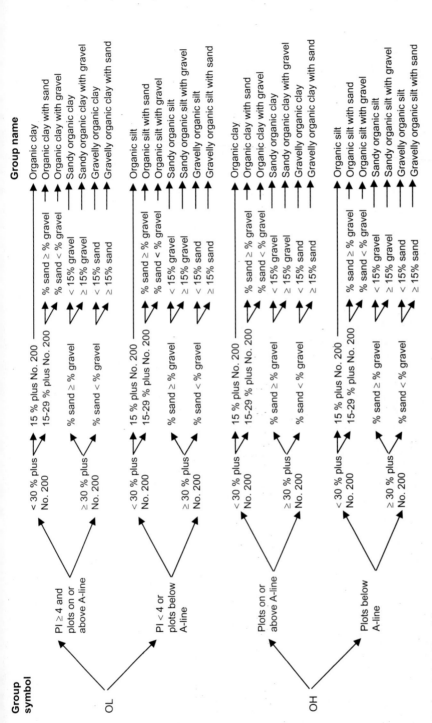

Figure 3.5, (Cont'd)

considerations as *bearing capacity* (the ability for the ground to support load without failing), *slope stability* (an evaluation of the degree of safety for a soil slope to resist failure), *durability* (resistance to freeze-thaw and wet-dry cycles, as well as leaching for some soils), and *liquefaction resistance* (the ability of a soil to withstand dynamic loads without liquefying, discussed in Section 4.2.5). The general equation for shear strength is

$$t_f = c' + \sigma' \tan \Phi' \tag{3.3}$$

where τ_f is the shear strength (shear stress at failure), c' the effective soil cohesion parameter, σ' the effective confining stress, Φ' the effective soil friction parameter.

As can be seen from Equation (3.3), shear strength is a function of the effective confining stress (σ'). Here effective stresses are used as opposed to total stresses. Effective stresses are the intergranular stresses that remain after pore water pressures are accounted for. These are the actual stresses "felt" between grains, adding to their frictional resistance (strength). Total stresses are the combination of intergranular and pore water pressure acting on soil grains. Figure 3.6 graphically depicts shear strength as a function of effective confining stress in terms of a shear strength failure envelope. In looking at this figure, the plotted line defines the failure envelope. Any state of stress described by a point below the line is a possible state of equilibrium. Theoretically, once a state of stress is reached which touches the failure envelope, the soil will fail. Stress states above the failure envelope are not theoretically possible. Evaluation of the shear strength parameters c' and Φ' may be obtained directly from laboratory tests or interpreted from in situ field tests performed as part of a site investigation.

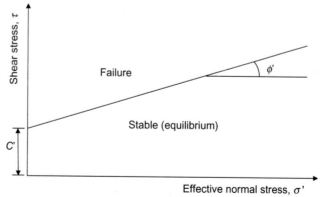

Figure 3.6 Graphical representation of the shear strength failure envelope.

Laboratory tests typically used include: direct shear tests (ASTM D3080), unconfined compression tests (ASTM D2166), triaxial tests (ASTM D7181), and simple shear tests (ASTM D6528). In each of these tests (except unconfined compression), effective stress can be varied so that the shear strength (and shear strength parameters) can be evaluated for the appropriate stress levels estimated for each field application. The unconfined compression test may actually be considered a special case of the triaxial test, where the lateral confining stress is equal to zero. This test is simple and is often used as a quick indicator of strength and for comparative strength purposes, but is limited to cohesive soils (or in some cases, cemented soils). More discussion regarding the use of these laboratory techniques will be addressed later in this chapter.

A variety of in situ field tests are also available to evaluate soil shear strength. These include simple handheld devices such as the pocket penetrometer and pocket vane, which can give a quick estimate of strength for cohesive soils in a freshly excavated cut, trench, or pit. In situ tests such as the standard penetration test, vane shear, dilatometer, pressuremeter, and shear wave velocity test can be performed in conventional boreholes as part of a field investigation. These techniques will be explained in more detail later in the section of this chapter called field tests.

The mechanism of *bearing capacity* failure is well documented and is described in detail in any text on shallow foundation design. While more detailed analyses address the finer aspects and contributions of irregular loads, footing shapes, slopes, and so forth, the fundamentals of foundation bearing capacity are dependent on size, shape, depth, and rigidity of a footing transmitting a level of applied stress to the supporting soil with respect to available resisting shear strength of the soil. A simplified schematic of a general soil-bearing failure beneath a spread footing is provided in Figure 3.7. Bearing failure occurs when the shear strength of the soil is exceeded by the stress imparted to the soil by an applied load. For the case of a shallow spread footing as depicted in Figure 3.7, shear failure occurs along a two- or three-dimensional surface in the subsurface beneath application of the load.

Slope stability may be simply described as the comparison of available resisting soil shear strength to the stresses applied by gravitational forces, and in more complicated situations, by water or seepage forces. Of course, there may be many more complexities involved, including geometry, soil variability, live or transient loads, dynamic loads, and so on, but in the context of soil improvement, any methods that increase the shear resistance of the soil along a potential shear surface beneath a slope will add to the stability. There are many applications of improvements and modifications that can solve a variety of slope stability issues. A simplified schematic of a slope stability failure is provided in Figure 3.8, which depicts a theoretical circular slip surface.

Liquefaction is an extreme and often catastrophic shear strength failure usually caused by dynamic loading, such as from an earthquake. When a soil loses shear strength as a result of liquefaction, a variety of related shear strength failures may occur, including bearing failures, slope stability failures, settlement, and *lateral spreading*. Examples of liquefaction-induced failures are presented in Figures 3.9–3.11. Several of the available soil and ground

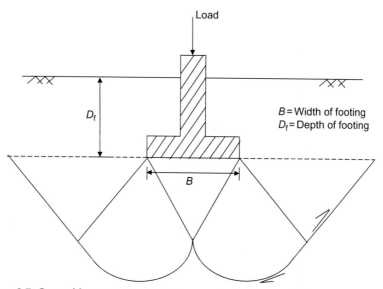

Figure 3.7 General bearing capacity failure mechanism.

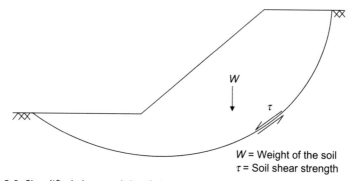

Figure 3.8 Simplified slope stability failure mechanism.

Figure 3.9 Liquefaction-induced bearing capacity failure. *Courtesy of GEER.*

Figure 3.10 Liquefaction-induced slope stability failure, San Fernando Dam. *Courtesy of EERC.*

improvement applications are intended to mitigate liquefaction that may result from seismic (earthquake) events. It therefore seems appropriate to provide an overview of liquefaction phenomenon and ground or soil conditions that provide susceptibility to this type of soil failure. Liquefaction is a soil state that occurs when loose, saturated, "undrained," cohesionless soil is subjected to dynamic loading or other cyclic loading that could result in the generation of pore water pressure. The conditions stated here show that a number of variables are involved, and all conditions are necessary to initiate liquefaction. To explain the phenomenon of liquefaction, consider the

Figure 3.11 Liquefaction-induced lateral spreading. *Courtesy of GEER.*

equation for effective stress given as Equation (3.3). We see that for a cohesionless soil ($c' = 0$), shear strength is directly proportional to the effective confining stress σ'. Given that effective stress is a function of the pore water pressure (u),

$$\sigma' = \sigma - u \qquad (3.4)$$

where σ' is the effective confining stress, σ the total confining stress, u the pore water pressure.

When the pore water pressure rises to near or equal to the total confining stress (σ), as can result from dynamic loading (or from other circumstances effectuating an undrained condition), then the effective stress approaches zero. At this point, the soil shear strength also approaches zero such that all soil response (or capacity) that relies on soil shear strength (i.e., bearing capacity, slope stability, lateral stability, resistance to uplift forces, etc.) is compromised. In fact, much of the resulting damage from earthquakes has been a direct result of soil liquefaction, especially in coastal areas, ports, and harbor facilities and properties. So, in order to improve soil to resist liquefaction, one or more of the necessary conditions (loose, saturated, undrained, cohesionless) must be eliminated. This can be done by (1) densifying the soil (shearing of dense soil tends to generate negative pore pressures), (2) draining the soil or providing for adequate drainage so that generation of positive pore pressures will be prevented, or (3) adding "cohesion" to the soil with a cementing agent.

Collapse due to saturation: When unsaturated, cohesionless soils become saturated due to submersion, surface infiltration, or rising groundwater, sudden settlement may occur. This type of behavior is most prominent in uniform sands and is often found in arid regions, windblown deposits (loess), and alluvial fans where soil grains are deposited in a low-energy environment. One mechanism postulated to cause this phenomenon is the loss of capillary tension ("apparent cohesion" caused by negative pore water pressures). As demonstrated by Equation (3.4), negative pore pressure increases the effective stress, thereby increasing strength. Additional water will reverse the negative pore pressure, leading to a rapid loss of strength. The amount of settlement may be as much as 5-10% in loose sands, but may be only 1-2% in dense sands (Hausmann, 1990). Understanding this behavior may be useful in designing improvement techniques to eliminate or reduce the impacts of this phenomenon. Collapse can also occur due to the loss of cementing action when salt solids are leached from certain soils. Additional loads can also cause the collapse of a soil structure with or without the presence of water (Budhu, 2008).

Permeability: The measure or capacity of a fluid to flow through a porous medium such as soil is known as permeability (or hydraulic conductivity). Permeability is typically evaluated as a two-dimensional rate of flow that is critical in designing for drainage (including pumping and dewatering), filtering, or hydraulic barriers. While certainly related closely with grain size and grain size distribution, permeability is also strongly affected by density, grain arrangement (structure), confining stresses, and other variables. Of notable interest is that the magnitude of permeability varies more than any other soil property, most often reported by including order of magnitude. Also, it is typically anything but uniform in the field due to its truly three-dimensional nature, and the resulting effects on flow prediction can be one of the most difficult soil phenomena to accurately assess.

Several common applications of ground improvement address "improvements" in permeability (or drainage) of a soil. Improvements may be intended to reduce or increase permeability, depending on the desired end results. Many improvement techniques, such as densification, grouting, and use of admixtures, result in reducing permeability while achieving other desired properties (such as increased stiffness, strength, reduced compressibility, and swell). These are generally desired for stable earth structures, slopes, foundation soils, and hydraulic structures. On the other hand, where drainage is important or can improve stability by reducing water pressures and water content, other approaches can increase permeability and drainage characteristics. These will generally be covered in the section on hydraulic modification.

Filtering, seepage forces, and erosion: When water flows through the ground, the flow generates a *seepage force* that is a function of the *gradient* (*i*) of the flow. The gradient at any point in the ground is calculated as the head loss (Δh) due to the frictional drag as the water flows through a length (Δl) of its flow path through the ground, given as

$$i = \frac{\Delta h}{\Delta t} \tag{3.5}$$

If the gradient is too high, then the seepage force may become greater than the static force holding the ground (soil grains) in place, resulting in an unstable condition. This is especially problematic where the water exits a body of soil, as it can dislodge soil particles without resistance downstream of the flow, but can also exist internally in a soil body. If allowed to go unchecked, this condition can lead to a condition known as *piping* (or *internal erosion*), which has been attributed to a number of major catastrophic failures. One high-profile example of piping was the catastrophic failure of the Teton Dam on June 5, 1976, during its initial filling (Figure 3.12). In this case, the time between first reported seepage through the compacted earth dam structure (approximately 9 a.m.) and full breach of the 100-m (305-ft) high dam (at 11:57 a.m.), was a mere 3 h. Once breached, the nearly full reservoir released approximately 308,000,000 m^3 (250,000 ac-ft) of storage over the next 5 h, flooding three towns, causing over \$1 billion in damage, and killing 11 people.

Two common approaches to mitigating high gradients and internal erosion are to either lengthen the seepage flow path, thus reducing the gradient

Figure 3.12 Failure of the Teton Dam, June 1976. *Photo by Mrs. Eunice Olsen.*

for the same head difference, and/or to filter the water as it progresses through the ground by retaining the "upstream" soil while allowing the water to freely flow towards the downstream or outlet or exit. These improvement methodologies will be discussed in the sections regarding hydraulic modification, which may include redirection of the flow to reduce gradients, or filtering (with either natural soil or geosynthetic filters).

Compressibility: When a load is applied to a soil, there will be a volumetric, contractive response of that material. If the amount of that response is significant, it may be critical to the functionality of a constructed project. Compressibility may be evaluated as a relationship between stress and deformation, and may include either or both elastic and inelastic components. The amount of deformation under an applied load is directly related to the amount of settlement that a constructed project may experience. For nearly all projects constructed in or on the ground, the expected amount of settlement (or ground deformation) is an important consideration. This response is often most critical for saturated clays, which may exhibit excessive settlement as water is expelled from the soil under pressure, a phenomenon known as *consolidation*. In fact, settlement is often one of the governing design criteria for a project. For consolidation settlement, it is also important to be able to estimate the rate of consolidation as well as the total amount of settlement. The difficulty in accurate prediction of time rate of settlement is an extended consequence of predicting the rate of three-dimensional fluid flow through the ground (permeability).

The acceptable amount of settlement that can be tolerated may vary greatly depending on the characteristics of the load or structure placed over the compressed soil. Extreme cases include very small tolerances of less than 0.25 mm (0.01 in) for the case of foundations for precision equipment, to several meters for certain storage tanks or earth embankments for which large settlement displacements will not adversely affect the functionality and performance of the structure or component of a project. Another factor that must be considered is *differential* settlement, where the vertical settlement of the ground varies over relatively short lateral distances. This can lead to excessive tilting or structural damage.

The total amount of settlement expected may be composed of three parts as expressed by

$$S_T = S_e + S_c + S_s \tag{3.6}$$

where S_T is the total settlement, S_e the elastic (immediate) settlement, S_c the (primary) consolidation settlement, S_s the secondary (consolidation) settlement.

Elastic settlement occurs "immediately" upon application of a load without a change in the water content. The amount of expected elastic settlement can be calculated given the soil parameters and an accurate representation of the load application (e.g., foundation stiffness, load distribution, etc.). The equation for elastic soil settlement in sand is

$$S_e = \Delta\sigma \times B \left[\frac{1 - \mu_s^2}{E_s} \right] I_p \tag{3.7}$$

where S_e is the elastic settlement, $\Delta\sigma$ the net vertical pressure applied, B the nominal (smallest) width of applied (foundation) load the net vertical pressure applied, μ_s the Poisson's ratio for the soil, E_s the (Young's) modulus of elasticity, I_p the influence factor (nondimensional).

Consolidation settlement occurs when the structure of a saturated soil is compressed as pore water is expelled over time from the low permeability soil. As a consequence of the low permeability of the soil, consolidation is very much time dependent and may take many years to be mostly complete. In the field, this phenomenon is actually a complex, three-dimensional problem. But as the basic input parameters are so varied and difficult to accurately evaluate, it usually does not make sense to attempt more complex estimation models that will not likely add to accuracy. In fact, because the value of permeability (k) can vary so widely and is difficult to accurately estimate, the difficulty of estimating time rate of consolidation is even more complex, as it is compounded by including the uncertainty of k. The traditional and still most widely accepted means of consolidation evaluation is based on the Terzaghi 1D theory and laboratory consolidation testing. This analysis assumes one-dimensional (vertical) pore water flow and settlement, and assumes a parabolically slowing rate of consolidation from instantaneous at 0% consolidation to infinite as consolidation approaches 100%. In the conventional, one-dimensional consolidation test (ASTM D2435), a saturated soil specimen is incrementally loaded in a "stiff" (horizontally resistant) ring so that all deformation is vertical (Figure 3.13). The vertical deformation is measured as a function of time for each load increment until the deformation rate becomes very slow. The time rate of consolidation is determined from the deformation versus time data. From a plot of this data, the *coefficient of (vertical) consolidation* (c_v) can be determined. As it has been recognized that stress–strain results may be strain rate dependent, a variation of the 1D test (ASTM D4186) provides for testing with limited strain rates to alleviate any introduction of errors due to high strain rates.

(a)

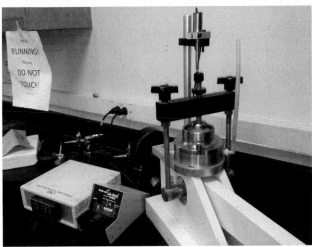

(b)

Figure 3.13 Consolidation test equipment for measuring soil compression. (a) Sample test ring. (b) Complete test apparatus setup.

So, there are essentially two major components of consolidation settlement evaluation: (1) How much total consolidation settlement will occur? and (2) How long will it take? Answers to these questions are not easy. We will examine each of these questions separately.

3.1.2.1 How Much?

To estimate the total amount of expected settlement, the total amount of deformation for each load increment is recorded. A plot can then be made of the deformation as a function of "load" or applied stress. If the soil horizon in question has been previously stressed to a level greater than the current stress level, known as the maximum past pressure (σ'_p), it is considered to be *overconsolidated*, such that the first portion of load will cause a smaller amount of settlement for a given load increase. This portion is evaluated by the reloading index (C_r). After the maximum past stress level has been reached, the settlement is calculated at the greater virgin compression index (C_c). Estimation of the total amount of "how much" settlement will occur is then calculated from coefficients $(C_c$ and $C_r)$, determined from the deformation versus load plot using actual initial stress (σ'_o), maximum past pressure (σ'_p), and net applied load $(\Delta\sigma')$, at the point where consolidation settlement is to be calculated. Total expected consolidation settlement can then be calculated as

$$S_c = \frac{H_o}{1 + e_o}\left[\left(C_r \times \log\frac{\sigma'_p}{\sigma'_o}\right) + \left(C_c \times \log\left(\frac{\sigma'_o + \Delta\sigma'}{\sigma'_o}\right)\right)\right] \qquad (3.8)$$

Secondary consolidation is compression that occurs in a soil after the completion of primary consolidation. The amount of settlement that may be contributed by secondary consolidation may range from less than 10% to nearly all of the total settlement, and is most prevalent in highly organic soils (i.e., peat). It can be measured from laboratory test data (and/or from field measurements) as the slope of the settlement versus log time curve and may extend for very long periods of time.

Some have attributed secondary compression to a plastic adjustment of soil fabric in cohesive soils or a delayed compression due to either release of previously bounded water or diffusion of water bound by organic membranes in organic materials (Mesri, 1973; Mesri et al., 1997).

3.1.2.2 How Long?

In order to evaluate the time to reach a certain percent of the total (or ultimate) amount of consolidation settlement, there is a dimensionless *time factor* (T) for all soils that is based on the soil-dependent c_v, and the maximum drainage length (L_{dr}) (maximum distance that dissipating water would have to travel to drainage). The relationship between the time factor and actual time to reach a certain percent consolidation is calculated as

$$T_n = \frac{t_n \times c_v}{L_{dr}^2} \qquad (3.9)$$

where T_n is the time factor for n% consolidation (dimensionless), t_n the time to achieve n% consolidation (year), c_v the coefficient of consolidation for a soil at appropriate stress level (m^2/year), L_{dr} the maximum drainage length (m).

The time factor, as described here, is technically the *vertical* time factor because drainage is assumed to be vertical for the 1D consolidation case. Values of T commonly used for analyses are $T_{50} = 0.197$ and $T_{90} = 0.848$, referring to 50% and 90% of total calculated consolidation. The coefficient of consolidation, c_v, typically lies between the values of 0.4 and 10.0 m^2/year. Actual values may vary between 0.1 and 20 m^2/year or more.

Example: A 30 m thick layer of clay over "impermeable" bedrock (single drained), with a $c_v = 1.0$, will achieve 90% consolidation in

$$t_{90} = \frac{T_{90} \times L_{dr}^2}{c_v} = \frac{0.848 \times 30 \text{m}^2}{1.0} = 763 \text{ year}$$

Some historically significant examples of buildings having experienced excessive consolidation settlement include the ("Leaning") Tower of Pisa in Italy and the Palacio de las Bellas Artes in Mexico City. In both cases, the architects' designs resulted in extremely high loads being applied to highly compressible clay soils. The infamous "tilt" of the Tower of Pisa is a result of differential settlement, where the amount and rate of settlement was different beneath opposing sides of the tower. This effect was exacerbated by attempts during construction to "correct" the tilting by adding height (and subsequently more weight) to the tilting side (where consolidation settlement was worst!).

The Palacio de las Bellas Artes has "sunk" more than 4 m, and is now more than 2 m below street level. Where the building was originally designed with steps up to the first floor, visitors now must walk down a set of steps to reach the same entrance to the building (Lambe and Whitman, 1969). Fortunately, the settlement of the building was extraordinarily uniform, such that no major damage occurred, and differential settlement was minimal. The building has been occupied continuously since its completion.

There are a variety of different types of ground improvement schemes to reduce and/or minimize the impacts of consolidation settlement. These would all be considered in situ deep improvements that may include deep densification, forced consolidation and drainage, gravel columns, or other ground "stiffening" schemes.

For a complete explanation of the procedure as well as the theories of consolidation settlement analyses, consult a fundamental soil mechanics or foundation textbook.

Volume stability (shrink/swell): One of the most destructive "quiet" phenomena is a result of the damage that can occur from soils that undergo significant volume change. This phenomenon is commonly evaluated by testing for swell potential of a compacted soil or, in some cases, an "undisturbed" sample taken from the field (ASTM D4546). Certain soil types and conditions can cause severe damage to constructed projects. For instance, damage to roadways as a result of soil expansion due to high swell clay soils (or frost heave susceptibility) has resulted in millions of dollars of necessary repairs each year in the United States alone. Even heavy structures may sustain critical damage when constructed in expansive soils, as the tendency to expand under high confining pressures results in the buildup of very high pressures capable of irreparable structural damage, sometimes requiring demolition of structures.

Stiffness/Modulus: For some applications (e.g., roadways), a higher stiffness (or soil modulus) is desired. This is generally evaluated by laboratory or field tests that measure deformation as a function of load. The measurements may be made for a single loading or for repeated loading cycles, and may be of interest for compacted samples that have been stabilized and/or cured after various periods of time. Soil stiffness may be desirable when deformations (particularly for applications expected to undergo repeated loading cycles) could be detrimental to long-term stability or durability. Soil stiffness, as expressed by elastic modulus, is used to predict immediate settlement of foundations.

These fundamental ground response characteristics are therefore the basis for many, if not most, geotechnical designs. Certain values for each must be achieved in order to provide adequate performance of the ground, whether in situ or constructed as engineered material. Fortunately, there are methods of soil and ground improvement that can dramatically alter each of these characteristics so that acceptable engineering property values and responses can be achieved.

In addition to the response of completed projects, other soil properties and site characteristics may be important either prior to or during construction. These include site accessibility, soil workability, (lateral) support and dewatering requirements (temporary and/or permanent), and so forth. For these cases, there are also ground improvement techniques that can be used to solve each situation depending on specific requirements and needs.

3.2 SITE INVESTIGATIONS, DATA, AND REPORTS

Before any attempt can be made to design construction with, on, or in the ground, certain information about the subsurface must be gathered, analyzed, and reported in such a fashion that it can be used to adequately design a project. The detail and depth of investigation and exploration will depend on the size and type of project, and the geotechnical parameters analyzed will be depend on the function(s) and purpose(s) of the proposed geosystem. There are several steps that must be taken for any project. Once data is gathered and analyzed, a geotechnical report is prepared, where all observations, data, test results, and analyses are presented along with preliminary recommendations and discussion of potential difficulties or problems.

The main purposes of conducting a site investigation are

(1) To evaluate the suitability of the site for the proposed project. This may include a study of the environmental impact of the construction methods and completed project.

(2) To obtain physical and mechanical properties of the subsurface materials, in order to determine their suitability as they may affect the construction and performance of the project.

(3) To enable safe and economical design of the project components.

(4) To obtain groundwater conditions as they may pertain to the project.

(5) To identify any potential problems or difficulties with the ground conditions that may affect construction or performance of the proposed project.

The extent and detail of a field investigation and subsequent geotechnical report are, of course, going to be relative to the size, type, tolerances, and critical nature of the project. In general, a geotechnical report will be prepared for all soil or site investigations, from small residential housing to large, critical civil engineering projects (emergency facilities, nuclear power plants, large bridges, etc.). There is almost always a tendency to attempt to trim costs on a project, and unfortunately, the site exploration is often seen as a target for cost cutting. In reality, hindsight has shown that even if initial costs are high, the intangible geotechnical information and data that may be omitted from a field investigation may later end up costing much more to the project due to overruns, redesign, remediation, or even possible distress or failure. The time, effort, and cost put in "up front" should be considered an investment that may be critical to construction costs and functionality of a project. *Rule of thumb #1*: "Don't skimp on field investigations."

A site investigation program may be thought of having a number of reasonably distinct phases: desk study and site visit, subsurface exploration, laboratory testing and analyses, and reporting. The detail to each phase will be dependent on the scale and critical nature of the project, and not all phases will be included in all investigations.

3.2.1 Desk Study and Site Visit

A *desk study* involves collecting all available existing information for the site (or adjacent or nearby property), including maps, reports, rainfall data, historical information (e.g., from newspapers or internet), photographs, and any previous site or construction plans. This information is compiled and sorted to provide an initial overview of the site conditions. If enough detail is available pertaining to the subsurface materials, the information can be used to accelerate and/or supplement the planning of subsequent investigation and subsurface exploration.

A site visit (or reconnaissance) allows the engineer to get a better "feel" for the topography, local geology, vegetation, and other aspects of the site and surrounding areas such as access, necessary site preparation, surface working conditions, available utilities, existing infrastructure, and so on.

3.2.2 Subsurface Exploration and Borings

Exploration of the subsurface other than field tests (described in the next section) may be undertaken in a number of ways. Shallow soils can be explored by means of open *test pits* typically created by a backhoe, shallow hand sampling with shovels, hand augers, or hand driven samplers. For investigation of deeper soil strata, depth to bedrock, or depth to the water table, exploratory drill holes (or "borings") are typically made. There can be several purposes of drilling exploratory borings at a site. During drilling of borings, careful observations can be made continuously by an experienced field engineer and entered into a *boring log*. Much can be ascertained by watching the cuttings coming from a boring, including change in soil type, change in moisture level (maybe as a result of intercepting the water table), or change in difficulty (or ease) of drilling. Ultimately, enough information should be gathered so that a reasonable estimate of the subsurface profile can be approximated.

There is always a question regarding the number of and depth to which exploratory borings should be made. The ideal scenario is that borings should be advanced to the depth below which there will be no significant loads applied. Unfortunately, this estimation is often difficult, as this depth depends largely on the soil profile and soil properties that exist. *Rule of thumb #2*: "Boreholes should go deeper than you thought." There are a number of published "general guidelines" on the minimum number of borings that should be made based on size and type of structure to be built.

3.2.3 Field Tests

As part of most field investigations, a variety of in situ tests may be performed to gather information and data regarding the ground response parameters at the site. In situ field tests alleviate some of the problems associated with sample disturbance, scaling effects, anisotropy, and 3D effects.

Exploratory borings will typically include one or more field tests. These may include the Standard Penetration Test (SPT; ASTM D1586), or Cone Penetration Test (CPT; ASTM D5778), flat plate Dilatometer Test (DMT; ASTM D6635), Pressuremeter Test (PMT; ASTM D4719), and Vane Shear Test (VST; ASTM D2573). Each type of test applies different types of loadings to the soil to measure corresponding soil response in attempts to evaluate various material characteristics. Boreholes are required for conducting SPT and common versions of the PMT and VST. No borings are required for CPT and DMT (and some versions of PMT), thus providing some obvious advantages in time, equipment, and costs. A disadvantage of the "push-in" methods is that they may not be applicable where hard or cemented layers are present, or where cobbles or boulders may prevent penetration.

SPT and CPT provide means of measuring the resistance of advancing the penetrometers through the soil strata. Penetration resistance can then be used to characterize certain soil parameters through well-developed correlations. Modern cones are now typically equipped to measure a number of other parameters. Dilatometers and pressuremeters are able to measure stress-strain parameters. Each of these tests is explained in more detail here.

The *Standard Penetration Test* (SPT; ASTM D1586) uses a standardized split-spoon sampler that is driven into the bottom of the boring, and measurements are made as it is advanced. The split-barrel type sampler (pictured in Figure 3.14) has a 50.8 mm (2 in) outside diameter, and has a 61 cm (24 in) split-center section, held together by a 76.2-mm-long (3 in) driving shoe at the tip and threaded connector behind. It is attached to the end of threaded driving rods, and struck with a standard impact hammer of 63.5 kg (140 lbs) falling 0.76 m (30 in) (Figure 3.15). Penetration resistance is measured as number of "blows" (N) with a theoretically standard level of energy

Figure 3.14 Split-spoon sampler empty and with a disturbed sample. *Courtesy of Prof. Horst Brandes.*

to advance a certain depth. The sampler is driven a total depth of 450 mm (18 in), with the number of blows counted and recorded for each 150 mm (6 in) interval (e.g., 13/15/17). As there is often disturbance and/or debris at the bottom of the open boring, the first increment is usually omitted and the "*N*-value" is reported as the total number of blows to penetrate the last 300 mm (12 in). SPT measurements are typically made at approximately 1.5 m (5 ft) depth intervals within a boring, sometimes more frequently near surface, and sometimes alternating with or substituted by "undisturbed" sampling at selected depths.

Figure 3.15 Standard penetration test hammer. *Courtesy of Prof. Horst Brandes*

While this N-value does not have any direct meaning, there are many correlations between SPT N-values and other soil parameters, such as density, effective friction angle, liquefaction resistance, undrained shear strength, and so forth (Tables 3.3 and 3.4). The split–spoon sampler also allows for retrieval of a disturbed "grab sample" of soil so that the material through which the sampler was driven can be identified. Unfortunately, there is a lot of variation in application of this "standard" test, which can make the accuracy of results from this field test somewhat suspect unless all appropriate corrections are made.

Given that the resistance to penetration is going to increase with overburden stress (effective confining stress), a correction must be made to normalize measured blow counts. The standard is to normalize to 96 kN/m^2

Table 3.3 Approximate Correlations for SPT N-values and CPT Cone Resistance for Coarse-Grained Soil (www.oce.uri.edu)

Density Description	Relative Density (%)	Corrected N-value	Static Cone Resistance, q_u (kg/cm^2)	Effective Friction Angle (°)
Very loose	<20	0–4	0–20	<30
Loose	20–40	4–10	20–40	30–35
Medium	40–60	10–30	40–120	35–40
Dense	60–80	30–50	120–200	40–45
Very dense	>80	>50	>200	>45

Table 3.4 Approximate Correlations for SPT N-Values with Undrained Shear Strength for Fine-Grained Soil (www.oce.uri.edu)

Consistency Description	Corrected N-Value	Undrained Shear Strength, S_u (kPa)	Undrained Shear Strength, S_u (psf)
Very soft	<2	<12	<250
Soft	2–4	12–25	250–500
Medium	4–8	25–50	500–1000
Stiff	8–15	50–100	1000–2000
Very stiff	15–30	100–200	2000–4000
Hard	>30	>200	>4000

(1 tsf) effective overburden pressure by applying an overburden correction factor, C_N. Often, the correction factor now uses a normalizing stress equal to one atmosphere, such that

$$C_N = \left(\frac{P_a}{\sigma'_{vo}}\right)^{0.5} \text{ not to exceed 2.0} \tag{3.10}$$

where P_a is the atmospheric pressure (101 kPa), σ'_{vo} the effective vertical overburden stress at a given depth (stress in atm).

Another of the major variables in the test is energy efficiency of the actual hammer blows. This variable is highly dependent on the equipment used and the operator. Different types of equipment deliver different amounts of the theoretical energy based on weight and fall height due to friction and eccentric loading. The standard is now to correct equipment with different amounts of the maximum theoretical free-fall energy to 60% of the maximum by applying an energy correction factor, C_E.

$$C_E = \frac{ER}{60} \tag{3.11}$$

where ER is the energy ratio (typically 60-85 for safety hammer, 30-60 for donut hammer, 85-100 for automatic hammer; ASTM D4633).

Additional correction factors that may be included (but are not always used) include corrections for oversized borehole diameters (>120 mm), short rod lengths (<10 m), use of the split-spoon sampler without a liner, and so on. So the complete, "corrected" blow count to be reported should be

$$N_{1(60)} = N_m C_N C_E C_B C_R C_S \qquad (3.12)$$

where N_m is the measured field blow count, C_N the overburden correction factor, C_E the energy correction factor, C_B the borehole diameter correction factor (1.0-1.15), C_R the rod length correction factor (0.75-0.95), C_S the sampling method (w/o liner) correction factor (1.1-1.3).

The *Cone Penetration Test* (CPT; ASTM D5778) uses a standardized probe with a 60° conical tip to measure the penetration "tip" resistance (q_t) or undrained cone resistance (q_u) in the subsurface. Cones now come in a variety of sizes and may be equipped with various instrumentation in addition to reading tip resistance. Figure 3.16 shows an assortment of CPT cones. The cone is typically hydraulically pushed at a steady rate of 2 cm/s into the ground from the surface. The CPT can be used in a variety of soil types from very soft clay to dense sands, but is not applicable for gravelly or rock soils. While the database of CPT correlations is growing, there is much greater abundance of data for SPT *N*-values. Because of this, CPT

Figure 3.16 Assortment of cones used for cone penetration test. *Courtesy of Prof. Paul Mayne.*

resistance is sometimes first correlated to SPT values, then SPT correlations are used. Some correlations for q_u are also provided in Table 3.3.

One distinct advantage of the CPT with respect to SPT is that it provides a continuous record with depth as it is advanced through the subsurface strata. In addition, the relatively high speed of CPT testing as compared to most all other in situ field tests allows for a large number of tests to be done, providing a much more complete record over a greater areal extent. A number of variations or modifications have been made to the CPT by adding sensors or transducers to the probe. These include means to measure side friction (f_s), pore water pressures (u_b), and even shear wave velocities (V_s). When pore pressure measurements are made, the test is referred to as CPTU. When the equipment is configured to receive shear wave measurements, it is referred to as a "seismic" CPT or SCPT. If both additions are included, it may be considered a SCPTU. A disadvantage of the CPT test is that no sample is retrieved, although results can also be correlated with those of the SPT test, so that a combination of the two tests may be desirable. The CPT test also requires overburden correction and modifications to tip and sleeve friction for pore water pressures, but is not influenced by many of the other issues that must be addressed for the SPT. Given tip resistance, sleeve friction, and pore pressure readings, soils can be classified by behavior (Figure 3.17).

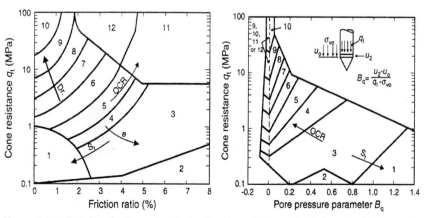

Figure 3.17 CPT classification of soil behavior types. (1) Sensitive fine grained, (2) organic material, (3) clay, (4) silty clay to clay, (5) clayey silt to silty clay, (6) sandy silt to silty sand, (7) silty sand to sandy silt, (8) sand to silty sand, (9) sand, (10) gravelly sand to sand, (11) very stiff fine grained (overconsolidated or cemented), (12) sand to clayey sand (overconsolidated or cemented). *After Campanella and Robertson (1988).*

The flat plate *Dilatometer Test* (DMT; ASTM D6635) is a rapid field test that involves the penetration of a flat steel plate with a sharp leading edge equipped with an expandable membrane that measures stress-strain of the soil in one dimension (Figure 3.18). There is no boring required and therefore no disturbance except for possible remolding of very soft soils resulting from penetration of the instrument. Test results are useful for evaluating stratigraphy, homogeneity of soil layers, identification of voids and/or cavities, and other discontinuities. The DMT allows for both penetration resistance *and* stress-strain measurements at selected depths that can be used for classification of soil types, estimation of compressibility, pore water pressure, and so forth.

The *Pressuremeter Test* (PMT; ASTM D4719) is another type of field test that provides stress-strain response measurements of in situ soils. The ASTM standard specifies placement of the measuring sensor cell in a predrilled boring between two guard cells (Figure 3.19). As opposed to the 1D expansion of the dilatometer membrane, the pressuremeter membrane expands radially in two dimensions. The test results are commonly used for foundation design and other geotechnical analyses requiring stiffness or stress-strain moduli. It has been noted that the accuracy of pressuremeter results are dependent on the degree of disturbance from drilling of the boring.

The *Vane Shear Test* (VST; ASTM D2573) provides a direct indicator of in situ undrained shear strength for silts, clays, and other fine-grained geomaterials such as mine tailings. The test is performed by pushing a mechanical vane into the bottom of a boring, and measuring the torque needed to rotate the vane (and thus shearing the soil). Vanes come in a variety of sizes to be used with different ranges of soil strengths. Figure 3.20 shows a photo of

Figure 3.18 Flat dilatometer.

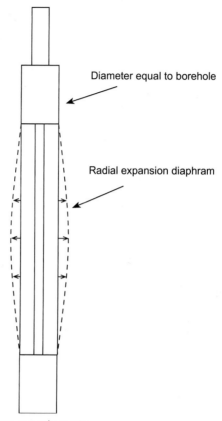

Figure 3.19 Pressuremeter schematic.

typical field vanes. The test is generally applicable to soils with undrained shear strengths of less than 200 kPa (approx. 2 tsf). The test results are applicable for designs that may experience rapid loading. Vane shear tests are usually conducted in combination with other field and lab tests to aid in accurate correlations. While manual vanes have been commonly used for over 30 years, automated operation and data acquisition is now available.

In addition to these common tests applied at single geographic points, a number of *geophysical methods* may be used to survey lateral expanses of subsurface materials across sites. These include surveys by ground penetrating radar, electrical resistivity, magnetic measurements, and gravity techniques. Seismic methods are another variety of geophysical measurements that will be addressed separately. Geophysical methods have a number of advantages: (1) they are (for the most part) nondestructive; (2) they tend to be fast, which

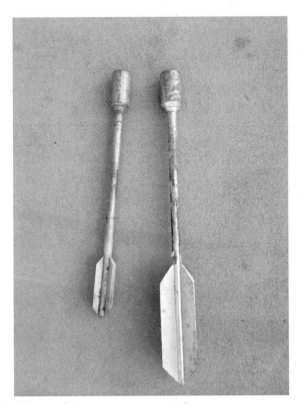

Figure 3.20 Field vane photo.

can be economical (once the equipment has been purchased); and (3) they are generally not limited to specific soil types and may be applied to both soils and rock. The disadvantages to using geophysical methods are that (1) no samples are collected; (2) there are no physical penetration measurements; (3) interpretation often relies on assumed models and experience; and (4) results may be influenced by water (especially salt water), clay content, and depth (Mayne et al., 2001).

Seismic methods measure the propagation of *mechanical waves* through the ground, which can be used to determine layering, density, stiffness, and damping. Considering the tests usually induce only very small strain levels, the measurements are essentially of elastic parameters. Typical seismic field tests include measurements of compression waves, V_p, and shear waves, V_s. These mechanical wave measurements are included in seismic refraction surveys, crosshole, downhole, and spectral analysis of surface wave tests. Compression waves vary between different densities of geomaterials from approximately 6000 m/s for intact rock to 1000 m/s for loose or soft soils. But V_s values are most valuable as indicators of stiffness and density for soils. Crosshole seismic testing allows direct measurement of individual layers at specified depths, but requires drilling of boreholes. As previously mentioned, some cone penetrometers may also be configured with seismic wave receivers, thus eliminating the need for borings, but wave travel paths will generally be from a surface source to the instrument location (depth).

Field permeability tests of in situ materials may be desired in some cases, where it is critical or important to performance of the project. There are a number of methods of performing field permeability tests. These include pumping tests—where the flow of water entering a boring is measured by pumping water from a boring, often in conjunction with monitoring the water level in adjacent observation wells—and infiltration tests (ASTM D6391) where water is pumped into the ground, either by gravity or under pressure, allowing permeability to be calculated from the quantity of flow that enters the ground from a boring well. While these procedures have a number of limitations, the advantage of in situ tests is that 3D effects, soil anisotropy, and sampling disturbance are summarily addressed.

3.2.4 Sampling and Laboratory Testing

Sampling of soils from a proposed construction site is often done in order to obtain material for identification of properties and testing of material behavior to be used in developing geotechnical designs. There are a wide variety of sampling techniques that range from very crude and inexpensive to painstakingly complex and expensive. The type of sampling done may be commensurate with the detail of accuracy needed for evaluation and designs, and/or the overall magnitude and critical nature of the project. It is always recommended that at least one boring should include continuous sampling if thin seams or lenses or variable material may be present that would affect the performance of the project construction.

Grab or bag samples are *disturbed* samples of material collected from a site that can be evaluated and tested for index properties such as grain size analyses, Atterberg limits, specific gravity, and so on. These types of samples can be obtained from shallow test pits opened by a backhoe or small excavator. Disturbed samples can also be obtained from split-spoon samplers used for the SPT test or from other downhole test samples not intended for "intact" lab tests. When samples are needed for testing of in situ field behavior, "undisturbed" samples can be obtained using thin-walled tubes typically pushed into the ground through the bottom of drilled borings (Figure 3.21). All of these types of samples will actually have some level of disturbance, but measures can be taken to minimize sample disturbance through added care (and cost). Thin-walled samples of cohesive materials may provide quality samples for laboratory testing, but undisturbed, cohesionless materials are more difficult to obtain, especially if dry or saturated. Very high quality, undisturbed samples and samples of cohesionless materials may be obtained either by "hand-carved" blocks or by freezing the soil and extracting the frozen mass of soil to retain in situ structure and density.

Depending on the type of project, subsurface soils encountered, expected loads, and geotechnical behavior tolerances required, a range of laboratory tests may be conducted on soil samples retrieved from the field site. Aside from classification and index property tests, which can be made on disturbed samples, strength, compressibility, permeability, and durability tests, and so on may be performed on collected, undisturbed samples in order to derive parameters needed for evaluation of expected performance or to enable proper design parameters through the use of engineering analyses.

3.2.5 Typical Geotechnical Soils Report

For each project where a soil or site investigation is performed, a geotechnical soils report is usually generated to present all of the information gathered and analyzed for the conditions encountered at the project site. A geotechnical report should include a detailed description of the site from the information collected from the desk study and site visit, subsurface

Figure 3.21 Thin-walled Shelby tube sampler.

exploration, field tests, laboratory tests, and any analyses made. All of the subsurface information for each soil boring, including observations made during drilling, field tests and subsequent lab tests, interception of a water table, and so forth, should be reported in a boring log (Figure 3.22). A written summary of the subsurface conditions should accompany the boring logs. The report should provide a general description of the project, make recommendations for geotechnical aspects of the project, point out any potential problems or difficulties, and describe all limitations of the field investigation, testing and analyses, and recommendations made.

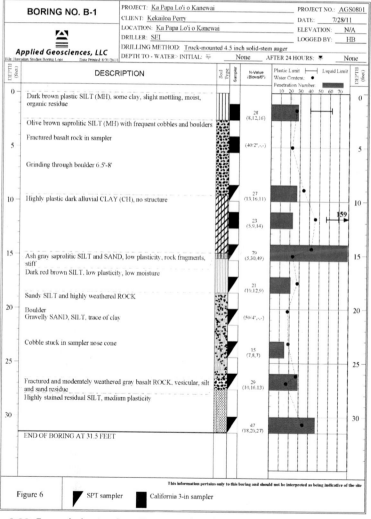

Figure 3.22 Example boring log. *Courtesy of Prof. Horst Brandes.*

3.3 PRELIMINARY MODIFICATION DESIGN EVALUATION

In order to conduct a preliminary evaluation to determine the suitability of a site for ground modification, one must consider a number of variables and must have adequate data with which to make informed decisions.

First, the geometric parameters and fundamental soil parameters of the subsurface strata must be adequately delineated. This includes classification of soil type(s) (including particle-size distributions and index properties as available or necessary), engineering properties, layer thicknesses, depth of groundwater, and so on. Strength, density, and stiffness may be determined by standardized field tests such as SPT, CPT, or PMT. Some geophysical tests may also prove to be useful tools for identifying certain properties. As previously described, the soil type or depth requirements alone may narrow the realm of available methods. Any available site history may also prove to be a source of valuable information to aid in the decision-making process or to improve on preliminary designs.

Effects of disturbance from noise, vibrations, or ground displacements from improvement processes on adjacent properties must be examined and considered. This is of particular interest and a concern in urban areas or adjacent to existing infrastructure or utilities. There may be restrictions that preclude the use of certain approaches or require additional attention to details or remediation.

Next, design requirements must be addressed. This may include maximum limitations on settlement, minimum required strength or bearing capacity values, and so forth. Finally, preliminary cost estimates must be made for the most promising improvement method candidates remaining after initial screening from the above assessments.

Once a preliminary choice (or two) have been selected, it is often prudent to perform a pilot study or test section to ensure the feasibility and/or effectiveness of the proposed improvement method. A full-scale field test can also aid in adjusting certain parameters and details of the improvement approach so that a more efficient improvement is made and quality can be enhanced.

RELEVANT ASTM STANDARDS

D1586-11 Standard Test Method for Standard Penetration Test (SPT) and Split-Barrel Sampling of Soils, V4.08
D2166/D2166M-13 Standard Test Method for Unconfined Compressive Strength of Cohesive Soil, V4.08

D2435/D2435M-11, Standard Test Methods for One-Dimensional Consolidation Properties of Soils Using Incremental Loading, V4.08

D2487-11 Standard Practice for Classification of Soils for Engineering Purposes (Unified Soil Classification System), V4.08

D3080/D3080M-11 Standard Test Method for Direct Shear Test of Soils Under Consolidated Conditions, V4.08

D3282-09 Standard Practice for Classification of Soils and Soil-Aggregate Mixtures for Highway Construction Purposes, V4.08

D4186/D4186M-11 Standard Test Methods for One-Dimensional Consolidation Properties of Saturated Cohesive Soils Using Controlled-Strain Loading, V4.08

D4318-10 Standard Test Methods for Liquid Limit, Plastic Limit, and Plasticity Index of Soils, V4.08

D4546-08 Standard Test Methods for One-Dimensional Swell or Collapse of Cohesive Soils, V4.08

D4633-10 Standard Test Method for Energy Measurement for Dynamic Penetrometers, V4.08

D4719-07 Standard Test Methods for Prebored Pressuremeter Testing in Soils, V4.08

D5778-12 Standard Test Method for Electronic Friction Cone and Piezocone Penetration Testing of Soils, V4.08

D6391-11 Standard Test Method for Field Measurement of Hydraulic Conductivity Using Borehole Infiltration, V4.09

D6528-07 Standard Test Method for Consolidated Undrained Direct Simple Shear Testing of Cohesive Soils, V4.08

D6635-01 Standard Test Method for Performing the Flat Plate Dilatometer, V4.08

D7181-11 Standard Test Method for Consolidated Drained Triaxial Compression Test for Soils, V4.08

REFERENCES

Budhu, M., 2008. Foundations and Earth Retaining Structures. John Wiley & Sons, Inc., Hoboken, NJ, 483 pp.

Campanella, R.G., Robertson, P.K., 1988. Current status of the piezocone test. Keynote paper. In: 1st International Symposium on Penetration Testing, ISSMFE, Balkema, pp. 93–117.

Das, B.M., 2010. Principles of Geotechnical Engineering, seventh ed. Cengage Learning, 666 pp.

Hausmann, M.R., 1990. Engineering Principles of Ground Modification. McGraw-Hill, Inc., 632 pp.

Lambe, T.W., Whitman, R.V., 1969. Soil Mechanics. John Wiley & Sons, 553 pp.
Mayne, P.W., Christopher, B.R., DeJong, J., 2001. Manual on Subsurface Investigations. National Highway Institute, Publication No. FHWA NHI-01-031, Federal Highway Administration, Washington, DC.
Mesri, G., 1973. Coefficient of secondary compression. J. Soil Mech. Found. Div., ASCE 99 (SM1), 123–137.
Mesri, G., Stark, T.D., Ajlouni, M.A., Chen, C.S., 1997. Secondary compression of peat with or without surcharge. J. Geotech. Geoenviron. Eng., ASCE 123 (5), 411–421.

SECTION II
Soil Densification

CHAPTER 4

Objectives and Improvements from Soil Densification

This chapter provides an overview of the objectives and improvements attained by densification of soil, including effects on fundamental soil engineering properties, basic geotechnical design, and special attention to liquefaction phenomenon. An explanation is made to discern the differences between shallow and deep densification. The fundamental differences between methodological processes used to densify different soil types is addressed along with an introduction to how different equipment can achieve these different densification processes. The ending sections of this chapter describe the effects of soil densification on each of the basic soil engineering behaviors that are important to design of various geotechnical projects.

4.1 OVERVIEW OF SOIL DENSIFICATION

Without much question, the most common method of soil and/or ground improvement is densification. Most fundamental, desired, engineering properties of soils can be achieved and/or improved by creating a denser packing of soil grains. These include soil shear strength (critical to foundation bearing capacity, slope stability, liquefaction mitigation, etc.), minimized compressibility and settlement, increased stiffness, resiliency and durability, reduced permeability, and so forth. Depending on the approach and equipment used, densification may be applicable to a wide range of soil types and site conditions, including soft fine-grained marine sediments, liquefiable sands, heterogeneous fills, sinkholes, municipal wastes, and even low-level nuclear waste (Schexnayder and Lukas, 1992a,b). In today's practice, "unusable" sites no longer exist.

In general, there are two fundamentally different categories of soil densification with important differences in mechanisms. *Compaction* is the process by which soil is densified by eliminating (or squeezing out) *air* from void space between grains. *Consolidation* is the process by which soil is densified by eliminating (or squeezing out) *water* from void space between grains. A big difference between the two is that compaction occurs almost

Soil Improvement and Ground Modification Methods
63

immediately after application of a load or densification process. Consolidation, on the other hand, is time dependent, and is a function of the soil permeability or rate that water will be expelled from the soil. Consolidation is also a function of length of travel path, in that the longer the travel path, the longer the time to consolidate.

4.1.1 Shallow vs. Deep Densification

There are many methods and technologies available for achieving increased soil density that will be described herein. But first it would be useful to differentiate between "shallow compaction" and "deep densification." *Shallow compaction* typically refers to processes where soil is worked at the ground surface or where material is placed and compacted in layers. This may involve compacting existing near surface soils in place, but more commonly refers to *engineered fill*, where soil is placed in controlled *lifts* (i.e., layers, typically 20-30 cm = 8-12 in thick) and compacted to achieve a minimum desired (usually specified) density. This type of application is often called earthwork construction and results in earth structures such as prepared soil foundations, engineered slopes and embankments (including earth dams and levees), transportation projects, and stable backfill to create level ground (e.g., behind retaining walls). *Deep densification* usually refers to in situ processes where existing subsurface soils are densified by a variety of methods such as *blasting, vibrocompaction, deep dynamic compaction, compaction grouting*, and others that will be described in Chapter 6. *Forced consolidation* techniques may be considered a method of dewatering, but for practical purposes may also be considered a form of in situ densification for fine-grained soils. One of the essential differences between deep densification of primarily granular soils and densification of saturated fine-grained soils by consolidation is the time required for consolidation, controlled largely by the permeability of the soils being densified. Ground improvement methods employing *preloading* or forced consolidation will be introduced here, but described and addressed in greater detail in Section 6.1.6 in Chapter 6 under the heading of Deep Densification and in Chapter 9, which is dedicated to this improvement application.

The approach taken in this text is to address each general category of densification separately. Shallow compaction will be covered in Chapter 5. Deep densification will be covered in Chapter 6. In each chapter, the objectives, methodological approaches, application and equipment choices, design specifications, and QA/QC will be described.

4.1.2 Processes and Equipment

When discussing or designing for densification applications, it is important to understand some of the basic processes of compaction techniques. One should know how the soil grains are physically rearranged during a compaction process and how the compaction energy is delivered. This will largely be a function of the soil type being densified and the equipment being used. There are a variety of densification processes that can be administered by means of different equipment and methods.

The most efficient way to compact granular, primarily cohesionless soils is with the assistance of vibrations. This is due to the fact that cohesionless soils attain all of their strength from friction between grains. The introduction of vibrational loads "shakes" the particles so that the frictional resistance between grains is overcome. When combined with static load and/or impact load, vibrations can help to attain high levels of compaction. A notable example of this is the use of a vibratory table for the maximum density test of cohesionless soils (ASTM D4253). Compaction equipment is available for both shallow and deep compaction processes that employ vibrations, principally through oscillatory motors with controllable frequencies of oscillation. For shallow compaction, vibratory rollers are available that apply both static load through their own weight combined with vibrations. The most effective vibration frequencies for clean sands have been found to be between approximately 25 and 30 Hz (Xanthakos et al., 1994). In situ densification of deep cohesionless deposits is often achieved through application of induced vibrations through specialized vibratory probes (vibroflot) or by other dynamic means. These will be further described in Chapter 6.

The application of static loads has been conventionally applied through a range of heavy, steel drum, tired, or tracked vehicles for shallow or surface compaction. An additional densification technique applying a static load is *preloading*, where large loads approximately equivalent to (or sometimes greater) than the final constructed project load are placed on a site to allow soil compression and settlement to occur prior to the actual construction, thus alleviating postconstruction distress. Static compaction is applicable for most soil types, but is most effective for use with well-graded and cohesive soils.

In addition to static and vibratory compaction (and combinations of the two), other loading methods that employ somewhat different or modified equipment include impact, tamping, kneading, and so forth. A more detailed description of shallow compaction equipment is provided in Chapter 5.

4.2 ENGINEERING IMPROVEMENT OBJECTIVES

As mentioned previously, several fundamental soil engineering properties may be enhanced by densification. Each of the major improvement objectives is addressed in the following sections.

4.2.1 Bearing Capacity, Strength, and Stiffness

One of the most important engineering properties of interest for design and performance of structures built on, of, or within the earth is soil strength. This generally refers to *shear strength*, as soils tend to fail in shear. Soil shear strength is fundamental to analyses and design for engineering use of earth materials. A background discussion of soil shear strength was provided in Chapter 3. Basic soil mechanics teaches us that the capacity of a soil to support bearing loads (bearing capacity), the ability of soil to stand up to lateral forces imposed by retaining walls, excavations, and so forth. (lateral earth pressures), and the stability of sloping ground or sloped earth structures (slope stability), all rely on the shear strength of soil.

As described in Section 3.1.2, soil shear strength is a limiting state of shear stress as a function of applied load. Theoretical values of strength can be determined as a combination of the combined components of cohesive and frictional strengths under a set of limiting stress conditions. Soil mechanics theory, further supported by laboratory tests, has shown that, for the same states of stress conditions, a soil with grains arranged in a tighter packing configuration (denser) will have higher frictional strength or greater frictional resistance. Thus, attaining a greater degree of density will generally result in increased shear strength, leading to greater bearing capacity, greater slope stability, ability to resist higher, lateral earth pressures, and so on.

Stiffness (or stress–strain behavior) will also generally increase with increased density. Stiffness is an important parameter for many engineering components where smaller tolerances on deformations are needed or desirable. Soil fabric or "structure" (arrangement of soil grains) may also account for variations in soil stiffness, especially for cohesive soils whose structure plays a critical role in response characteristics. For granular soils with more rounded or "bulky" grains, the soil "structure" is essentially just a matter of grain packing. Figure 4.1 depicts how a soil made up of rounded grains can be arranged into a denser state by simply packing grains in a closer configuration. For fine-grained cohesive soil, moisture (or water content) at the time of compaction and the compaction method can be vitally important in controlling the arrangement of soil grains. This is described in some detail

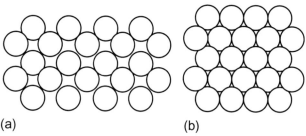

Figure 4.1 Packing arrangement of rounded "bulky" soil grains: (a) loose, (b) dense.

in Chapter 5. Research has indicated that some variation in stiffness may also occur for granular soils at low confining stresses compacted at different moisture levels (Carrier, 2000).

4.2.2 Compressibility and Settlement

As densification of a soil will reduce void space within the soil mass, it will also reduce the *compressibility* of the ground, as the soil structure will already be in a compressed state. The fundamental goal of reduced compressibility is to reduce future settlement of the ground under the load of built structures or of a prepared engineered site. Depending on soil type, depth, structure, and stress–deformation characteristics, soil compressibility (and hence the amount of expected settlement or deformation under load) may be comprised of elastic and/or inelastic components.

The compressibility of granular soils is directly related to soil density (or degree of compaction). For the most part, the settlement of granular soil is essentially "immediate" and for practical purposes is often considered elastic. Settlement of structures founded on granular soils can be estimated using relatively simple equations (Hooke's law) employing such soil parameters as the elastic (Young's) modulus (E_s), and Poisson's ratio (μ_s), as described in Chapter 3. Denser sands will have higher E_s and μ_s values and will therefore exhibit less settlement, as can be seen by examining Equation (3.4).

Thus, densification of sands and other mostly granular soils is therefore an "immediate" result unless saturated. As will be described later in Chapter 6, there are a number of deep densification techniques for granular soil with a maximum of approximately 15-20% "fines" (minus #200 material). But as will be discussed, some percentage of fines can actually improve the overall density and stiffness.

Compressibility and settlement of cohesive materials, particularly saturated clays, is often a controlling design parameter for many structures. As described in Chapter 3, the dominant portion of settlement in saturated clays occurs as a result of consolidation. Consolidation settlement can be significantly reduced by densifying these types of soils. This involves techniques that will preconsolidate the ground prior to construction and application of final load. Due to the typically low permeability of clay and the significant time required to consolidate these materials, especially when significant depths are encountered, densification techniques for these soil types often employ methods to expedite consolidation.

4.2.3 Permeability and Seepage

With the decrease in void space when a soil is densified, it intuitively follows that the permeability of the soil will be decreased. All else being held equal, this assumption is essentially true. But for cohesive soils or soils containing appreciable amounts of clay, the permeability and seepage rate will also be heavily dependent on soil structure. Fortunately, the structure of clayey soils can be controlled to a great degree by the compaction conditions and method (and equipment) of compaction used. These attributes will be discussed further in Chapter 5.

4.2.4 Volume Stability (Shrinking and Swelling)

Volume stability is an important parameter, as it has been noted that excessive shrinking, and particularly swelling, has been known to cause millions of dollars of damage each year to roadways, airfields, and foundations. Repeated shrinking and swelling of expansive clayey soils in alternating wetting and drying cycles of certain soils has also been attributed to downward slope movement. Volume stability is not easily achieved by merely densifying soil. In fact, as will be discussed in Chapter 5, soil densification may actually aggravate swelling potential in some soils dependent on compaction conditions.

4.2.5 Liquefaction Phenomenon and Mitigation

Several of the available soil and ground improvement applications are intended to mitigate liquefaction that may result from seismic (earthquake) events. An overview of liquefaction phenomenon and ground/soil conditions that provide susceptibility to this type of soil failure was presented in Section 3.1.2. As noted in Chapter 3, three fundamental conditions must be present for initiation of liquefaction:

(1) The soil must be essentially cohesionless, such that all of its shear strength results from intergranular friction and shear strength is a direct function of effective stress.

(2) The soil must be in a "loose" condition in that applied shear stress will cause a tendency for compression or contraction of the soil mass.

(3) The soil must be saturated and effectively undrained so that any increase in loads will tend to generate positive water pressures, thereby decreasing effective stress.

If any of these conditions are not met, then soil liquefaction is unlikely.

Some general methodological approaches to mitigate liquefaction occurrence is presented throughout discussions of the ground modification techniques in this text. These mitigation approaches can be fulfilled by several different (or combination of) ground improvement techniques. In principle, to mitigate liquefaction potential, one must eliminate one or more of the causative or susceptibility factors. This simply means that if (1) density is increased, or (2) water saturation is eliminated, or (3) the material is made to be "cohesive" by means of additional intergranular strength, the soil deposit would be rendered less likely to liquefy under dynamic (earthquake) loads. Each of these variables can be addressed by means of ground improvement methods. Densifying soil is one of the most accepted and well-defined means to achieve this goal while enjoying several other gains in engineering performance.

RELEVANT ASTM STANDARDS

D4253-06 Standard Test Methods for Maximum Index Density and Unit Weight of Soils Using a Vibratory Table, V4.09.

D7263-09 Standard Test Methods for Laboratory Determination of Density (Unit Weight) of Soil Specimens, V4.09.

REFERENCES

Carrier III, W.D., 2000. Compressibility of a compacted sand. J. Geotech. Geoenviron. Eng. 126 (3), 273–275.

Schexnayder, C., Lukas, R.G., 1992a. Dynamic compaction of nuclear waste. Civil Engineering. ASCE, New York, pp. 64–65.

Schexnayder, C., Lukas, R.G., 1992b. The use of dynamic compaction to consolidate nuclear waste. Grouting, Soil Improvement and Geosynthetics. ASCE, Geotechnical Special Publication No. 30.

Xanthakos, P.P., Abramson, L.W., Bruce, D.A., 1994. Ground Control and Improvement. John Wiley & Sons, Inc, 910 pp.

CHAPTER 5

Shallow Compaction

This chapter provides coverage of the topics related to common practices of compacting (densifying) shallow surface soils, or more commonly, placed layers (lifts) of soil as *engineered fill*. This includes efforts utilized to construct roadways, airfields, other transportation facilities, compacted backfill behind retaining walls, prepared material for slab construction, support of spread footings, embankments, earthfill dams, and so forth.

The principles of shallow compaction theory, control of compacted soil engineering properties, and, finally, a discussion of field applications are provided. Various compaction processes and equipment available for implementing these processes for field applications are described in order to provide an understanding of the different physical manner in which soil materials are densified, along with the effect on different soil types. A description of soil properties that can be achieved by controlling field compaction parameters is presented with construction specifications and tests that can be used to assure that desired engineering properties are attained.

5.1 METHODS OF SHALLOW COMPACTION

The concept of shallow compaction (introduced in Chapter 4) is the conventional method of densifying surface soils, new fill, or constructed earthworks such as embankments and transportation facilities. This type of compaction is usually carried out using a variety of commercially available rollers or tampers. These compactors may apply static load, vibrations, impact loads, or kneading to the soil. In some cases, a combination of applied compaction loads may provide the best results. The choice of applied loading method is primarily a function of soil type and desired outcome. Other types of methods and equipment used for shallow compaction of soils will also be described.

Static compaction generally refers to applications that apply a load without dynamic, vibratory, or impact components. This is done in the field by means of heavy rollers, stacking large weights, filling tanks with water, or simply piling up soil. Static loads will compress the soil structure of materials with relatively low frictional resistance. In the laboratory, static compaction is sometimes applied by compressing a known amount of soil into a prescribed volume.

Soil Improvement and Ground Modification Methods 71

As presented in Chapter 4, induced vibrations will aid in compaction of primarily granular soils by overcoming frictional resistance. Clean sands can be densified to as deep as 2 m using *vibratory compaction*, with the highest degree of compaction within about 0.3 m with diminishing densification at greater depths. The most effective vibration frequencies have been found to be between 25 and 30 Hz (Xanthakos et al., 1994).

Kneading compaction is a process by which the soil is "worked, formed, and manipulated. . . as if with the hands" (www.thefreedictionary.com), not unlike kneading bread dough. In this process, the equipment imparts a shearing force to the soil, which can contribute to better compaction in some soils. Kneading compaction is most commonly achieved in the field by *sheepsfoot* compactors or other similar types of compactors with protrusions (or feet/tampers). This equipment will be discussed in the next section. Kneading compaction can also be performed in laboratory tests to simulate the type of compaction achieved by the field equipment.

Dynamic or impact methods are also used for shallow compaction, involving loads that are applied dynamically by mechanical tampers. These methods of compaction can be applied in both laboratory and field applications, as will be described.

5.2 PRINCIPLES OF COMPACTION/COMPACTION THEORY

When compacting a soil at shallow depths or compacting new material placed at the surface, there are a number of variables to consider in order to achieve the desired degree of compaction and associated engineering properties. In many cases, the desired outcome is simply the highest density achievable with a set of given equipment. But in other cases, there are more subtle goals that can be achieved by carefully controlling other variables that may affect the properties and characteristics of the compacted soil. The main variables that will affect the degree of compaction of a soil are:

- Type of soil being compacted
- Method of compaction
- Compactive effort
- Moisture content of the soil being compacted

It is generally well known that for a given *compactive effort* (often noted as compactive energy per unit volume of soil) and compaction method, the density that a soil will achieve will vary with change in water content. Compaction theory tells us that from a relatively low water content, density will increase with increased water content up to a point and will then decrease with additional water. To measure the degree of compaction, geotechnical engineers use *dry unit weight* (γ_d). This alleviates possible ambiguity, as compacted samples with the same dry unit weight would have different moist weights at different moisture contents. The use of γ_d also helps with clarity and uniformity of construction and design specifications. Dry unit weight can be calculated by the equation

$$\gamma_d = \frac{G_s \gamma_w}{1 + e} = \frac{\gamma}{1 + (w\%/100)} \tag{5.1}$$

where G_s is the specific gravity of soil solids, γ_w the unit weight of water, e the void ratio, γ the moist unit weight of soil.

When water is added to a relatively dry soil, it acts to soften and "lubricate" the soil so it becomes easier to compact. This effect continues to allow the soil to be compacted to higher unit weights so that the dry unit weight (density) increases with an increase in water content until a certain point, the optimum water content (w_o). Beyond that level of moisture, the air voids attain approximately a constant volume but the water takes up additional space, resulting in an increase in total void space (air plus water), therefore reducing the dry unit weight.

A generalized compaction curve, as shown in Figure 5.1, represents the relationship for "as compacted dry unit weight" as a function of "as compacted water content," sometimes referred to as the *moisture-density relationship*. An exception to the typical curve is found for some soils. At very low moisture levels, the as compacted unit weight of uniformly graded sands

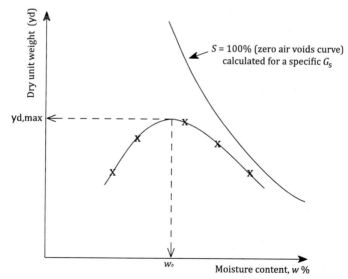

Figure 5.1 Moisture density (dry unit weight) relationship for a soil.

actually drops with increased water content. This has been explained by a phenomenon known as *bulking*, where capillary tension resists the effort of compaction at low moisture levels. As capillary tension builds, compacted unit weights are lower. Addition of water at this point "breaks" the capillary bonds and allows for more of the compactive effort to achieve higher unit weights, and the curve then resumes an upward trend until a peak is reached.

5.2.1 Laboratory Tests

In order to evaluate compaction parameters for a particular soil, prepare specimens for testing of engineering properties, and prepare design specifications, laboratory tests are generally utilized. There are a number of different types of compaction tests that have been designed to simulate various types of field compaction. To assure uniformity and alleviate ambiguity, tests are usually standardized. ASTM provides testing standard specifications that are recognized internationally. Other organizations, such as AASHTO, state DOT's and other regulatory (and governmental) agencies, also have various test standards. The results of compaction tests and specimens tested for engineering behavior under controlled compaction conditions can be used to optimize field placement and compaction of soils, and assist in compaction design parameters.

The most common types of laboratory tests are the Standard Compaction (or Proctor) Test (ASTM D698; AASHTO T-99) and the Modified Compaction (or Proctor) Test (ASTM D1557; AASHTO T-180). The equipment and procedure are similar to those originally proposed by R. R. Proctor in 1933 to simulate the compactive effort achievable by typical equipment of that era (Figure 5.2). In these tests, a free-falling steel rammer is dropped a fixed height repeatedly on loose soil placed in a mold. The

Figure 5.2 Standard and modified laboratory compaction hammers and molds.

diameter of the rammer is approximately half the diameter of the mold. The compaction with this equipment actually employs a dynamic or impact load as opposed to a static load or kneading. These types of tests may be appropriate for evaluating compacted soils used for earth fills, foundations, and road bases, for example. A uniform procedure is used to compact samples over a range of moisture contents to obtain the relationship between moisture and dry unit weight for a soil by the specified procedure. In the standard test, soil specimens are compacted in 101.6 mm (4 in) or 152.4 mm (6 in) diameter molds, depending on maximum grain size of the soil used. Each of three approximately equal amounts of soil are then compacted in layers with a 24.5 N (5.5 lb) rammer dropped from a height of 305 mm (12 in). For the 101.6 mm diameter mold, 25 blows of the hammer are applied to each layer. For the 152.4 mm diameter mold, 56 blows of the hammer are applied to each layer. By multiplying the fall height, hammer weight, and total number of blows, a total compactive effort of 600 kN m/m^3 (12,400 ft lbf/ft^3) is achieved. The Modified test was developed by the U. S. Army Corps of Engineers in response to the development of larger and more efficient compaction equipment which could deliver a higher degree of compaction, and greater compaction requirements for airfields (Holtz et al., 2011). The Modified test (ASTM D1557) uses the same molds, but an increased fall height of 457.2 mm (18 in), a larger hammer weight of 44.48 N (10.0 lbf), and five layers. This gives a compactive effort of 2700 kN m/m^3 (56,000 ft lbf/ft^3). With either test, all of the major variables affecting compaction are held constant except for moisture content. While many industry and research laboratories still use the labor-intensive standard test equipment, automated compactors are available and can significantly increase production (Figure 5.3). They can typically perform both Standard and Modified effort tests and are accepted as an ASTM standard as long as properly calibrated according to ASTM 2168.

Different laboratories and/or different projects may use one or the other of the standard Proctor-type tests. Based on examining many data sets (including different soil types), a reasonable approximation can be made of the compaction curve that would result for one compactive effort (standard or modified), given the results of a compaction test from the other compactive effort. Typically, the maximum dry unit weight (dry density) of a soil compacted using the Modified test will be approximately 5–10% higher than achieved by the standard test effort, while the optimum water content will be approximately 2–5% lower (in actual percent less moisture). The actual difference in $\gamma_{d,max}$ will depend on soil type, with smaller differences for well-graded granular soils (i.e., SW, GW) and greatest differences for high

Figure 5.3 Automated laboratory compactor.

plasticity cohesive soils (i.e., CH, MH). Estimation of the optimum water content may be aided by assuming that the peak values from each curve will fall on a *line of optimums* that would connect the peaks of compaction curves on a soil compacted at different efforts. As described earlier, the peaks of compaction curves occur at approximately 80% saturation (ranging from 75% to 90%). Therefore, the line of optimums will be subparallel to the 100% saturation ($S = 100\%$) or *zero air voids* line (ZAV). It is important to note that only the peak of a curve should be estimated in this manner, not all data points from a test so that curves from different efforts should (theoretically) not cross. In addition, compaction curves generated for a single soil should have roughly the same "shape" at different compactive efforts. An example of a set or "family" of compaction curves for different compactive efforts is provided in Figure 5.4.

Kneading compaction may be simulated in the laboratory by use of a California Kneading Compactor (ASTM D1561), used for preparation of 102-mm diameter and 127-mm high cylindrical specimens to be tested in a stabilometer. Another popular test used to compact soils with the characteristics of kneading compaction is the Harvard miniature compaction test.

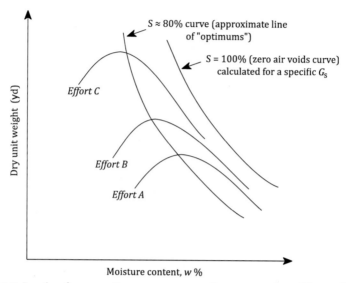

Figure 5.4 Family of compaction curves on a soil compacted at different levels of compaction effort (Effort A < Effort B < Effort C).

Due to its miniature size of 25.3 mm (1 in) diameter, it is only suitable for fine-grained soils. The equipment used for the Harvard miniature compaction test is shown in Figure 5.5. The size and ease of using the Harvard miniature equipment allows for a large number of specimens to be produced in a short amount of time. Extruded specimens can be tested quickly for strength, permeability, stiffness, and so forth. It has been suggested that the compaction achieved is most representative of that in the field by sheepsfoot compactors. The test results have been suggested to be similar to standard Proctor test results with regard to maximum dry unit weight, while results on some soils have been shown to underestimate maximum Proctor densities (Demars and Chaney, 1982). The obvious advantage is the ability to more closely duplicate the compaction process and thereby better replicate compacted conditions in the field. The effects of the compaction procedure (Harvard vs. Proctor) have been shown to be significant (D'Onofrio and Penna, 2003). This test is no longer a recommended standard by ASTM, but is still widely used in research and industry, including state DOTs and consultants (www.igesinc.com; www.nevadadot.com).

Most laboratory compaction tests are performed on a soil sample deemed to be representative of the material to be compacted in the field. Oversized soil particles are typically removed to eliminate possible effects of too large a ratio of grain size to sample size. This may affect the compaction test results, for example, by proportionally changing the amount of fine-grained to

Figure 5.5 Harvard miniature compactor equipment.

coarse-grained material in the sample. For material with a significant portion of gravel and/or cobbles, certain corrections and provisions can be made, including use of larger sample mold sizes, mathematical corrections prescribed by ASTM D4718, or by simple methods of "scalping and replacement" described in the literature (Hausmann, 1990; Houston and Walsh, 1993; Lin et al., 2001). The simple approach, repealed as an ASTM standard in 1991, is to add an amount of material between the maximum useable size (typically 19 mm = ¾ in) and the next sieve size smaller (e.g., 4.75 mm or No. 4 standard sieve), that is equal in dry weight to the amount of oversized grains removed.

Static compaction is not very common in the laboratory for general practice, but has been used for research when accurate moisture levels and unit weights are required. Static compaction is generally performed by a steady motorized or hydraulic load that compacts a known amount of soil into a

prescribed volume without any effects of dynamic load, impact, or kneading. Stress path simulation using a triaxial apparatus is another variation
sometimes utilized in the laboratory (primarily for compaction research).

5.2.1.1 Presentation of Laboratory Compaction Test Results

In the compaction test, individual specimens are compacted over a range of
water contents, with each specimen being compacted under "identical"
conditions, with a specific method and effort. Test specimens are usually
prepared from lower to higher water contents over a range that includes
the optimum water content (w_o). (Note: The optimum water content is
sometimes referred to as the *optimum moisture content*, or OMC.) While water
content can only be estimated at this time, water content samples are taken
to later determine actual "as compacted" water contents for each specimen.
Each compacted specimen is trimmed to a standard volume and then carefully weighed. As the specimens increase in weight, the moist (total) density
(weight/volume) is increasing. Once the measured weight decreases for an
increase in water content, the specimen density has decreased, therefore
indicating that the optimum water content has been exceeded. Once the
as compacted water contents of each specimen is determined, the dry
unit weight of each compacted specimen can then be calculated by using
Equation (5.1). The data collected for each prepared specimen is then plotted on a graph of dry unit weight versus (as compacted) water content, or
compaction curve (a.k.a. moisture–density relationship). All compaction curves
should be clearly labeled indicating the particular compaction method/effort
used, and should also include a curve representing the theoretical maximum
density for a given specific gravity (G_s). This curve is called the *zero air voids*
line (ZAV or $S = 100\%$), as this would represent the condition if *all* air was
expelled from the sample. As a theoretical maximum, the ZAV also provides
a boundary for the test data, which cannot be crossed (or even reached). This
curve can be calculated given (or assuming) G_s for the material by plotting
the dry unit weight for the ZAV (γ_{ZAV}) over a range of water contents as

$$\gamma_{ZAV} = \frac{\gamma_w}{w + (1/G_s)} \tag{5.2}$$

The peak in dry unit weight for most soils occurs at approximately 75–
90% saturation. This peak is the *maximum dry unit weight* ($\gamma_{d,max}$) for the soil as
compacted at a specific compactive effort/method. The corresponding
water content at which the maximum dry unit weight occurs is the *optimum
water content*, w_o. These two parameters of compaction will be important for

use with designs and construction specifications, as will become apparent when the relationship between engineering properties and anticipated behavior to compaction conditions is described later in this chapter.

5.2.2 Compaction of Different Soil Types

Different soil types will exhibit a wide array of properties and characteristics that will play a major role in many of the improvement methodologies and approaches described in this book. These variations are a function of both physical and chemical differences, including size, shape, intergranular forces, chemical charge, mineralogy, and so on.

Due to the differences and variety of characteristics for different soil types, it should not be expected that compaction curves should be similar. In fact, except for some well-documented and common soil types, estimation of compaction curve relationships may be difficult without actually performing (standardized) tests.

Figure 5.6 shows some typical compaction curves for different soil types. This is just an example of the variability that may be expected. One trend that seems to follow is that, in general, optimum water content and maximum dry weight will both increase with increasing plasticity of soil (as defined by Atterberg limits). There's an exception to this general trend: for "free-draining" (poorly graded) granular material, peak densities achieved by standard laboratory (Proctor) compaction tests are often low

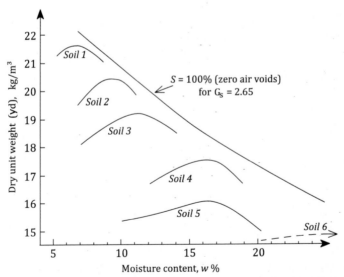

Figure 5.6 Typical compaction curves for various soil types. Soil 1, SW-SM; Soil 2, SM; Soil 3, SC; Soil 4, CL; Soil 5, SP; Soil 6, MH (volcanic ash).

and optimum water content high (as seen in Figure 5.6). In some cases, it has been reported that free-draining granular soils never attain a clear peak density. The difficulty with compaction test results for these types of soils has been attributed to the inability of the laboratory sample to maintain a uniform water content and to the lack of confinement in the laboratory test. Some have suggested that the compaction method used in the laboratory may be in appropriate for comparison to field compaction of these materials.

Some soils require special treatment, as the standard test methods may not provide results that accurately represent field compaction. An example of this is with certain tropical and residual soil with amorphous minerals that undergo irreversible changes upon drying, such as is encountered for soils with the mineral halloysite. For such soils that are typically moist in the field, compaction of samples may be more representative if performed from wet to dry. Irregular variability is also important for other properties of these special soils, such as Atterberg limit tests. Calcareous soils (derived from ocean coral or calcium deposits) must also be carefully evaluated, as the soil grain mineralogy is very different than most terrigenous soils. The materials are softer, often resulting in crushing or greater than expected compressibility, and specific gravity tends to be relatively high, sometimes reaching $G_s = 3.0$.

5.3 SHALLOW FIELD COMPACTION EQUIPMENT

A variety of equipment is available for shallow compaction of soil. The differences between equipment choices are principally related to the compaction method, coverage, uniformity of results, compactive effort, and effectiveness for different soil types. An overview of the readily available equipment is presented here with some comments on uses and advantages of each. Figure 5.7 shows photographs of some of the common shallow compaction equipment.

Smooth drum rollers (Figure 5.7a) are probably the most traditional type of equipment used for compaction of soils and asphalt pavements (which are actually just soils stabilized with bituminous admixtures). This type of equipment applies a uniform static load over the width of the drum and has the advantages of providing 100% coverage and a smooth finished surface. Smooth drum rollers can apply a modest static pressure (typically about 300-380 kPa = 45-55 lb/in^2), which may be adequate to compact thin layers of aggregate base coarse but may not apply sufficient pressure for other soil types or greater layer thicknesses. These types of rollers have been found to be ideal for compacting paving mixtures. Smooth rollers have also been found to be useful for *proof rolling*, which is a means of confirming

(a)   (b)

(c)   (d)

Figure 5.7 Photographs of typical shallow compaction rollers. (a) Smooth drum roller (Courtesy of Bomag). (b) Pneumatic tire roller (Courtesy of Bomag). (c) Combination roller photo (Courtesy of Dynapac). (d) Sheepsfoot/padfoot roller (Courtesy of Bomag).

uniform compaction or identifying "soft" spots that may require additional compaction.

Pneumatic (rubber tire) rollers (Figure 5.7b) are designed to apply very high static loads that are effective at compacting a wide range of soil types, and have been widely used for compaction of roadway bases, subbases, and asphalt mixes. Due to the configuration of alternating high–pressure tires and gaps between tires, these compactors also contribute some kneading action that can enhance the compaction. These machines may have up to seven or nine wheels, and larger versions can apply pressures up to 1000 kPa (145 lb/in^2) (Murthy, 2003). The individual tires can move up and down a small amount independently, which enables them to find small soft spots that may be missed by other types of drum rollers, providing better uniformity for uneven lifts.

Combination rollers (Figure 5.7c) are hybrid compaction rollers with both pneumatic tires and a smooth drum. The principle is that this equipment can utilize some advantages of both the smooth drum, with complete (100%) coverage, and the greater degree of compaction offered by pneumatic rollers.

Vibratory rollers are similar in appearance to static roller compactors, with the addition of oscillating motors that apply eccentric loads, providing impact and vibrations to the soil. As mentioned earlier, the vibratory action (along with static load and sometimes impact) can provide better compaction of granular soils, particularly cohesionless materials, by overcoming the frictional resistance inherent to granular soils. Vibrations provided by these types of equipment tend to be most effective at frequencies of 1000-3500 cycles per minute (approximately 17-60 Hz).

Sheepsfoot, padfoot/tamping foot, and wedgefoot rollers (Figure 5.7d) are essentially drum rollers with protrusions (knob-headed spikes of various sizes and shapes, see Figure 5.8) that can apply very high static load (up to 2000-7000 kPa; approx. 300-1000 lb/in^2) due to the concentrated load over small contact areas of as little as 8-12% of the roller area for sheepsfoot rollers (Holtz et al., 2011). The larger pads of tamping or padfoot rollers, which apply about 40% coverage, are used for wetter and softer soils. These types of compactors are also effective at breaking up cohesive soils by a process of *kneading* the soil. This type of equipment is the most effective for compaction of clayey cohesive soils, as the kneading action helps to break bonds within the soil mass, enabling better compaction. The kneading effect will provide the most uniform and highest degree of compaction for clays and other cohesive soils. The roller protrusions first compact and manipulate the soil below the surface, and then as the soil becomes more compact, the feet "walk out" on top of each layer.

This type of compactor was reportedly first introduced in the United States in the early 1900s (Hausmann, 1990; www.contrafedpublishing.com). Stories attribute the origins of the sheepsfoot roller to successful compaction of soft clays by herding sheep across soft ground. The high contact pressure and

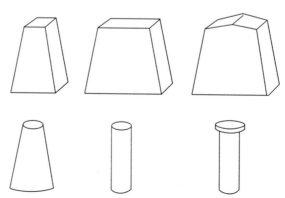

Figure 5.8 Various typical "pads" for sheepsfoot/padfoot rollers—schematic.

manipulating action of the sheep hooves have been attributed to the design seen in sheepsfoot compactors. Use of sheepsfoot rollers also leave a "pock-marked" surface, which is important in providing good bonding between compacted layers. This has been found to be especially important in construction of hydraulic structures and embankments, as it minimizes the chances of a preferred shear plane or seepage plane within a constructed earthwork. The kneading compaction also tends to render the soil in a more dispersed structure, leading to lower permeability and greater ductility (lower stiffness), which are also advantageous for hydraulic structures.

Grid rollers (a.k.a. mesh rollers) are another version of roller that applies a high contact pressure through concentrated contact (Figure 5.9). These rollers have approximately 50% coverage and can apply pressures in the range of 1500-6500 kPa (200-900 psi). These rollers are ideally suited for breaking up and compacting rocky soils, gravels, and sands.

Trailer rollers describe any of the drum roller types that are towed equipment rather than self-propelled driven equipment (Figure 5.10). The concepts and mechanics of compaction are the same, but these towed rollers have some advantages in that they may often be used at faster speeds and can be used with existing, nonspecialized equipment. They may, however, be lighter and apply lower effort than self-propelled compactors.

Impact rollers are significantly different than conventional rollers that use smooth drums, tamper feet, or pneumatic tires. Impact rollers are designed much like a "square wheel" (actually, impact rollers typically have three or five "sides") with "rounded" corners (Figure 5.11). As the corners roll over, the weight of the roller (up to 15 tons; www.impactor2000.com) provides impact compaction with up to 100 kJ of kinetic energy (www.broons.com) at rates of 90-130 blows per minute (www.landpac.co.za). Figure 5.12

Figure 5.9 Grid roller. *Courtesy of Broons, LLC.*

(a)

(b)

Figure 5.10 Trailer combination roller. (a) Smooth drum mode, (b) pneumatic mode. *Courtesy of Broons, LLC.*

depicts the dynamic action of the impact drums. This imparts dynamic compaction loads on the order of two to five times that of conventional shallow compaction equipment. The higher compaction energy allows achievement of high densities over a wider range of water contents so that moisture control may not be as critical. This is an important advantage when compacting in situ soils, or placed lifts of over 1–2 m (3–6.5 ft). This method of compaction, sometimes referred to as "rolling dynamic compaction,"

Figure 5.11 Impact roller. *Courtesy of Landpac Technologies.*

Figure 5.12 Dynamic action of the impact drum. *Courtesy of Landpac Technologies.*

or high-energy impact compaction, can be effective to moderate depths of 2-3 m (6.5-9 ft) due to the surface impact force and transmission of dynamic waves into the ground, with the best compaction occurring in the top 2 m (Jaska et al., 2012; www.impactor2000.com; www.landpac.co.za; www. broons.com). With this zone of influence, these compactors can effectively improve near-surface soils to greater depths than any other surface compactor roller, enabling construction of a thickness often sufficient for many projects without the need of layered engineered fill. In other cases, lift thickness of up to 1.5 m (approx. 5 ft) have been successfully compacted to required specifications (www.impactor2000.com; www.landpac.co.za). In an example case study, 12 million m^3 of fill compacted in four 1.2-m lifts consistently met the minimum requirement of 95% modified maximum dry density. These rollers are often towed behind conventional, nonspecialized equipment. This can be a tremendous economic advantage with very high volumetric production rates, as they can travel at speeds typically in the range of 10–13 km/h (6–8 mph). Some manufacturers are also marketing stand-alone, self-propelled equipment. This type of compactor was first introduced in the 1950s, but was not widely used until modified and marketed to a wider international audience. This type of compaction has been demonstrated to be successful at compacting a wide range of soil types, including dredged marine fills, deep loose sandy materials, unsaturated and saturated silty and clayey materials, collapsible soils, and rockfill (www.landpac.co.au). In recent years, the impact compactor has gained significant interest and has been used for compacting roadways, port facilities, airport runways, landfills, mine and quarry waste, heterogeneous (mixed) fills, and for development of reclaimed land. This equipment has also been very successful in roadway/runway rehabilitation, as existing pavement layers up to 0.5 m can be broken up and recompacted as part of the new base material all in one step. Recent applications include projects in the United States, Australia, New Zealand, Africa, Europe, Asia, and the Middle East. As with any high-energy impact loading, there are a few disadvantages, including disturbance (actual loosening) of the top 0.5 m (1.5 ft) and generation of moderate vibrations within 10 m of the application. In general, the top surface layers will be finished with conventional compactors and/or paved/repaved.

Small *portable compactors* may be very useful and efficient for small and difficult locations, including corners, edges against walls and abutments, backfill in trenches, around utilities and pipes, and so on. There are several types of portable compactors, including vibratory (impact) tampers and rammers, vibratory plates, and heavy remote control (RC) rollers. An obvious advantage of RC compactors is that they may be used in hazardous or potentially unstable situations. These types of portable compactors come in a wide variety of weights, power, and so forth. Some examples of these are shown in Figure 5.13.

5.4 PROPERTIES OF COMPACTED SOILS

Engineering properties and soil behavior may be heavily influenced by how soils are compacted. Because of this, control of compaction conditions (as compacted moisture and density) can aid in achieving the desired properties for a given soil.

5.4.1 Soil Structure

One of the characteristics that can play a critical role in achieving desired soil properties is the *soil structure*, or arrangement of soil grains.

(a)

(b)

(c)

(d)

Figure 5.13 Small portable and hand-held compactors. (a) Vibratory rammer, (b) vibrating plate compactor, (c) remote controlled compactor, (d) field application of hand operated compactor. *Photos courtesy of Wacker Neuson.*

Cohesionless (granular) soils: The interaction between cohesionless soil grains is essentially all frictional, and because the grains are more "bulky" or rounded than clay particles, there is not much significant difference between "structures" other than the density of packing. One exception to this is for granular soils at very low moisture levels. The low moisture may provide enough "apparent cohesion" between grains due to water surface tension to form a very loose *honeycombed* structure. But this structure is relatively unstable and tends to collapse with any manipulation or applied load. Some minor differences have been noted in certain properties of cohesionless soils compacted at different moisture levels, but in general, density of packing is by far the dominant factor that controls engineering properties and behavior of these soils. In general, the higher the density of a cohesionless soil, the stiffer and stronger (higher shear strength) the compacted material will be. Higher density will also result in a reduction of permeability in cohesionless (granular) soils. Hence, for controlling desired engineering properties in cohesionless soils, "as compacted" density (reported as dry unit weight) is usually the only requirement. Knowing the optimum water content for a specific compaction method can aid in achieving the desired densities, but will have little effect on engineering properties or behavior.

Cohesive (clayey) soils: The wide variety of soil structure present in cohesive (clayey) soils plays an important and often critical role in achieving desired engineering properties and behavioral characteristics. The structure of cohesive materials may be somewhat controlled as a function of compaction conditions and may have much more to do with water content than density. As opposed to rounded granular soil grains, clay particles tend to be very thin and flat or "platy" in shape. This results in a very high ratio of surface area to volume such that the "physio-chemical-electrical" properties of a clay particle's surface plays a vital role in the properties, characteristics, and behavior of clay soils. The details of these "physio-chemical-electrical" interactions, along with a discussion of the importance of clay mineralogy, will be addressed in the discussion of Admixture Soil Improvement provided in Chapter 11.

The structure of clayey soils may be a state where clay particles are configured with edge-to-edge or edge-to-side contact. This is called a *flocculated* structure, referring to the "flocs" (or knits) that are created by the attraction between soil grains that occurs when compacted at lower moisture (water content) levels. At higher moisture levels, water forms "bonded" layers around the clay particles known as diffuse double layers. These water layers create a natural repulsion between soil grains, thus keeping soil grains apart (i.e., no edge contact). This type of structure is called *dispersed*. Figure 5.14 schematically depicts possible arrangement of grains found in flocculated and dispersed structures. Higher compaction energy will also tend to orient

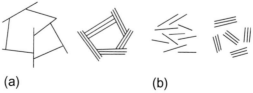

(a)   (b)

Figure 5.14 Clay particle structure: (a) flocculated and (b) dispersed.

groups of grains in a more subparallel to parallel configuration. Lambe (1958a,b) described the effect of compaction on the structure and properties of cohesive soils. His studies showed that, in general, clayey soils compacted to the wet of optimum (above the optimum water content, w_o) for a given compactive effort would render a more dispersed structure. As we can now see, control of the compaction conditions (water content and density) can induce different structures in clayey (cohesive) soils. The difference in structure along with compacted density (dry unit weight) will result in different soil properties and behaviors, including strength, compressibility, permeability (hydraulic conductivity), stiffness/ductility, and swell. For cohesive soil compacted at the same relative density but with different structures, some noticeable differences can be seen. In general, compacted samples with a more flocculated structure will exhibit lower compressibility, higher peak strength, and higher stiffness, while samples with a more dispersed structure would be more ductile (less brittle), have a lower permeability, and may have a higher residual strength. An exception to the general rule for compressibility exists for soils compacted dry of optimum (above the optimum water content, w_o) such that a highly flocculated structure is achieved. In this case, subsequent saturation may cause "collapse" of the structure, leading to additional settlement. Compressibility may also be greater for soil compacted dry of optimum if subjected to high applied stresses (Murthy, 2003).

An example of the variation of shear strength and stiffness characteristics for a cohesive soil is shown in Figure 5.15. The specimen compacted dry of

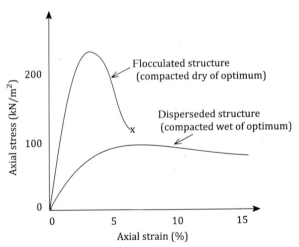

Figure 5.15 Variation of strength and stiffness for a silty clay compacted at different moisture levels.

optimum has a flocculated structure and exhibits a high peak strength and stiffness (stress–strain ratio), but this soil then fails in a brittle manner with low residual strength. This is in comparison to the specimen compacted wet of optimum at roughly the same density (dry unit weight), with a lower peak strength and lower stiffness (more ductile), but with significant residual strength after peak to a relatively large strain.

5.4.2 15-Point Method

In order to relate the differences in engineering properties to compaction conditions, a battery of tests may be conducted on soil specimens densified to "as compacted" conditions. A procedure known as the "15-point method" is utilized in this way. An example of a 15-point method plot generated from test data is shown in Figure 5.16. The procedure is used for determining the variation in test values for an individual property of interest for a soil:

(1) Approximately five specimens of a representative soil are prepared by compacting with a uniform effort over a range of water contents that span the optimum, much as would be done for a compaction test.

(2) Two more sets of approximately five specimens are prepared in the same manner, but at two additional and different compactive efforts.

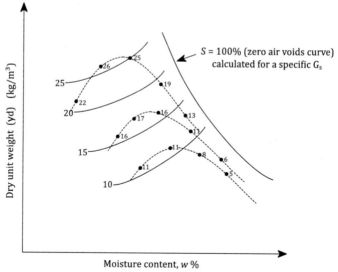

Figure 5.16 Example of a "15-point" plot showing the trend of an engineering property with compaction conditions.

(3) Each of the approximately 15 specimens is tested for some property of interest (e.g., strength, compressibility, permeability).

(4) The resulting test values are plotted on a graph of density (dry unit weight) versus "as compacted" water content *at the compaction conditions* for each specimen.

(5) The plotted values are then contoured to show the numerical variations of the tested soil property as a function of compaction conditions.

As shown by the results of the 15-point method for different properties, there are three general "trends" of improved properties that can be achieved through the control of compaction conditions. These are depicted in Figure 5.17. In the first scenario (Figure 5.17a), properties improve with

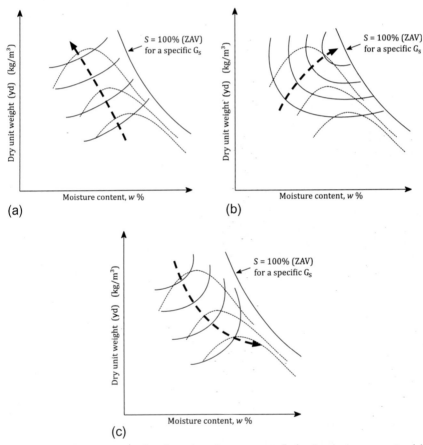

(a)

(b)

(c)

Figure 5.17 Common trends of engineering property/behavior improvements. (a) Improvement with increased density and decreased water content. (b) Improvement with increased density and increased water content. (c) Improvement with decreased density and increased water content.

increased density and decreased water content. These soils will tend to have a more flocculated structure. The trend shown is representative of improved stiffness, "as compacted" strength, and reduced compressibility. Data supporting these trends are available in the literature (e.g., Hausmann, 1990; Seed and Chan, 1959). In the second scenario (Figure 5.17b), properties improve with increased density and *increased* water content. These soils will tend to have a more dispersed structure. The soil compacted under these conditions will exhibit lower permeabilities (important for hydraulic structures) (Lambe, 1958b; Mitchell and Hooper, 1965), typically higher residual strength (as might be expected given the discussion accompanying Figure 5.15), and higher *strength after soaking*. This last property has not yet been discussed, but plays an important role for compacted soils and earth structures that may get "soaked" (or submerged) subsequent to compaction. The strength after soaking (sometimes referred to as "soaked strength") is tested for compacted samples that have been subsequently submerged for 24-48 h after being compacted and prior to testing. This sample preparation is similar to that for California bearing ratio (CBR; ASTM D1883) tests. Strength after soaking is often considerably less than the "as compacted strength." This strength should be used for designs where soaking is expected. A third scenario (Figure 5.17c) seeks to minimize swell (a major problem for roadway and structural foundations) and improve ductility (important for earth structures susceptible to damage due to brittle failure). These properties are optimized when the soil is compacted at lower densities and at higher moisture levels.

It is important to recognize that the plots depicted in Figure 5.17 are typical trends only and may vary considerably with individual soils. For actual design data to be used in developing specifications, property tests should be performed on representative samples of the actual soils to be used.

A few soil properties deserve particular attention due to their variability and often critical importance to the success of a project. Damage resulting from the swell of expansive soils costs millions of dollars each year. *Swell potential* is strongly affected by soil structure and compacted density, as well as clay mineralogy. Because of these variables and the importance of soil expansion to certain projects, swell tests may need to be performed on specimens compacted to expected "as compacted" conditions in order to evaluate the actual anticipated swell potential for particular soils. An example of variation in swell is shown in Figure 5.18. Permeability is another important property that can vary widely over a range of compaction conditions. With proper compaction, permeability can be reduced by a factor of more than 100 if compacted wet of optimum and at relatively high density.

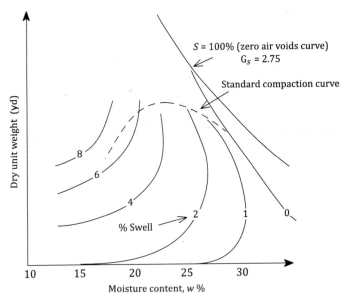

Figure 5.18 Example of swell as a function of compaction conditions. *Redrawn after Holtz and Gibbs, 1956.*

Obviously, when more than one property or characteristic is of importance, the engineer must evaluate what is most critical to the project and sometimes combine (or compromise) the best overall compaction characteristics of one property to achieve the desired results. This will have to be taken into consideration when design specifications are prepared for compaction of a soil for a project.

5.5 FIELD COMPACTION AND SPECIFICATIONS

5.5.1 Field Compaction Variables

A number of variables that may affect the field quality of the compacted material are not all present or as important in laboratory tests. These variables include:

- Controlled water (moisture) content during compaction
- Size or weight and the number of passes with the equipment
- Type of the compactor and compaction method
- Lift (layer) thickness
- Uniformity of the source (borrow) material

5.5.1.1 Water Content

The importance of water content has been described as having a profound influence on the density achieved, and water content is a fundamental part of compaction theory. Water content during the compaction of clayey soils has also been demonstrated to have significant (even controlling) effects on soil engineering properties as a result of variations in the soil structure achieved. In field applications, control of water content may be difficult, especially in very arid regions or where there is heavy and/or irregular rainfall.

5.5.1.2 Weight and Number of Passes

The combination of weight and number of passes are analogous to compactive effort delivered by a standardized laboratory test. While a variety of combinations of weight and number of passes can theoretically give the same mathematical solution for compactive effort, experience has shown that additional passes may provide diminishing returns of increased density, and that if more compactive effort is needed after 5-8 passes, then a larger (i.e., heavier) compactor may be appropriate. Other factors may have even greater effects and must be monitored and controlled in actual field applications. Generally, a minimum of 4-8 passes are normally needed to economically achieve desired density results. An exception may exist for saturated sands, which have been shown to improve with increased numbers of passes, up to 15 or 20 (Hausmann, 1990).

5.5.1.3 Type of Compactor and Compaction Method

As previously described, the method of compaction can influence the degree of compaction, uniformity of compacted fill, and sometimes soil structure. It was discussed that different types of equipment (e.g., static vs. vibratory or impact) would be more effective with different categories of soil. It was also noted that typical laboratory tests do not always apply the same method of compaction as applied by certain types of field equipment. Certain types of compactors may be more beneficial in achieving the desired degree of compaction and/or characteristics than others. For example, a sheepsfoot roller applies very high static pressure and kneading compaction that is beneficial for achieving a high degree of density and uniformity, while promoting better bonding between layers of cohesive soils, and achieving a more dispersed structure. Some compactors may be able to adequately compact to greater depths than others, and are therefore able to compact thicker layers of material.

5.5.1.4 Lift Thickness

Lift thickness is the measured thickness of each layer of compacted soil. It must be made clear as to whether lift thickness refers to the thickness before or after compaction of a soil layer. For most specifications, lift thickness refers to the "as compacted" thickness, as this is less ambiguous and more meaningful to evaluating the compacted soil. The effective depth of compaction is indirectly proportional to the pressure applied to compact the soil and is also a function of soil type. As previously described, vibrations can be very effective in compacting granular (cohesionless) soils so that less pressure may be required to compact the same thickness. But it also needs to be considered that sandy soils will not be well compacted near the surface due to lack of confinement and, in fact, may be looser at the surface than prior to compacting. To demonstrate the relationship between compacted density and lift thickness, Figure 5.19 depicts the relationship between compacted density and depth for a typical sandy soil. Looking at Figure 5.19a, note that the density increases with depth to a point and then decreases at greater depths. When compacted lift thickness is limited, then the minimum density within the compacted fill will be at the juncture of the overlap between density curves for each successive lift (Figure 5.19b). Based on this scenario, the limiting (maximum) lift thickness should be specified to assure that the entire fill is compacted at or above the minimum required value. A balance must be made between the geotechnical engineer, who would prefer thinner lifts to assure higher and more uniform densities, and the contractor, who would

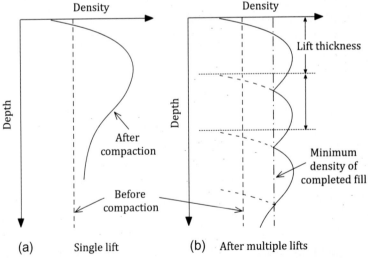

Figure 5.19 Example of density vs. depth for compacting lifts of sandy soil.

have higher volume productivity if afforded thicker lifts. Experience has resulted in some general guidelines on requirements for maximum lift thicknesses that may be appropriate for construction specifications. Typical maximum required lift thicknesses range between about 15 and 30 cm (6-12 in), with greater thicknesses allowed for larger, heavier, and more dynamic equipment, and for greater particle sizes (e.g., rockfill). As described earlier, some very heavy equipment and those delivering high dynamic loads are able to compact lift thickness up to 1-1.5 m (3-5 ft).

5.5.1.5 Uniformity of Source Material

In many cases (especially for large projects), unless the soil is prepared from a supplier such as an industrial quarry, the source material (borrow) for earthwork construction will vary in properties as it is excavated for use as fill. This is a natural and expected occurrence and should be considered by the design engineer. As the construction continues, variations may be addressed by periodic reevaluation of the material. This is often included in specifications for quality control of the engineered fill by requiring systematic testing and adjustments to compaction efforts.

5.5.2 Shallow Compaction Specifications

In order to achieve the desired results and to assure the engineering properties that the design engineer is expecting, certain requirements must be made for the compaction of soils for each application. Very often, the contractor performing the compaction work in the field is not trained in engineering and needs only to be contracted to perform the compaction to meet parameters set by the geotechnical engineer, along with any other necessary construction guidelines. Fortunately, as previously described, most needed soil design properties can be adequately achieved by controlling the soil compaction conditions. In order to avoid any confusion or controversy that could potentially lead to future performance problems or contractual issues, the compaction requirements must be clearly specified in written documentation that is agreed to by all responsible parties, including contractors, owners, developers, and others.

Shallow field compaction requirements (specifications) are most often based on the results of standardized compaction tests, such as ASTM D698 and D1557, and/or any other associated tests of important properties of compacted samples under controlled compaction conditions. These tests are most commonly done in the laboratory, but sometimes can be performed on site in the field for larger and more critical projects, especially when large volumes of potentially variable borrow material are used as the fill or construction material. Some compaction specifications (e.g., for some transportation facilities) may have other test requirements such as performance tests described at the end of Section 5.6.

5.5.2.1 Density Requirements

For most applications, a minimum density is required so as to assure an adequate degree of compaction is accomplished. As previously described, higher degree of compaction will improve many desirable engineering properties and response behaviors.

There are a few different ways that minimum density can be specified. Geotechnical engineers tend to refer to density of soils in terms of *relative density* (D_r). This definition, while useful in describing soil mechanics terms, may be problematic in that the maximum and minimum values $(\gamma_{d,max}, \gamma_{d,min}, \text{ or } e_{max}, e_{min})$ used to determine the value of D_r are often difficult to determine accurately.

Sometimes *procedural specifications* are used. This is where the construction details are defined, such as, "Soil to be placed in 20 cm (8 in) lifts, at a water content between 10% and 14%, and compacted with 10 passes of a 12-ton vibratory compactor." While procedural specifications are sometimes appropriately used, there are a number of issues with using this type of specification. First, the contractor must be monitored to be sure he/she is following the prescribed procedure. Second, even if the procedural specifications are followed completely, there is a real possibility that the desired results may not be obtained.

In order to avert confusion, ambiguity, and possible legal/contractual issues, specification with respect to *relative compaction* (RC) is recommended. Relative compaction is defined as

$$RC = \frac{\gamma_{d,\text{field}}}{\gamma_{d,\text{max}}} \qquad (5.3)$$

This type of specification requires only determination of $\gamma_{d,max}$ by a specified method (i.e., modified Proctor test; ASTM D1557), and comparing it to the measured field dry density (dry unit weight). The important key to avoiding problems with this kind of specification is to clearly specify the reference compaction test (e.g., ASTM D698 or D1557). For example, use of subscripts such as RC_{mod} of $\gamma_{d,max,mod}$, for reference to the modified Proctor test, will usually be sufficient to avoid any confusion. As pointed out earlier, if necessary, the expected $\gamma_{d,max}$ of one compaction test could be estimated from the results of another test, but this type of estimation should be avoided if possible.

5.5.2.3 Cohesionless Soils

As essentially *all* important soil properties and behavior characteristics of cohesionless soils are a function of density alone, specifications for cohesionless soils will generally have a minimum density requirement but no water content requirement. As long as a contractor can meet the density requirement, then the water content is usually immaterial to achieving the desired results. Other requirements may also be needed.

Given that approximately 98% modified relative compaction can be achieved for cohesionless soils with typical equipment, a reasonable expectation would be to require 95% or 96% compaction as a minimum specification. With specialized "heavy" equipment, RC_{mod} of over 100% (such as needed for major airport runways) may be achieved. Depending on the type of project component being considered, higher or lower required minimum densities may be appropriate (or needed). For instance, when only basic stability of a backfill is required without any need to support additional loads, 80% RC_{mod} may be sufficient. On the other hand, saturated cohesionless soils in high seismic hazard areas may warrant a minimum of 97% RC_{mod} or more to mitigate liquefaction potential. Generally, cost is commensurate with degree of compaction with up to 90% RC_{mod} easily and economically achieved with smaller equipment, while $RC_{mod} > 96\text{-}97\%$ can get expensive.

So, in conclusion, a "good" written specification for compaction of cohesionless soils should read something like:

The soil shall be compacted to not less than ____% of the maximum dry unit weight as achieved by (specify test method, e.g., ASTM D1557, Modified Proctor) *compaction test.*

Rather than specify actual dry unit weight values, this type of specification continues to be valid even when there are changes in soil conditions or variability in borrow source material. The specified minimum RC (percent of the maximum dry unit weight) should be chosen considering that most of the soil will be compacted to a greater degree in order to pass field compaction control (inspection) testing. Some additional specifications may also be desirable, such as maximum compacted lift thickness (as per the discussion of Section 5.5.1) and/or maximum particle size (usually required to be no greater than one-half of the maximum lift thickness or some smaller size).

5.5.2.3 Cohesive Soils

As described in some detail in Section 5.4, the compaction water content is critical in achieving the desired engineering properties and behavior of cohesive soils. As a consequence, specification of water content in addition to degree of compaction density (dry unit weight) is almost always required for compaction of cohesive soils. Based on the desired engineering properties and values obtained from testing (such as from a 15-point method or similar), density and water content ranges can be specified. As a general rule, water content requirements should only be limited to values that will affect the desired levels of engineering properties. For example, if there is no need to limit the maximum water content, then only a minimum should be specified. When both upper and lower bounds are important to control the combination of all desired soil properties, a minimum range of at least 3% should be provided to the contractor, as control of water content in field applications of cohesive soils is difficult, at best. Therefore, a "good" written specification for compaction of cohesive soils should read something like:

> The soil shall be compacted to not less than ___% of the maximum dry unit weight as achieved by (specify test method, e.g., ASTM D1557, Modified Proctor) compaction test, and must be compacted at a water content (w) greater than (lower bound, e.g., $w_o - 1\%$) and less than (upper bound, e.g., $w_o + 3\%$), where w_o is the optimum water content as determined by (specify test method, e.g., ASTM D1557, Modified Proctor) compaction test.

As the writing of the water content limits can be somewhat confusing, it is always recommended that the written specifications be drawn onto a compaction test plot to be certain that they make sense. Some additional details may also be added to specifications for cohesive soils, such as maximum compacted lift thickness and/or maximum particle size, and sometimes compactor type (i.e., sheepsfoot roller for hydraulic structures).

5.6 COMPACTION CONTROL/FIELD INSPECTION

As noted in earlier sections of this chapter, careful control of compaction conditions and meeting of required specifications may be critical in assuring that the compacted soil performs as expected. It is also important that compaction specifications are designed within reasonable limits and achievable ranges.

Traditional monitoring of compaction moisture contents and testing of compacted soil density provide the most direct means of field compaction quality assurance. These tests are the ultimate tool for field compaction inspection and are often required as part of specifications. These will be discussed in more detail later in this section. But there are a number of methods and tools now available to assist the contractor in understanding how good a job he is doing or where there is a possible problem that can be addressed while still early in the construction process. Some of these methods directly address the compaction specification parameters of moisture content and density, while others may indirectly provide indicators of the compaction quality.

Proof rolling is a qualitative method of identifying "soft" spots or areas that may need further densification or greater degree of uniformity. It may also be used to identify where pumping may be a concern for subgrades. Proof rolling is typically carried out on a subgrade or at the completion of compaction of a layer of engineered fill. The general premise is that a heavy roller traverses the prepared area and a note should be made of where there is an irregularity or excessive deformation. High resolution, onboard GPS instrumentation can keep track of locations where irregularities occur. Some compaction specifications may include proof rolling as an interim quality control tool, which is usually followed by one or more quantitative tests.

Many of the new vibratory compaction rollers are now equipped with *intelligent compaction* (IC) control systems such as the *compaction meter* for arguably better and more efficient quality control monitoring. A compaction meter uses a frequency domain accelerometer sensor mounted on the drum that continuously emits signals that are processed and displayed on the operator's instrument panel. The displays show a compaction meter value (CMV) and or color coding, which indicates the stiffness/density of the compacted material to effective depths of 0.3-1 m (approx. 1-3 ft). An operator can watch in real time for the CMV to increase and ultimately reach a peak whereby the maximum compaction is achieved under the particular effort for the specific soil conditions. Readings can also give an indication of the uniformity of the compacted material. Recordings of CMVs along with GPS coordinates allow for accurate documentation, which can be useful for review or for revisiting problematic locations. Use of this type of monitoring during construction can result in significant increase in efficiency for the operator or contractor, as wasted time or additional effort can be addressed immediately during the process. In comparison to traditional moisture and density measurements made only at point locations, IC provides complete coverage of all areas compacted. Furthermore, the measurements provide performance data comparable to that measured by deflectometers, plate load tests, and dynamic cone penetrometers (DCPs). A number of state DOTs are pushing to have IC included in compaction specifications for combined advantages of efficiency, personnel and time cost savings, and the belief that the results more accurately portray mechanistic design values.

Landpac Technologies has developed instrumentation on its impact compactors to provide continuous impact response (CIR) and continuous impact settlement (CIS) measurements in real time during compaction (www.landpac.co.za). The CIR system uses an accelerometer mounted in the compactor drum, which measures the deceleration for each impact while GPS records the position for each reading. Slower deceleration indicates softer ground. Locations where low readings are recorded can then be further compacted. The CIS uses GPS to accurately measure the settlement deformation as compaction proceeds to indicate where no further compaction can be achieved with that effort (Black Geotechnical, personal communications).

5.6.1 Compaction Control Tests

There are a number of standard tests used for compaction control (inspection) in the field. Depending on the specifications for a particular project, variables that need to be tested may include density, moisture, compacted lift thickness, maximum particle size, and compactor type. These last three variables simply require physical dimension measurement and observation, but density and moisture must be measured by carefully controlled and regulated tests.

5.6.1.1 Density Control Tests

The two components required to calculate density are volume and dry weight. Traditionally, compacted samples could be taken by means of thin-walled sample tubes and transported to the laboratory, where they could be tested for moist (wet) density (weight/volume), and one or more samples of the test cylinders would then be tested for moisture. Dry density (dry unit weight) calculations could then be made. A couple of problems are inherent to this methodology. For one thing, *sample disturbance* can skew the results of dry density calculations. In some cases, there is an inability to obtain a good "undisturbed" representative sample. This is especially true for granular soils and for very coarse materials. And finally, the time required to obtain test results, which may be overnight, may create a hardship for the contractor. It seems unreasonable that a contractor would have to rip out perfectly good compacted layers to redo a buried layer completed a day prior that did not meet specifications. To alleviate these problems, a number of tests have been derived for performing density tests in the field.

5.6.1.2 Volume Tests

One approach to determine volume is to excavate a small hole into the compacted fill material and measure the volume of the hole. The compacted material excavated from the hole is weighed, providing the "wet" weight from which dry weight can be calculated once the moisture content has been determined by other means. There are a few common tests used to measure the volume of the excavated material.

The *sand cone* method (ASTM D1556) is one of the most trusted conventional tests for field compaction. The test involves pouring a standardized (uniform, dry, 20-30 Ottawa) sand from a jar through a standard valve and cone that regulates a uniform flow of sand passing through. This provides a deposit of sand that is repeatable at a constant density. In this test, a hole approximately 10 cm in diameter is made in the compacted fill, over which the jar and cone are placed on a base plate (Figure 5.20). The valve is then opened so that sand pours into the excavated hole until the hole and cone are completely filled. At that time, the valve is closed. The apparatus is calibrated so that the weight of the sand that fills the hole below the base plate is determined. Knowing the G_s of the sand used in the test, the volume of the hole can be calculated as

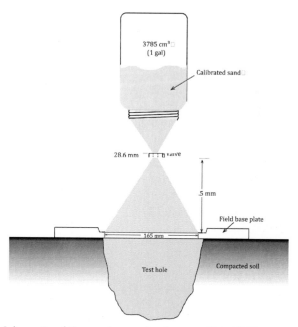

Figure 5.20 Schematic of the sand cone density test. *After ASTM.*

$$V = \frac{W_s}{G_s \times \gamma_w}$$ (5.4)

The *rubber balloon* method (ASTM D2167) is another test used for determining the volume of an excavated hole. It utilizes a liquid-filled (usually water), calibrated cylinder attached to an expandable rubber membrane (balloon) (Figure 5.21). The fluid is pumped into the balloon until the excavated hole is filled. This allows the volume of the hole to be read directly on the graduated cylinder. Once the volume of the hole is determined, the dry unit weight can be calculated in the same manner as for the sand cone method. An advantage of the rubber balloon method is that no sand can flow beyond the limits of the excavated hole, as may occur in relatively clean coarse (e.g., gravel) backfill material.

Some other simple methods to evaluate the volume of a hole excavated into a compacted fill include the so-called water or oil methods, where the excavated hole is lined with a membrane and filled with a measured volume of liquid. This type of method is used for evaluation in very coarse material for which the sand cone and balloon tests are too small to provide accurate results.

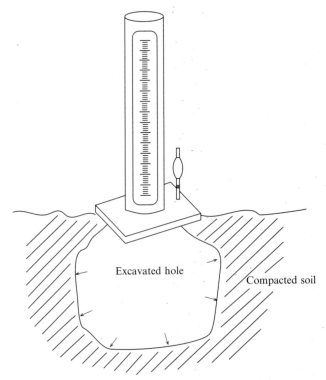

Figure 5.21 Balloon density test schematic.

5.6.1.3 Moisture Control Tests

As is now quite clear, the moisture content of a compacted material is itself very often a compaction specification parameter. It is also needed to calculate the dry unit weight (dry density) of a soil being tested for compaction control. In order to expedite determination of the moisture content, a few methods (some with limitations) are available.

ASTM D4959 provides for a method to determine moisture content rapidly by "direct heating," where the water in moist soil is evaporated by heating the soil on a hotplate, stove, or blowtorch, for example. This test is sometimes substituted for conventional oven drying (as per ASTM D2216) to expedite other phases of testing, although results may be somewhat less accurate for certain soil types with hydrated materials or appreciable organics.

Speedy/Instant moisture testing (ASTM D4944) is another method to rapidly determine moisture content of soils with moisture ranges from 0% to 20%. This test procedure involves using calcium carbide as a chemical reagent, which reacts with the water in a soil. The available testing apparatuses are approved by ASTM D4944, but are not recommended for high plasticity clays, or soils with appreciable organics.

ASTM D4643 allows for drying in a *microwave oven*. One of the biggest concerns with this method is overheating the soil so as to give incorrectly high water content determinations. This issue can be mostly alleviated by incremental drying or heating at lower power levels. This method may not be appropriate if high accuracy results are required, and should not be used for soils with hydrated materials, high plasticity clays, or appreciable organics.

The *Proctor needle* method (ASTM D1558) was developed for indirect rapid determination of compacted soils *in situ*. The test uses a spring-loaded needle calibrated to measure the penetration resistance of the compacted soil. Various sized needle tips are available for different soils and strengths. The penetrometer is pushed into the soil at a uniform rate of approximately 1.25 cm/s to a depth of 7.5 cm. Moisture content is determined from calibration curves generated from samples of the same material compacted and tested in the laboratory by test methods D698 or D1557. Accuracy is reasonably good for fine-grained soils, but not reliable for very dry or very granular soils.

ASTM D5080 allows for rapid determination of percent compaction and variation from optimum moisture content of soils used in construction without knowing the value of field moisture content at the time of the test.

It does not require determination of accurate moisture content, and results can be ready in 1-2 h (Holtz et al., 2011). The test is normally performed on soils with more than 15% fines, and may not provide accurate results on "clean" granular soils. For the test, a representative sample is compacted in accordance with standard Proctor test parameters: one at in place moisture, two more at other moisture contents. Assume a parabolic curve for the three field test samples and compare to the *moist* sample curve prepared in the laboratory.

5.6.1.4 Combined Tests

Nuclear gauges offer the ability to quickly determine both density and moisture content. Requiring no physical or chemical processing of the material being tested, these compact, lightweight gauges (Figure 5.22) meet ASTM D6938 for compaction control acceptance testing. Updated models provide a direct readout of wet density, dry density, percent moisture, and percent compaction (if a specified level is input) (www.troxlerlabs.com). They have additional advantages of data storage and download capabilities, along with stored GPS coordinates for all collected data. Measurements are made by transmission of gamma rays into and through the soil and counted by a detector located in the same device. The standard "direct" method of measurement (Figure 5.23a) involves inserting a probe into the compacted soil from which gamma rays are transmitted to detectors and counted by the device at the ground surface. This method is suitable for testing compacted

Figure 5.22 Field operation of a nuclear gauge. *Courtesy of Troxler Laboratories.*

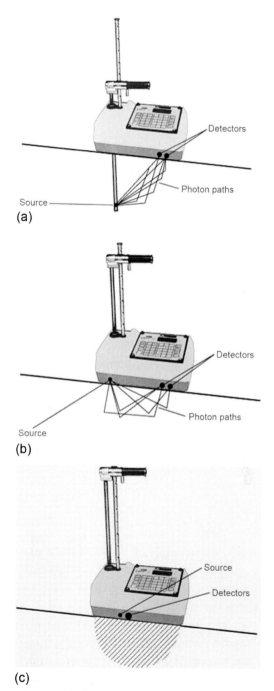

(a)

(b)

(c)

Figure 5.23 Operation modes for a nuclear gauge. (a) Direct method, (b) backscatter (indirect) method, (c) moisture reading. *Courtesy of Troxler Laboratories.*

lifts of up to 30 cm (12 in) in thickness. A completely nondestructive version of the measurement uses a *backscatter* "indirect" mode (Figure 5.23b), where photon paths are scattered into the ground from the device at the surface and then travel through near-surface soils to the detectors. This limits the use primarily to density determinations in shallow ($<$10 cm $=$ 4 in) layers. The moisture measurement is also nondestructive, where neutrons are emitted into the soil by the device at the surface (Figure 5.23c). Fast neutrons are slowed when encountering hydrogen atoms present in the soil water. A helium3 detector in the gauge detects and counts the number of slowed (thermalized) neutrons, which relates directly to the amount of moisture in the soil (www.troxlerlabs.com). In most cases, the results for *in situ* dry density and moisture content are very accurate. The equipment has a moderate up-front cost, and certification for use is required. Some precautions are also necessary, as the equipment is technically a nuclear device.

5.6.1.5 Accept/Reject Criteria

In theory, *all* field control test data should meet or exceed specification limits. But in reality, it is generally acceptable to allow for a small number (or percentage) of data points to fall below or outside of the specifications depending on the critical nature of the project, or portion of a project dependent on performance of the compacted material. While there is no set limit for this allowance, a general rule of thumb in the industry is that up to 5% (or sometimes more) of the compaction control test data points may be allowed to "fail" meeting the specifications, especially if failed points are relatively close to passing. This "unofficial" allowance takes into consideration that the average compaction values are likely to be well above the specified limits.

5.6.1.6 Performance Tests

In addition to field control tests to assure moisture and density specifications have been met for compaction conditions, some other tests may be appropriate to assess the performance of compacted soils. A few of the most common performance tests are described here.

Field California bearing ratio (CBR) tests (ASTM D4429) have been commonly used as a strength parameter to test various components of pavements such as subgrades, subbases, and base coarse layers, or for unpaved roadways. The test is essentially a penetration test, whereby a value is determined by

measuring the load required to penetrate a standard probe into the compacted soil surface. Several correlations are available to estimate engineering
properties. As strength can be a function of water content, ASTM recommends that direct evaluation or design with field CBR should only be
performed when (a) the degree of saturation is greater than 80%, (b) when
the material is coarse-grained and cohesionless (so that water content is not
significant), or (c) when the soil has not been modified by construction
activities for 2 years preceding the test.

The DCP is a low-cost tool designed to provide a measure of *in situ*
strength for a range of granular and fine-grained natural and modified
near-surface soils. It has been widely used for compaction control and roadway work, and test results have been correlated to determinations of
required base coarse thickness needed over a subgrade material, or to indicate where (additional) soil modification may be needed. This penetrometer
test simply involves counting the number of blows needed to drive a standard 20-mm diameter, 60° cone into the soil with a manually dropped hammer. There are a few different versions of the test. ASTM D7380 uses a
2.3 kg (5 lb) weight, falling 508 mm (20 in) and measures the number of
blows required to drive the cone 83 mm (3.25 in) into the soil. ASTM
D6951 is a version of the DCP that uses an 8 kg (17.6 lb) hammer, falling
57.5 cm (22.6 in), and measurements are made for a penetration depth of
15 cm (6 in). The length of the drive stem allows recording penetration
resistance up to a depth of 90 cm (36 in).

Plate load tests provide a direct measure of load bearing capacity
(strength), as well as deformation under load. The plate load test basically
consists of loading a steel plate placed at the level where the actual load is
to be applied, and recording the deformation (settlement) corresponding
to successive load increments. The test load is either gradually increased until
the plate starts to sink at a rapid rate (failure), or loaded with typically two to
three times the design load to establish a suitable margin of safety. Data is
usually plotted as load–deformation curves, which very clearly provide visual
indicators of the soil response to loading. If the plate is loaded to "failure,"
then the total value of load on the plate divided by the area of the steel plate
gives the value of the ultimate bearing capacity of soil, from which the value
of safe bearing capacity of the soil can be derived. From the load–deflection
measurements, subgrade modulus can also be determined. For some applications, a maximum permissible amount of settlement may control design.
The load capacity corresponding to amount of settlement can be easily

obtained from the load settlement curves by reading the value of load intensity corresponding to the limiting settlement of the test plate. Obviously, because the short duration of the test will not be representative of time-dependent consolidation settlements or for the settlement of loaded areas other than the shape and size of the test plate, the following formula can be used to relate settlement of the test plate to settlement of the actual footing dimensions:

$$Sf = Sp \left\{ \frac{B(Bp + 0.3)}{Bp(B + 0.3)} \right\}^2 \tag{5.5}$$

where B is the width of footing in mm, Bp the width of test plate in mm, Sp the settlement of test plate in mm, Sf the settlement of footing in mm.

Bearing plates are typically round or square, but may also be configured in other shapes. For concentrated foundation loads, these tests are ideally performed with a plate of the actual dimensions of the design footing along with the full design load to avoid scaling effects and limited testing depth. Loads are typically applied either by gravity (i.e., dead weights), or by means of a securely anchored "stiff" reaction frame for higher loads. ASTM D1195 and D1196 are primarily designed for testing of pavement components. ASTM D1195 applies repetitive static loads, while D1196 uses only a single load.

RELEVANT ASTM STANDARDS

D698-12 Standard Test Methods for Laboratory Compaction Characteristics of Soil Using Standard Effort $(12,400 \text{ ft lbf/ft}^3 (600 \text{ kN m/m}^3))$, V4.08; similar to AASHTO T-99

D1195/D1195M-09 Standard Test Method for Repetitive Static Plate Load Tests of Soils and Flexible Pavement Components, for Use in Evaluation and Design of Airport and Highway Pavements, V4.03

D1196/D1196M-12 Standard Test Method for Nonrepetitive Static Plate Load Tests of Soils and Flexible Pavement Components, for Use in Evaluation and Design of Airport and Highway Pavements, V4.03

D1556-07 Standard Test Method for Density and Unit Weight of Soil in Place by the Sand-Cone Method, V4.08

D1557-12 Standard Test Methods for Laboratory Compaction Characteristics of Soil Using Modified Effort $(56,000 \text{ ft lbf/ft}^3 (2700 \text{ kN m/m}^3))$, V4.08; similar to AASHTO T-180

D1558-10 Standard Test Method for Moisture Content Penetration Resistance Relationships of Fine-Grained Soils, V4.08

D1561-92(2005)e1 Standard Practice for Preparation of Bituminous Mixture Test Specimens by Means of California Kneading Compactor, V4.03

D1883-07e2 Standard Test Method for CBR (California Bearing Ratio) of Laboratory-Compacted Soils, V4.08

D2167-08 Standard Test Method for Density and Unit Weight of Soil in Place by the Rubber Balloon Method, V4.08

D2168-10 Standard Test Methods for Calibration of Laboratory Mechanical-Rammer Soil Compactors, V4.08

D2216-10 Standard Test Methods for Laboratory Determination of Water (Moisture) Content of Soil and Rock by Mass, V4.08

D2844/D2844M-13 Standard Test Method for Resistance R-Value and Expansion Pressure of Compacted Soils, V4.08; equivalent to AASHTO T-190

D2974-07a Standard Test Methods for Moisture, Ash, and Organic Matter of Peat and Other Organic Soils, V4.08

D4253-00(2006) Standard Test Methods for Maximum Index Density and Unit Weight of Soils Using a Vibratory Table, V4.08

D4254-00(2006) Standard Test Methods for Minimum Index Density and Unit Weight of Soils and Calculation of Relative Density, V4.08

D4429-09a Standard Test Method for CBR (California Bearing Ratio) of Soils in Place, V4.08

D4643-08 Standard Test Method for Determination of Water (Moisture) Content of Soil by Microwave Oven Heating, V4.08

D4718-87(2007) Standard Practice for Correction of Unit Weight and Water Content for Soils Containing Oversize Particles, V4.08

D4914-08 Standard Test Methods for Density and Unit Weight of Soil and Rock in Place by the Sand Replacement Method in a Test Pit, V4.08

D4944-11 Standard Test Method for Field Determination of Water (Moisture) Content of Soil by the Calcium Carbide Gas Pressure Tester, V4.08

D4959-07 Standard Test Method for Determination of Water (Moisture) Content of Soil by Direct Heating, V4.08

D5030/D5030M-13 Standard Test Method for Density of Soil and Rock in Place by the Water Replacement Method in a Test Pit, V4.08

D5080-08 Standard Test Method for Rapid Determination of Percent Compaction, V4.08

D6938-10 Standard Test Method for In-Place Density and Water Content of Soil and Soil-Aggregate by Nuclear Methods (Shallow Depth), V4.09

D6951/D6951M-09 Standard Test Method for Use of the Dynamic Cone Penetrometer in Shallow Pavement Applications, V4.03
D7380-08 Standard Test Method for Soil Compaction Determination at Shallow Depths Using 5-lb (2.3 kg) Dynamic Cone Penetrometer, V4.09

REFERENCES

D'Onofrio, A., Penna, A., 2003. Influence of compaction variables on the small strain behaviour of a clayey silt. In: Benedetto, D. et al., (Ed.), Deformation Characteristics of Geomaterials. Swets & Zeitlinger, Lisse, 1433 pp.

Demars, K.R., Chaney, R.C. (Eds.), 1982. Geotechnical Properties, Behavior, and Performance of Calcareous Soils. ASTM, Special Technical Publication 777, 415 pp.

Hausmann, M.R., 1990. Engineering Principles of Ground Modification. McGraw-Hill, Inc, 632 pp.

Holtz, R.D., Gibbs, H.J., 1956. Engineering properties of expansive clays. Trans., ASCE 121, 641–677.

Holtz, R.D., Kovacs, W.D., Sheahan, T.C., 2011. An Introduction to Geotechnical Engineering. Prentice Hall, 853 pp.

Houston, S., Walsh, K., 1993. Comparison of rock correction methods for compaction of clayey soils. J. Geotech. Eng. 119 (4), 763–778.

Jaska, M.B., Scott, N.L., Mentha, A.T., Symons, S.M., Pointon, P.T., Wrightson, P.T., Syamsuddin, E., 2012. Quantifying the zone of influence of the impact roller. In: Proceedings of International Symposium on Ground Improvement IS-GI, Brussels, p. II(41–52).

Lambe, T.W., 1958a. The structure of compacted clay. J. Soil Mech. Found. Div., ASCE 84 (SM2), 1654-1–1654-35.

Lambe, T.W., 1958b. The engineering behavior of compacted clay. J. Soil Mech. Found. Div., ASCE 84 (SM2), 1655-1–1655-35.

Lin, D.-F., Chang, M.-K., Luo, H.-L., 2001. Study of scalp and replace equation in compaction specifications for soil with high gravel content. Transportation Research Record, vol. 1772. National Academies, Washington, DC, 55–61.

Mitchell, J.K., Hooper, D.R., 1965. Permeability of compacted clays. J. Soil Mech. Found. Eng., ASCE 94 (4), 41–65.

Murthy, V.N.S., (2002). Geotechnical Engineering: Principles and Practice of Soil Mechanics and Foundation Engineering. CRC Press, 1056 pp.

Seed, H.B., Chan, C.K., 1959. The structure and strength characteristics of compacted clays. J. Soil Mech. Found. Div., ASCE 85 (SM5), 87–127.

Xanthakos, P.P., Abramson, L.W., Bruce, D.A., 1994. Ground Control and Improvement. John Wiley & Sons, Inc, New York, 910 pp.

http://www.blackgeotechnical.com.au/index.html (accessed 05.07.13.).

http://www.broons.com (accessed 05.06.13.).

http://www.contrafedpublishing.co.nz/Historical/The+history+of+the+compactor.html (history of the compactor) (accessed 04.04.13.).

http://www.dot.il.gov/materials/research/pdf/ptat4.pdf (accessed 05.29.13.).

http://www.dynapac.com (accessed 04.05.13.).

http://www.fhwa.dot.gov/engineering/geotech/pubs/05037/08.cfm (accessed 05.17.13.).

http://www.igesinc.com (accessed 4/28/13).

http://www.impactor2000.com/soil.html (Impact Roller Technology, accessed 05.07.13.).

http://www.intrans.iastate.edu/research/documents/research-reports/High%20Energy%
   20Impact.pdf (accessed 05.07.13.).
http://www.landpac.co.za/Index.html (accessed 05.07.13.)
http://www.nevadadot.com/uploadedFiles/NDOT/About./MTM_T101F.pdf   (accessed
   04.28.13.)
http://www.theconstructioncivil.org/plate-load-test-determine-bearing-capacity-of-soils
   (accessed 05.29.13.).
http://www.thefreedictionary.com/knead (accessed 05.15.13.).
http://www.troxlerlabs.com (accessed 05.28.13.).

CHAPTER 6

Deep Densification

Densification of soils at depth is by its very nature an in situ process. There are a number of methods that can be utilized depending on project-specific variables. In particular, the soil type will play a major controlling factor in the choice of method(s) applicable. The objectives are principally the same as for other densification applications. In addition, some of the methods are applicable to irregular fills and variable ground conditions. One of the major uses of deep densification techniques is for liquefaction mitigation. This has contributed greatly to the widespread use of deep densification worldwide.

The material in this chapter covers an array of available methods for densifying soil in situ to significant depths. The choice of method or application will depend on several variables, including soil type, uniformity, fines content, saturation, pretreatment density, degree of improvement needed, required uniformity of improved ground, location (proximity to existing and critical structures), and other specific project requirements. Available techniques and equipment are described along with some general guidelines on uses and quality control (QC) parameters, including design specifications. While not purely a densification process, related construction of gravel or stone columns is included here, because that method mostly uses the same equipment as some deep densification applications, and can often include a significant densification component. Special techniques, such as compaction grouting, are required where there is existing infrastructure or where access in difficult.

6.1 DEEP DENSIFICATION APPLICATIONS AND TECHNIQUES

A number of very different techniques have been developed for in situ densification of soils at depth. Each particular method will have advantages and disadvantages depending on the variables previously mentioned (i.e., soil type, soil variability, depth requirement, uniformity requirement, etc.). Costs associated with deep densification techniques are somewhat difficult to state a priori, as they will vary by size, depth, and other specifics of each project. What can be approximated are general relative costs between different deep densification alternatives. Following some relative cost

Soil Improvement and Ground Modification
Methods

115

guidelines proposed by Xanthakos et al. (1994), some rough approximations can be made between some alternative methods:

Deep dynamic compaction (DDC) = 1-6

Vibrocompaction (VC) = 2-14

Stone columns = 10-22

Compaction grouting = 30-200

6.1.1 Blasting

Blast densification, also known as explosive compaction, or deep blasting, has been used to densify loose sandy soils since the 1930s (Narsilio et al., 2009). This method fundamentally involves setting off explosive charges at prescribed depths, generating shock waves through the ground. Many case histories have shown its effectiveness at densifying uncemented granular deposits to significant depths (up to 35 m = 105 ft or more!). Applications have included dam sites in Canada, India, Nigeria, Pakistan, and the United States, transmission towers, power plants, airport projects, highways, bridges, mines, offshore platforms and man-made islands, as well as liquefaction and earthquake experiments.

Blast densification is typically most effective for deposits with relative density less than about 50-60%, and for saturated, free-draining soils (Narsilio et al., 2009). It can achieve relative densities on the order of 70-80%. This technique is limited to soils that contain little clay content (generally less than 5%) with a total of no more than 15-20% fines (minus #200 sieve). It is also important that the moisture condition is such that there will be little or no surface tension forces (e.g., best if dry or saturated). These limitations are principally due to the need to overcome internal strength and allow dissipation of pore pressures generated from the dynamic energy released. Blasting works by generating radial shock waves, which initially causes compression (P-waves) in the soil mass, followed by a rarefaction wave front. The cycling of compression and expansion creates a shear force that assists in collapsing the soil structure (Dowding and Hryciw, 1986; Narin van Court and Mitchell, 1998). The compression of a loose, saturated soil creates generation of an excess positive pore pressure that may reach the initial effective stress, thereby creating a state of transient liquefaction. The effects can be seen at the surface by transient surface "jump" and expulsion of excess water pressure (Figure 6.1). Under these conditions, the soil will rearrange into a denser packing as the soil grains resediment. Densification is expected to be significant, with greater densification in initially looser deposits, demonstrated by rapid settlement after blasting of up to 2-10% of the treated layer thickness. Penetration resistance is

Figure 6.1 Example of field blasting showing expulsion of excess pore water. *Courtesy of Explosive Compaction, Inc.*

commonly used to evaluate the degree of densification, although it should be noted that an increase in penetration resistance may take weeks or even months to be fully observed. In fact, in some cases the penetration resistance measured shortly after blasting has been found to decrease even though significant settlement has taken place. The reasoning is that some light cementation or resistance of the initial soil structure may be broken down by the blasting, while at the same time pore water pressures generated by the blasting may result in lower than expected resistance. With time, pore water pressure dissipates and soil grains sediment into a tighter configuration, ultimately resulting in higher resistance measurements. In most cases, penetration resistance values increase by as much as 50-200%.

Design of blast densification applications involves a number of variables, including (1) mass (weight) of explosives per unit volume of soil, (2) location of charges (lateral spacing, patterns, depths, and vertical distribution), and (3) number and sequence of events. Designs have generally been developed by experience rather than from analytical theory (Narsilio et al., 2009). Usually the explosives are arranged in a lateral grid pattern with typical spacing of 3-8 m. The radius of influence is a function of the size (weight) of the charge and has been estimated by the following relationship (Mitchell and Soga, 2005):

$$W = 164CR^3 \tag{6.1}$$

where W is the weight of explosive (N), C the coefficient (approximately 0.0025), R the radius of influence (m).

Explosive charges are typically placed at 2/3 of the depth of the layer to be treated for deposits of up to 10 m. When the depth of soil to be improved is greater than about 10 m, multiple charges have been prescribed at different depth horizons (Raj, 1999). Generally, charges are detonated in time-delayed sequence, from bottom upward and to take advantage of residual, transient shear waves and loosening of the soil structure from detonation of previous charges. Experience has indicated that repeated blasting of smaller charges with interim "rest periods" is more effective at achieving desired results than single, larger charges (Murthy, 2002).

Some of the greatest advantages of the blasting technique are the lack of any special construction equipment needed, minimal labor, and the speed of application. One only needs to get the explosive charges in place at the desired depths; this is done typically through conventional borings, but in some cases can be achieved by hydraulic pushing similar to advancing a cone penetrometer. An obvious disadvantage is the possible disruption of adjacent property due to vibrations and displacement, and there is sometimes a perceived danger associated with use of explosives, although this has little real merit. Thus, use of this method is usually limited to development and/or redevelopment of sites not immediately adjacent to sensitive properties. Also, as with other vibratory densification techniques, blasting may disrupt or loosen the near surface soils, which must then be densified by conventional equipment.

6.1.2 Vibrodensification

Vibrational loading is most effective at densifying cohesionless or mostly granular soil materials. The vibrations overcome the frictional resistance in granular soils, rearranging loose, cohesionless grains into a denser packing. This was described in Chapter 4 and again in Chapter 5 for shallow compaction of granular soils. With this understanding, equipment was originally devised (and patented) in Germany in the 1930s for in situ deep densification of granular soil deposits. Using the same basic equipment, a few different construction tools have been developed. The benefits provided to ground modified with the use of "vibro" systems may be considered to fall into three categories: (1) improvement of material properties (i.e., shear strength, stiffness, dynamic shear modulus, reduced compressibility, etc.), (2) drainage, and (3) reinforcement (Lopez and Hayden, 1992). As vibratory methods have been shown to be effective at densifying loose granular soils, it should be no surprise that these methods have been widely used for mitigation of liquefaction and earthquake-related deformations. Vibrodensification is now commonly used worldwide for a vast range of projects.

6.1.2.1 Vibrodensification Equipment

Most vibrodensification systems utilize downhole variable frequency vibratory probes (or *vibroflots*) that come in a variety of sizes and configurations. The probes can range in size from approximately 30 to 45 cm (12-18 in) in diameter, and 3-5 m (10-16 ft) in length. They are now manufactured by a number of different companies around the world. These probes are typically suspended from a standard crawling crane. Vibrations are generated by motor-driven, rotating eccentric weights mounted on an internal vertical shaft. The rotating action generates vibrations that travel laterally and propagate radially away from the vibrator. Vibratory compaction generates lateral stresses, which result in imparting permanent increases in lateral stresses. The vibrator penetrates the ground as it is lowered vertically under its own weight, typically assisted by high-pressure water/air jetting. A schematic of a typical vibroflot is shown in Figure 6.2, and a photo is shown in Figure 6.3.

Modern VC equipment is now most often instrumented with onboard computers capable of monitoring construction in real time. Typical parameters of energy consumption (amperage), lift rate, and so on, can be monitored and compared to target values, allowing the operator to make adjustments as construction progresses. Data is recorded and so can also be reviewed later for quality assurance (QA).

Until the 1970s, the vibroflot was the only vibrodensification tool available. Since then, a number of other variations have been developed. A theoretically less expensive alternative to the vibroflot that gained some popularity is the terraprobe, which works on much the same principles as the vibroflot, but with some important differences. First, there are no specialized equipment or water/air jets involved. The terraprobe is essentially a hollow, rigid, open-ended pipe, typically about 0.75 m (30 in) in diameter, driven by a vertical vibrating hammer, similar to those used for driving sheet piling. The major attractions of the terraprobe were that field studies showed that densification rates were approximately four times that of VC and generally did not require water jetting to reach maximum depths. However, in most cases, this method, along with other variations, has not shown much advantage because the spacing required to get the same densities requires at least four times as many probe holes, and maximum densities achieved by the vibroflot are still greater (Brown and Glenn, 1976). A resurgence of this type of method incorporates a variety of probe designs, including the use of an "H" pile probe with significantly higher horsepower vibratory hammers. More recent implementation of such equipment has shown promise for deep densification improvements in gravelly sands, particularly when saturated or below the water table. This lends itself well to liquefaction

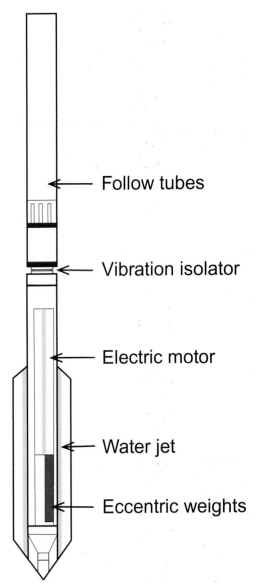

Figure 6.2 Schematic of a typical vibroflot.

mitigation. Case studies report density increases of more than 250% as measured by standard penetration test N_{60} values (Nottingham, 2004). For one example case, the average N_{60} blow counts increased from 26 to 66. Some other purpose-built probes of various geometric designs (e.g., Vibrorod, Y-probe, Vibro Wing, MRC compaction probe) have been designed and

(a)

(b)

Figure 6.3 Photographs of vibrocompaction (VC) probes in the field: (a) with vertical water jets (Courtesy Earth Tech, Inc.); (b) with vertical water jets (Courtesy Earth Tech, Inc.).

implemented with some variations in results (Massarsch and Fellenius, 2005). A limitation of these "waterless" vibratory probes is the inability to reach depths much greater than about 10-15 m (30-45 ft).

VC usually refers to the densification of sandy soils with generally less than 15% fines. It was found that the deep densification vibratory equipment would more easily penetrate the ground and provide better densification with the addition of water or air jets integrated into the vibrator assembly. This equipment was found to be able to readily penetrate not only mostly granular soils, but many additional strata, including dense gravelly soils, as well as a wide range of fine-grained soils and heterogeneous fills. The probe penetrates the ground to the depth of the bottom of the treatment zone. The vibratory energy (and water jetting if equipped) laterally densifies the soil around the probe. During the process, additional "similar" fill material is added to the annulus created by the vibratory probe to compensate for the reduction in volume and compacted to create a uniform densified stratum (Figure 6.4). Relative densities of 70-85% can typically be achieved, improving the soil strata both above and below the water table, and achieving allowable bearing pressures of up to about 480 kPa (10 ksf) (www. haywardbaker.com). This allows economical shallow spread footings, which may otherwise be insufficient. While most applications require a treatment

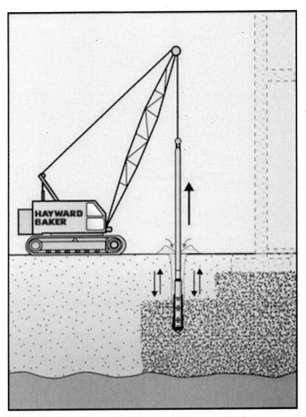

Figure 6.4 VC installation schematic. *Courtesy of Hayward Baker.*

depth of around 5-15 m (15-50 ft), successful applications using vibroflots have reached depth of up to 50 m (approx. 160 ft). Improvements will depend on the initial in situ conditions. In unsaturated zones, the additional water provided by vibroflot-type equipment aids in collapsing any structure and lubricating the soil grains, allowing them to be rearranged in a more closely packed configuration. Below the water table, the water jets increase pore pressure, effectively creating a state of transient liquefaction, which allows rearrangement of soil grains into a denser configuration as they settle during dissipation of pore water pressures.

It has been demonstrated through experience and analyses that vibration frequency plays an important role in the densification process. While relatively high frequencies (above 30 Hz) can aid in penetration of probes, lower frequencies of about 15-20 Hz tend to be close to the natural frequency of the ground so that more energy is transferred to the surrounding soils as the probe and soil achieve resonance (Massarsch and Fellenius, 2005). Degree of improvement of soil characteristics by VC is also dependent on spacing between penetration points and time spent (duration of) compacting. Typical VC spacing is between 2 and 5 m (6 and 14 ft), with compaction centers arranged in a triangular or square pattern. Closer spacing typically results in increased density and uniformity.

Vibroreplacement refers to the process used in fine-grained soils or soils otherwise unsuitable for VC (due to excessive fines or other deleterious materials), whereby the existing soil materials are replaced with coarse aggregate (gravel or crushed stone) to form *stone columns*. The aggregate is compacted in incremental lifts through vertical and horizontal forces resulting from the equipment weight and induced vibrations to form well-compacted, tightly interlocked stone columns surrounded by the adjacent densified soil (Figure 6.5). Stone columns, generally constructed with 0.6-1 m (2-3 ft) diameters, provide substantial load-bearing capacity as well as offer reasonably good drainage. As a general rule that has withstood the test of time, granular drains are considered to be satisfactory if their permeability is at least 20 times that of the soil being drained. A concern when combining use of materials with such disparate permeabilities then becomes whether the hydraulic gradient will be so high as to promote internal erosion and/or clogging of the "drain." A more detailed discussion of drainage and filtering guidelines and requirements will be addressed in the chapters concerned with hydraulic modification.

Construction of stone columns results in a composite foundation system with stiff, strong elements that can also be considered as reinforcement

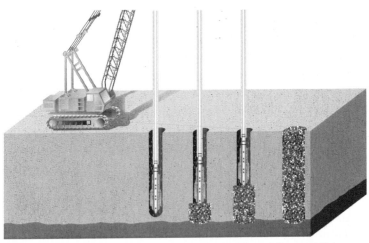

Figure 6.5 Stone columns installation schematic. *Courtesy of Hayward Baker.*

components and, as such, have also been used for slope stabilization or to resist lateral deformations due to earthquake-related loads. Stone columns have also been used in saturated fine-grained soils. They assist and expedite consolidation by both exerting an increased lateral confining load on the preexisting fine-grained soils, while at the same time providing a greatly reduced drainage path for dissipation of generated pore water pressures. This greatly speeds up consolidation times. Often, a layer of aggregate is placed across the surface of the gravel/aggregate columns to provide load distribution and also to provide lateral drainage when the columns are functioning in a drainage capacity. Some studies have been made to assess the value of using *composite stone columns* along with prefabricated vertical drains (PVDs, a geocomposite to be discussed later in Chapter 8) to speed up improvement and provide reinforcement of fine-grained soils.

Vibrodisplacement is a term sometimes used to describe the use of stone columns installed with vibrator probes in primarily cohesionless soils. In these cases, the existing soil remains in the ground and is densified in part by the vibratory probe and then by further lateral *displacement* by the compacted stone column. This method can provide added capacity to sandy soils that cannot be achieved by VC alone. Along with densification of the cohesionless soils, the drainage capacity of stone columns may be so large that they can be utilized as a means of liquefaction mitigation in loose sandy and silty soil deposits. For effective use as liquefaction mitigation, drainage needs to dissipate excess pore pressures generated by dynamic loads, and so permeability of the drain materials should be at least 200 times that of the soil being drained.

6.1.2.2 Construction Methods

(Wet) Top feed method (replacement and displacement): With this method, water is jetted under high pressure from the nose of the vibroflot to assist with penetration of the probe. Additional water jets are also sometimes located along the side of the probe to loosen and remove soft material with the upwelling ejected fluid, and ensure that surrounding soils will be stabilized by induced horizontal forces invoked by the vibrator and jets. This keeps open the space created by the vibrator so that added material introduced at the ground surface will be able to reach the nose cone of the probe. Fill material, typically either sand or stone aggregate, is continually added at the surface through the annulus created around the probe (Figure 6.6). This is the most commonly used and most cost-effective of the deep vibratory construction methods. The backfill is typically densified in 0.7-1 m (2-3 ft) lifts by repeatedly raising and lowering the vibroflot. Wet spoil generated by using this method (particularly for vibroreplacement), must be carefully managed, especially when working on confined sites or in environmentally sensitive areas. Dry top feed construction is also possible to alleviate problems with wet spoil. This works for cases when an open hole can be drilled to depth at a diameter of between approximately 75-90% of the finished column diameter.

(Dry) Bottom-feed method (displacement): This method is used when the annulus around the vibratory probe has a tendency to close around the probe (such as unstable soils and/or soils below the water table) so that the backfill material must be introduced at the nose of the probe. A hopper system (Figure 6.7) with a supply tube feeds stone backfill directly to the nose cone of the vibroflot. The vibrating probe is then used to compact the fill material in

Figure 6.6 Stone column field application. *Courtesy of Hayward Baker.*

Figure 6.7 Hopper system for bottom-feed vibroreplacement. *Courtesy of Hayward Baker.*

subsequent, incremental lifts. Bottom-feed vibroreplacement is generally a dry operation with little spoil, enabling its use to a greater range of sites (www. haywardbaker.com). Often, the vibrating probe can penetrate to its full design depth, either under its own weight or with the assistance of air jets. In some cases, planned stone column locations are predrilled to facilitate the penetration of the vibrator. Dry bottom feed of well-graded gravel may also be assisted with air jetting (Xanthakos et al., 1994; www.earthtech.net).

6.1.2.3 Compacted Aggregate Piers

Other versions of stone columns have been developed and are known as *Rammed Aggregate Pier*® *systems (RAPs or Geopiers) and vibropiers*. These versions are generally shorter (shallower in depth) than stone columns, to intermediate depths of typically 3-10 m (10-33 ft), but have been successfully installed to depths of 15 m (46 ft). The difference between vibropiers and RAPs is that the first are compacted using a vibrating probe similar to that used for VC and vibroreplacement, while the RAP columns are tamped in lifts with an impact hammer or rammer tool. Both methods densify or compress the surrounding soil by expanding the annulus of the initial cavity.

The RAP provides an efficient and cost-effective solution for intermediate foundation depths. They may provide up to a 20-50% savings over traditional deep foundations, and may be installed at a rate of 30-60 piers a day (www.geopier.com). Originally developed in 1989, RAPs can provide strengths and stiffnesses reportedly 5–10 times that of stone columns (www. farrellinc.com). In soil that will remain stable without caving or collapse, the holes are first drilled, which allows inspection of the subsurface where the piers will be installed. This method is applicable to clays, silts, organic soils, and variable fills. Aggregate is then placed by a top feed process (or if the soil has a tendency to cave or collapse, then a bottom-feed method must be utilized) and compacted in lifts (Figures 6.8 and 6.9). An alternative type of Geopier is a full-displacement method, where the impact rammer is pushed into the ground without predrilling a hole. Compaction material (e.g., natural aggregate or recycled concrete) is supplied by bottom feed using a patented hollow mandrel/tamper (Figure 6.10). This type of method is best for liquefaction mitigation in conditions of high water table in sands and silts. Bearing capacity of compacted aggregate piers has been reported to be as

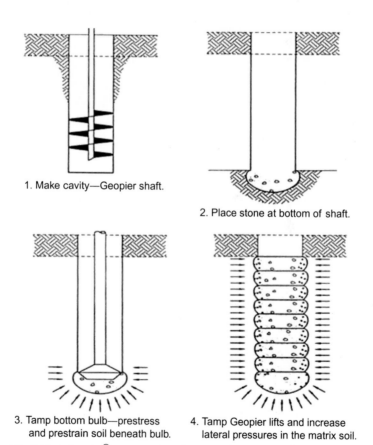

1. Make cavity—Geopier shaft.

2. Place stone at bottom of shaft.

3. Tamp bottom bulb—prestress and prestrain soil beneath bulb.

4. Tamp Geopier lifts and increase lateral pressures in the matrix soil.

Figure 6.8 Drilled Geopier® construction. *Courtesy of Farrell Design-Build, Inc.*

Figure 6.9 Equipment used for installation of a Rammed Aggregate Pier® system. *Courtesy of Geopier Foundation Company, Inc.*

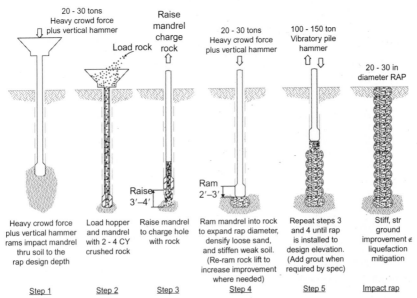

Step 1	Step 2	Step 3	Step 4	Step 5	Impact rap
Heavy crowd force plus vertical hammer rams impact mandrel thru soil to the rap design depth	Load hopper and mandrel with 2 - 4 CY crushed rock	Raise mandrel to charge hole with rock	Ram mandrel into rock to expand rap diameter, densify loose sand, and stiffen weak soil. (Re-ram rock lift to increase improvement where needed)	Repeat steps 3 and 4 until rap is installed to design elevation. (Add grout when required by spec)	Stiff, str ground improvement ℓ liquefaction mitigation

Figure 6.10 Construction sequence of a full-displacement impact pier. *Courtesy of Farrell Design-Build.*

much as 220-530 kN (50-120 kip) (www.haywardbaker.com) or up to 68,000 kPa (10,000 psf) per pier (www.geopier.com). RAPs have also been demonstrated to have significant uplift capacity when fitted with a confining anchor assembly, allowing uplift capacities nearly that of bearing capacity per pier. With a coefficient of friction of 0.4-0.5 (internal friction angles of up to 50°), RAPs also provide significant shear resistance, which is important for the resistance of lateral loads induced by wind, earthquakes, or slope forces.

Compacted aggregate piers have been used for many of the same types of applications as for gravel columns, including support for traditional shallow foundations and slabs, embankments, stabilized earth retaining walls (Chapter 14), industrial and storage tanks, slope stabilization, and liquefaction mitigation. An example of a RAP-reinforced site is shown in Figure 6.11.

6.1.3 Dynamic Compaction

Dynamic compaction (DDC, heavy tamping, dynamic consolidation, etc.) is a cost-effective method of soil compaction whereby a heavy weight is repeatedly lifted and dropped from a height, impacting the ground surface with a readily calculated impact energy (Figures 6.12 and 6.13). Costs are reportedly about 2/3 that of stone columns, with up to 50% savings over other deep densification alternatives (www.wsdot.wa.gov). Dynamic compaction is one of

Figure 6.11 Geopier® reinforced site. *Copyright by Geopier Foundation Company, Inc. Reprinted with permission*

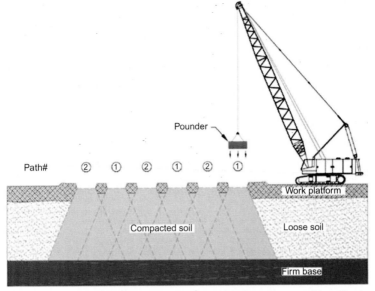

Figure 6.12 Schematic of deep dynamic compaction (DDC). *Courtesy of Densification, Inc.*

Figure 6.13 Photos of DDC field applications. *Courtesy of Hayward Baker (top) and Densification, Inc. (bottom).*

the oldest soil improvement methods known, reportedly used by the Romans prior to 100 AD and in the United States as early as the 1800s (Welsh, 1986). The name does not accurately portray the actual loading and energy transmission processes. One of the greatest misnomers regarding dynamic compaction is that it is a surface ground treatment, as loads are applied at the surface. But dynamic compaction, as opposed to conventional shallow compaction of controlled fill, is a process of densifying soils at significant depths by applying a large impact energy at the ground surface. Upon impact, craters as deep as six feet or more are created, which must then be backfilled prior to additional compaction passes and ultimately at the completion of the compaction process. But the densification at depth occurs as a result of the dynamic wave energy that is transmitted through the ground.

The main objectives of dynamic compaction are to improve strength and compressibility characteristics by either creating a uniform raft of densified material, or by compacting at locations where concentrated loads (e.g., column loads) will be applied. Improved soil properties result in increased bearing capacity and reduced settlements, including differential settlements. Dynamic compaction often allows for construction of conventional spread footings by providing bearing capacity of typically as much as 100–150 kPa (2000–3000 psf).

Dynamic compaction is suitable for densification of loose sand deposits such as those typically occurring in coastal, glacial, and alluvial deposits, as well as for dredged or hydraulically placed fills. This method has also been successfully applied to mine tailings, landfills, collapsible soils, sites underlain by sinkholes, and so forth (Zekkos et al., 2013). It is one of the better alternatives to densification of heterogeneous fills, and fills containing large debris that may create obstructions for other remediation techniques, such as stone columns or rigid inclusions (www.menard-web.com). Results are best for well-drained, high permeability soils with low saturation, although some satisfactory results have been reported for improvement of silty soils with the aid of PVDs or stone columns (or composite stone columns employing supplemental PVDs), and by providing time delays to allow for the dissipation of generated pore pressures (Dise et al., 1994; Shenthan et al., 2004). In certain conditions, saturated soils will be temporarily liquefied, allowing easier rearrangement and ultimately a tighter, denser packing upon dissipation of pore pressures. Because of this phenomenon and the benefits it can provide, "rest periods" between drop phases are sometimes specified, during which pore pressure dissipation can be monitored with piezometers to assure completion. This method is not appropriate for saturated clay soils.

Applications consist of dropping a heavy tamper (weight) from a specified height a calculated number of times at precisely determined locations in a pattern at the site. Drop patterns usually consist of primary and secondary (and occasionally tertiary) grids such as depicted in Figure 6.14. Grid spacing is typically about 3-7 m (9-21 ft). The weights typically range from 6 to 30 tons (up to 40 tons), and the drop heights typically range from 10 to 30 m (30-100 ft), sometimes more.

Effective densification is typical to depths of 10 m (or more with very big rigs and weights). The greatest improvement usually occurs between 3 and 8 m (10-25 ft) below the ground surface, with diminishing degrees of improvement at greater depths. The surface layers (surface to approximately 1-3 m) must be recompacted due to the disruption by the impact loads and lack of sufficient confinement. In order to estimate the required compaction effort using dynamic compaction, the Menard formula is generally followed:

$$Z = n\sqrt{MH} \qquad (6.2)$$

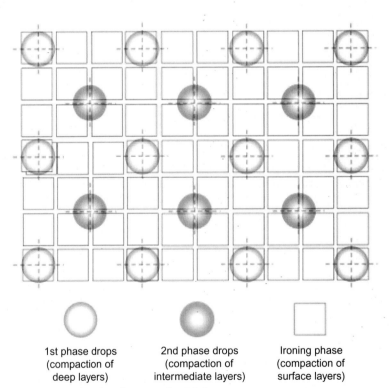

1st phase drops	2nd phase drops	Ironing phase
(compaction of	(compaction of	(compaction of
deep layers)	intermediate layers)	surface layers)

Figure 6.14 Example of grid pattern for DDC.

Figure 6.15 Menard's "Giga" compactor drops a 200 ton weight. *Courtesy of Menard.*

where Z is the (required) treatment depth, M the tamping mass (tons), H the fall height, n the (soil dependent) constant, typically between 0.3 and 0.6 for sandy soils.

Greater depths have been effectively densified using a system known as *high energy dynamic compaction*, where maximum efficiency is achieved with the complete free fall of the weight through the use of a specially designed weight release system (www.menard-web.com). As an extreme case, Menard developed a "Giga" compactor for deeper densification at the Nice airport in France (Figure 6.15).

Designing a dynamic compaction project application requires determining the most efficient application of energy at the site. This may be initially determined based on data from site investigations. Actual DDC program applications are typically fine-tuned, or modified, based on test sections or after field testing of preliminary applications (i.e., after an initial phase of drops). Field measurements of penetration (or "crater depths") and pore pressures are continuously monitored to allow for adjustments to the field program. Measurements of crater depths are also used in a manner similar to proof rolling in that deeper crater depths indicate "softer" or "weaker" locations that may require further attention.

6.1.4 Rapid Impact Compaction

Rapid impact compaction (RIC) is a ground improvement technique developed in England in the 1990s. It densifies moderately shallow depth, loose

Figure 6.16 Rapid impact compaction field application. *Courtesy of Hayward Baker.*

granular soils, using a hydraulic hammer, which repeatedly strikes an impact plate at rates of 40–80 blows per minute (Figure 6.16). Typical RIC equipment consists of a 7- to 9-ton weight dropping approximately 1 m (3.3 ft) onto a 1.5 m (5 ft) diameter plate. This impact load transmits approximately 61 kN m (45,000 ft lb) of energy directly to the ground surface. The energy is transferred to the ground by direct impact at the surface, but also by transmission of dynamic "shock" waves traveling in the ground as described previously for DDC. This compaction method allows suitable compaction of layers 4–7 m (13–23 ft) thick. Improvement of 6.1 m (20 ft) of uncontrolled variable fill was reported to provide a minimum allowable bearing pressure of more than 190 kPa (4000 psf) (www.haywardbaker.com). Some have reported effective depths of up to 10 m (33 ft) (www.farrellinc.com). Compaction results are highly dependent on soil conditions and are more effective for granular materials containing less than 15% fines.

An advantage of the RIC method is that the foot remains in contact with the ground, providing a safe, accurate, controlled compaction point. Also, the low headroom and relatively small equipment size provides access to difficult locations where other deep densification techniques may not be appropriate or possible. Continuous monitoring of GPS location, settlements, and applied energy are used to adjust to site conditions, resulting in more efficient and more uniform densification. With the relatively small space requirements, accurate control, speed of application, and onboard, real-time monitoring, RIC is gaining popularity for improvement at many locations.

6.1.5 Compaction Grouting

Compaction grouting consists of injecting low-slump (generally less than 5 cm = 2 in) soil cement mortar or "low mobility grout" under high pressures (3500-700 kPa = 500-1000 psi) to compact and displace surrounding soils (Xanthakos et al., 1994). The grout does not permeate into the soil pore space, but rather creates "grout bulbs" that expand at the injection point around the grout pipe tip (Figure 6.17). This application has been used most often for remediation of settlement problems, soil loss due to tunneling activities, and slab or foundation jacking (releveling). It has also been successfully used for treatment of sinkholes, to mitigate liquefaction susceptibility beneath existing structures, and in sensitive urban sites where other surface access treatments such as vibro methods are not feasible due to excess vibrations, access, or other concerns (Boulanger and Hayden, 1995; Wakeman et al., 2010; Xanthakos et al., 1994).

While grouting is typically an expensive proposition as compared to many other densification techniques, it may be an economical solution for certain difficult conditions—for example, where thin, loose, deep strata exist beneath dense layers, existing construction, utilities, or other infrastructure. In fact, the cost of compaction grouting may be an order of magnitude greater than other deep densification methods, but may be the only alternative, and still be less expensive than using drilled shafts or driven piles.

Typical compaction grouting projects for areal coverage are designed with grout pipes installed and injected on a square or triangular grid of primary and secondary (and sometimes tertiary and quaternary) spacing, with a final grid spacing of between 2 and 4 m (6.5-13 ft), and vertical spacing between 0.3 and 1 m (1-3.3 ft) (Wakeman et al., 2010; Welsh et al., 1987). Compacted grout columns may also be formed for localized bearing support by creating "columns" of compaction grout bulbs.

6.1.6 Consolidation Methods

Preloading is effectively a deep densification method applied to soft saturated clays, which involves the time-dependent expulsion of water to allow consolidation to take place. Preloading has been shown to be effective at improving large-scale project sites with a variety of compressible and non-uniform soils, including weak silts and clays, organic and marine deposits,

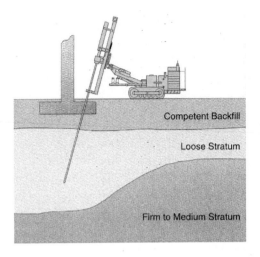

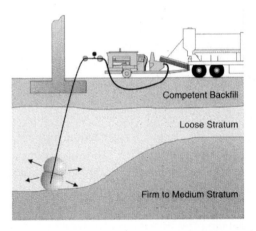

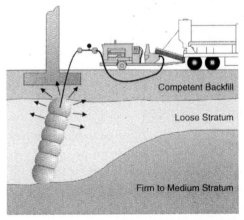

Figure 6.17 Schematic of compaction grouting process. *Courtesy of Hayward Baker.*

sanitary landfills, and so on. As this technique involves a time-dependent, geohydraulic process, a detailed discussion of variations of the method will be reserved for Chapter 9. The subject will also be mentioned in the section on stabilization techniques based on drainage (Chapter 7). Most of the variations of preloading techniques involve alternative modifications or approaches to speeding up the consolidation process. But as far as the fundamental goal of densification is concerned, only the basic philosophy of the approach will be introduced here.

When a project is constructed and applies a net load on a compressible soil, the soil will compress and settle under the application of that load. As previously described, settlement can be a controlling factor of design, especially for construction over soft cohesive soils. The approach taken here is to *preload* the site prior to construction of the actual project components, so that the compression (i.e., consolidation settlement) takes place before construction. The preload may be in the form of earthfill, temporary water tanks, or any other load that can be left in place long enough to cause the soil to consolidate. Once the soil has achieved the degree of consolidation prescribed by design, the preload can then be removed and the project construction loads applied with a greatly reduced settlement. Future differential settlement will also be greatly reduced as the softer, more compressible locations will undergo greater settlement, and the site will then be rendered more uniform. With consolidation of the soft cohesive soils, comes an increase in strength and stiffness, which can be as advantageous as the reduction of settlement for many projects.

A common modification that has been very successful in making this method economically and temporally feasible has been to provide vertical drainage to greatly speed up the consolidation process. Historically, this was accomplished by means of installing vertical *sand drains* in a grid pattern through the compressible layer. More recently, the use of PVDs made of geosynthetic materials has all but taken the place of sand drains except for smaller jobs. It has also been found that gravel columns installed through vibroreplacement can also provide this kind of drainage expedient. Vertical drains provide a much shorter drainage path for generated excess pore pressures to dissipate (Figure 6.18). From time rate of consolidation theory (Section 3.1.2), it can easily be seen that the speed of consolidation is directly proportionate to the square of the maximum drainage path, and that the theoretical time required may be increased many times with vertical drains. More detailed discussion of this concept and application is contained in Chapter 9.

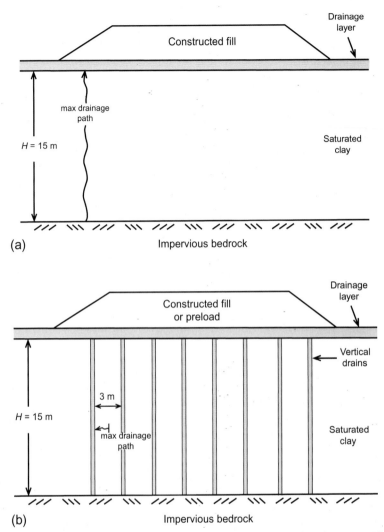

Figure 6.18 Example of maximum consolidation drainage path. (a) 15 m without vertical drains; (b) 1.5 m with vertical drains.

6.1.7 Combined Methods

It is often feasible to combine methods for a project to enhance the densification process. A popular combination previously mentioned is to use vertical drains with dynamic compaction or with stone columns. This is particularly useful in speeding up dissipation of pore pressures when the target materials are saturated and/or have low permeabilities (Shenthan et al., 2004).

In other cases, variations in soil characteristics between subsurface layers may warrant the use of a combination of techniques where certain methods are applicable to one soil layer but not to another. In this manner, the attributes of a particular method can be applied to the appropriate soil strata. An example of this would be for a site consisting of relatively thick layers of sand and clay. In the sand, VC may be appropriate for densifying and strengthening the ground; however, the clay would be better served by the installation of stone columns. In this scenario, a design may use a combination of VC for the sand and change to more expensive vibroreplacement with stone columns in the clay. This would still be more economical than simply using stone columns throughout the treatment depth as long as the capacity of the combination is adequate for the design.

6.2 DEEP DENSIFICATION QC, MONITORING, AND SPECIFICATIONS

6.2.1 Field Control for Deep Densification

Depending on the type of deep densification method employed, there may be large differences and variations in the types of quality control (QC) monitoring used. The methodologies range from simple observation of construction practices, physical measurements of size, spacing, and depths, to vertical displacement monitoring, standard field tests, laboratory testing of samples, energy monitoring, geophysical methods, and others.

6.2.1.1 Onboard Monitoring for Vibratory Densification

For vibratory densification construction, monitoring is now commonly included as part of the vibrator probe system. These systems collect real-time data, including start, finish, and completion times, penetration depths, energy consumption, and so on. Collection of this data provides a record of the construction details of each compaction point. Monitoring of the energy consumption data also allows the operator to adjust for additional time during construction, if necessary, to assure good compaction. The energy consumption is also used as an indicator of density.

6.2.1.2 Displacement/Volumetric Measurements

Simple physical measurement of displacements can be used to "see" the collapse/settlement or heave of the soil, as well as any lateral movements. Markers can be placed at the ground surface or at depth within a deposit (or fill), but care must be taken to ensure a stable datum point is established.

In some blasting cases, the ground surface was seen to drop by more than 1.5 m (www.wsdot.wa.gov). Displacement markers are also common tools used in monitoring settlement for preloading applications and for heave in displacement compaction applications. For cases where a high degree of accuracy is required or where movement tolerances are small, such as for projects in urban environments, laser levels and slope inclinometers may be appropriate.

Where volumes of material are added to the ground for displacement (i.e., VC, vibroreplacement, rammed stone columns/piers, compaction grouting, etc.), the volume of material added can be documented to assist in evaluation of degree of densification.

In dynamic compaction applications (DDC and RIC), crater depths are measured after compaction of the surface to estimate volume change and density increase. Similarly, "cones" created by vibrotechniques are also visually monitored.

6.2.1.3 Piezometers

For a number of deep densification applications (e.g., dynamic compaction, stone columns/RAPs, preloading), pore water pressures are generated. It is typically desirable to have these pressures dissipate before either continuing construction or preparing for QA testing of specifications. Pore pressure data can also be used to indicate the onset of liquefaction, which is desirable for blasting to aid in subsequent densification. Or, measurement of pore pressures can help adjust blasting design following preliminary tests. Basically, there are two types of piezometers used for water pressure monitoring. The simplest is an observation well or open pipe piezometer. This consists of an open pipe placed in a boring that allows water to rise to the groundwater or water pressure level outside of the pipe. The open pipe allows monitoring by lowering a sensing device to read the water level. The second type is closed to the atmosphere and senses water pressure through a diaphragm or electronic strain gage (Figure 6.19). These may be placed and packed into drilled boreholes or pushed in from the surface if conditions permit. The use of electronic "push in"-type piezometers have become common for these types of monitoring.

6.2.1.4 SPT and CPT

Density of in situ deposits, as well as liquefaction potential, are commonly evaluated with SPT and/or CPT penetration tests. As such, they are often used for before and after testing to evaluate densification improvements (Figure 6.20). These common field tests (described in Section 3.3.2, Field

Figure 6.19 Examples of some electronic piezometers. *Courtesy of Slope Indicator.*

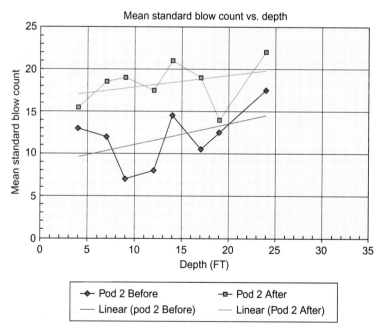

Figure 6.20 Example of improvement in penetration resistance before and after DDC. *Courtesy of Densification, Inc.*

Tests) are standard tools for field investigations and provide a wealth of data that allows reasonably good correlations to densities, strength parameters, liquefaction potential, and so forth. In this light, penetration tests (particularly CPT for cost and efficiency) are also sometimes used during construction to evaluate the densification progress. These tests can also ensure quality in construction specifications as discussed in the following section.

Piezocone (CPTU; ASTM D5778) testing enables pore pressure measurements to be made in conjunction with penetration resistance measurements. While these are only single temporal measurements of pore pressures, they may be useful for interim analyses.

6.2.1.5 Geophysical Measurements

Geophysical measurements can be helpful during certain types of deep densification projects. Blasting, dynamic compaction, and even vibrodensification can sometimes have restrictions on "vibrations," especially as they may generate a possible concern for adjacent properties. Vibrations are usually monitored by measuring peak particle velocities traveling through the ground at calculated distances from the application using conventional seismographs or *geophone* receivers. Other types of surface geophysics can be used to monitor densification, but are not commonly used for typical densification projects due to the expense and lack of accuracy stemming from the need to make interpretations. More often, geophysical methods are used for site exploration (Section 3.2.2) and assessing QA parameters (Section 6.2.2). An exception is the use of seismic CPT (or SCPT), which has advantages of providing multiple sets of data including V_s, q_c, f_s (and u_b if pore pressure transducers are included, SCPTU).

6.2.2 Specifications for Deep Densification

As for any engineering construction project, QA is typically included within design specifications. Deep densification applications will almost always have specified "densification criteria" or "acceptance criteria." Given the diverse methods and types of applications, these criteria will vary greatly. Many specifications will require that a *qualified* field QC representative be assigned specific inspection tasks and responsibilities to ensure proper QC/QA. Typical performance criteria may contain one or more requirements, including minimum bearing capacity, maximum settlement displacement under a specified load (now included in many LRFD designs), minimum stiffness or rigidity, minimum density (or relative density), and so on. While some

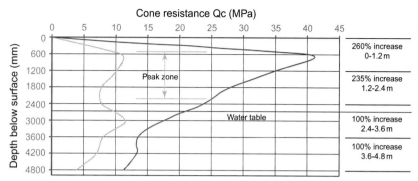

Figure 6.21 Increase of CPT measurements after compaction with high energy impact rollers. *Courtesy of Landpac Technologies.*

of these parameters can be measured directly, others are met through tests with accepted correlations.

As mentioned previously, *SPT and CPT* penetration tests have proven to be very useful in comparing "before and after" treatment improvements and are often used for specification criteria (Figure 6.21). As an example, for liquefaction mitigation, SPT blow counts of 20 or 25 have often been used as minimum benchmark criteria (Youd and Idriss, 1997), although some case studies have shown that SPT and CPT did not adequately predict the liquefaction resistance of mechanically improved soils. One needs to consider that most of the data that has been collected for correlations were developed from natural sites where no ground improvement had been performed (Lopez and Hayden, 1992). In these cases, it was shown by results of piezocone testing before and after treatment that although SPT and CPT penetration values had not significantly improved, pore pressure generation during pushing of the cone indicated that the soil had changed from a contractive tendency to dilative behavior, suggesting improved liquefaction resistance. Laboratory triaxial tests of thin–walled tube samples also showed dilative behavior, greatly increased undrained shear strength, and an approximate doubling of the critical stress ratio for treated samples (Lopez and Hayden, 1992). It may be prudent then to include additional piezocone or laboratory strength tests in QA programs when the penetration test values alone do not meet specified levels. SPT and CPT measurements have also been used to evaluate several other properties and predicted soil response behaviors through additional correlations. For coarse gravels of soil containing cobbles, the SPT and CPT are inappropriate due to their relatively small size and potential to be damaged. For these cases, a Becker penetration test, essentially a large-scale SPT-type penetration test, may be used, but is not standardized and is not as well correlated to liquefaction resistance or other penetration tests (Youd and Idriss, 1997).

Shear wave velocities (V_s) have become more commonly used as a tool for evaluating a range of soil properties, including liquefaction potential and soil density through correlations with penetration tests or with other site data. These measurements can be made through a number of different geophysical means, such as nondestructive surface wave (seismic) applications, downhole or crosshole measurements, or by seismic CPT if the soil conditions are appropriate. In the same manner as for SPT and CPT, certain minimum values for V_s may be used as a required specification (Youd and Idriss, 1997).

The *flat dilatometer* or *pressuremeter* tests are also sometimes used for before and after testing for evaluation in improvement to bearing capacity and settlement. As described earlier in Chapter 3, these types of tests provide evaluation of different soil parameters or responses such as stiffness, compressibility, or bearing capacity.

Full-scale load tests (ASTM D1143 or similar) may be employed to test the adequacy of completed deep densification applications, particularly for stone columns and compacted aggregate piers. In these tests, loads are applied to represent actual expected loading conditions. Often, criteria specify that test loads be applied to some level well above design loads to assure an adequate factor of safety and that displacements will be within tolerable limits. Figure 6.22 depicts a field load test on a completed Geopier® system.

Figure 6.22 Field load test of a Geopier® system. *Copyright by Geopier Foundation Company, Inc. Reprinted with permission.*

Many other details may be included in a specification for deep densification applications. Type, size, energy, or other specifics of equipment may be governed. Materials added to the ground will typically have requirements and limitations on grain size and or gradations.

6.2.2.1 Accept/Reject Criteria

If all goes well, most, if not all, test points will pass the specified minimum criteria. There may be some allowance for a small percentage of data points that come close, but do not meet a minimum specification (as described in Section 5.5.1). If a gravel column or other critical point fails to meet the specified criteria, it is almost always the responsibility of the contractor to either recompact the failed location(s) or to provide other evidence that the intended design parameters are met through performance of other testing methods.

RELEVANT ASTM STANDARDS

D1143/D1143M-07e1 Standard Test Methods for Deep Foundations Under Static Axial Compressive Load, V4.08

D1586-11 Standard Test Method for Standard Penetration Test (SPT) and Split-Barrel Sampling of Soils, V4.08

D3441-05 Standard Test Method for Mechanical Cone Penetration Tests of Soil, V4.08

D4428-07 Standard Test Methods for Crosshole Seismic Testing, V4.08

D5778-12 Standard Test Method for Electronic Friction Cone and Piezocone Penetration Testing of Soils, V4.08

D4719-07 Standard Test Methods for Prebored Pressuremeter Testing in Soils, V4.08

D5777-11 Standard Guide for Using the Seismic Refraction Method for Subsurface Investigation, V4.08

D7400-08 Standard Test Methods for Downhole Seismic Testing, V4.09

REFERENCES

Boulanger, R.W., Hayden, R.F., 1995. Aspects of compact grouting of liquefiable soils. J. Geotech. Eng., ASCE 121 (12), 844–855.
Brown, R.E., Glenn, A.J., 1976. Vibroflotation and Terra-Probe comparison. J. Geotech. Eng. Div. 102 (10), 1059–1072.
Dise, K., Stevens, M.G., Von Thun, J.L., 1994. Dynamic compaction to remediate liquefiable embankment foundation soils. In: In Situ Deep Soil Improvement. ASCE, pp. 1–25, Geotechnical Special Publication No. 45.

Dowding, C.H., Hryciw, R.D., 1986. A laboratory study of blast densification. J. Geotech. Geoenviron. Eng., ASCE 112 (2), 187–199.

Lopez, R.A., Hayden, R.F., 1992. The Use of Vibro Systems in Seismic Design. ASCE, Geotechnical Special Publication No. 30, pp. 1433–1445.

Massarsch, K.R., Fellenius, B.H., 2005. Deep vibratory compaction of granular soils. In: Indranatna, B., Jian, C. (Eds.), Ground Improvement-Case Histories. Elsevier, pp. 633–658 (Chapter 19).

Mitchell, J.K., Soga, K., 2005. Fundamentals of Soil Behavior, 3rd Ed., John Wiley & Sons, Inc., 577 pp.

Murthy, V.N.S., 2002. Geotechnical Engineering: Principles and Practices of Soil Mechanics and Foundation Engineering. Marcel Dekker, Inc.—CRC Press, 1035 pp.

Narin van Court, W.A., Mitchell, J.K., 1998. Investigation of Predictive Methodologies for Explosive Compaction. ASCE, Geotechnical Special Publication No. 75, pp. 639–653.

Narsilio, G.A., Santamarina, J.C., Hebeler, T., Bachus, R., 2009. Blast densification: multi-instrumented case history. J. Geotech. Geoenviron. Eng., ASCE 135 (6), 723–734.

Nottingham, D., 2004. Improvements in deep compaction using vibratory pile hammers. In: Proceedings of the 29th Annual Conference on Deep Foundations, Vancouver, Canada, DFI6 pp.

Raj, P.P., 1999. Ground Improvement Techniques. In: Laxmi Publications Ltd, New Delhi, 272 pp.

Shenthan, T., Nashed, R., Thevanayagam, S., Martin, G.R., 2004. Liquefaction mitigation in silty soils using composite stone columns and dynamic compaction. J. Earthquake Eng. Eng. Vib. 3 (1), 205–220.

Wakeman, R.C., Evanson, A., Morgan, T., Pastore, J., Blackburn, J.T., 2010. Compaction grouting for seismic mitigation of sensitive urban sites. In: Fifth International Conference on Recent Advances in Geotechnical Earthquake Engineering and Soil Dynamics, San Diego, pp. 1–9.

Welsh, J.P., 1986. In situ testing for ground modification techniques. In: Use of In Situ Tests in Geotechnical Engineering. ASCE, New York, pp. 322–335 Geotechnical Special Publication No. 6.

Welsh, J.P., Anderson, R.D., Barksdale, R.P., Satyapriya, C.K., Tumay, M.T., Wahls, H.E., 1987. Densification. In: Soil Improvement—A Ten Year Update. ASCE, New York, pp. 67–97, Geotechnical Special Publication No. 12.

Xanthakos, P.P., Abramson, L.W., Bruce, D.A., 1994. Ground Control and Improvement. John Wiley & Sons, Inc., New York, 910 pp.

Youd, T.L., Idriss, I.M. (Eds.), 1997. Proceedings of the NCEER Workshop on Evaluation of Liquefaction Resistance of Soils. National Center for Earthquake Engineering Research, Technical Report NCEER-97–0022, 310 pp.

Zekkos, D., Kabalan, M., Flanagan, M., 2013. Lessons learned from case histories of dynamic compaction at municipal solid waste sites. J. Geotech. Geoenviron. Eng., ASCE 139 (5), 737–751.

http://www.betterground.com/index.php?option=com_content&view=article&id=191&Itemid=194 (accessed 01.17.14.).

DDC, Robert Lukas, and Bill Moore, http://cbsink.lfchosting.com/brandenburg/dynamic/images/Dynamic%20Compaction%20Article_06.03.2005.pdf (accessed 03.16.13.).

http://www.earthtech.net/services-and-solutions/ (accessed 01.17.14.).

http://www.explosivecompaction.com/videos/Blast.wmv (accessed 05.20.13.).

http://www.farrellinc.com (accessed 05.15.13.).

http://www.haywardbaker.com/WhatWeDo/Techniques/GroundImprovement/VibroCompaction (accessed 05.15.13.).

http://www.landpac.com.au (accessed 05.10.13.).

https://mceer.buffalo.edu/publications/resaccom/06-sp04/pdf/09nashed.pdf   (accessed 05.15.13.).

http://www.menard-web.com (accessed 05.16.13.).

http://www.menardusa.com (accessed 05.15.13.).

http://www.penninevibropiling.com/Library/equipment_datasheets/Guide_to_Pennine_Vibroflots.pdf.

http://www.wsdot.wa.gov/research/reports/fullreports/348.1.pdf (accessed 05.11.13.).

http://civilprojects.wordpress.com/2007/03/27/ground-improvement-techniques/ (accessed 03.14.13.).

SECTION III
Hydraulic Modification

CHAPTER 7

Objectives and Approaches to Hydraulic Modification

The subject of hydraulic modification includes a variety of soil and ground improvement methods that can be achieved by altering the flow, presence, and pressures of water in the ground. This may involve any change or "improvement" in the ground that has to do with drainage, dewatering, seepage, or groundwater flow. On several occasions, Dr. Ralph Peck commented that the presence of water in the ground made for "most of the geotechnical engineering problems of interest." Therefore, it seems reasonable that if the presence or action of water in the ground can be controlled, the engineer may be able to affect the behavior of the ground in a positive manner. Some of the most serious engineering consequences caused by the presence, introduction, or change in concentration of water in the ground include foundation distress/failure, slope failure, excessive volume change (i.e., shrink, swell, or heave), liquefaction, piping failure, and total/differential settlement. Construction dewatering is also a common application where the water table must be drawn down to allow excavation with a dry working area.

This chapter provides an overview of a number of objectives for modifying water conditions at a site, along with some of the basic approaches to achieve those objectives. While some of the concepts are relatively simple, realizing the goals and desired results may be sometimes challenging. For many applications, permanent drainage or redirection of groundwater may be the primary objective. There are a number of methods available to attain these goals. The complexity of each approach will depend on several factors, including initial water and flow conditions, drainage capability of the particular soils and ground, and ability to adequately discharge unwanted flows. Modifying hydraulic conditions can provide means to reduce pressures behind retaining walls or beneath excavations, improve slope stability, and reduce risk of internal erosion or "piping." One of the principle causes of landslides and slope stability problems is a direct result of added water (or persistent high groundwater levels) in a slope. Because of this fundamental geotechnical issue and importance of water to slope stability, slope stabilization by drainage is addressed by itself in Section 7.5. For other cases, temporary or permanent lowering of the initial groundwater levels is needed for construction or to provide mitigation of future flooding for certain aspects of some projects.

7.1 FUNDAMENTAL OBJECTIVES AND IMPROVEMENTS

Altering the hydraulic properties of the soil/ground is a fundamental approach to making ground engineering improvements. This may be done by means of physical and/or chemical modification of the earth materials to alter permeability values or by dewatering target soil masses. Depending on the desired outcome, which may range from increased flow capacity for "free drainage" to creating a nearly "impermeable" barrier or boundary condition, improvement approaches will be very different. Compaction and other densification methods described in Chapters 4–6 can be effective ways to reduce permeability and groundwater flow. Admixture stabilization can also be effective at altering soil hydraulic properties. This will be described in more detail in Chapter 11.

It is obvious that standing water and/or flooding are intolerable for many projects. Even a very high water table may be unacceptable if it creates difficult working or construction conditions, especially if any excavation or earthwork is required. The depth of the water table is typically well documented from prior site exploration and therefore can be anticipated or planned for in design. Rainfall (especially if heavy or irregular) can cause flooding both during and following construction, and measures to handle these inflows should be included in design if expected or possible. Often, water is introduced during certain construction activities, and its proper handling, filtering, and removal must be addressed as well.

The main objectives to modifying hydraulic parameters in the ground include:
- Temporary lowering of the water table over a site area (construction dewatering)
- Permanent lowering of the water table (for permanent subsurface structures)
- Providing drainage to relieve hydrostatic and seepage pressures (reducing lateral earth pressures, upward gradient forces)
- Providing drainage to alleviate ponding or pumping
- Providing drainage to alleviate dynamic pore pressures (liquefaction mitigation)

- Redirecting flow to reduce seepage and exit gradients
- Creating low permeability "barriers" to retain or convey water
- Creating low permeability "barriers" to prevent water migration (shrink/swell and heave control)
- Increasing slope stability
- Increasing bearing capacity
- Reducing soil compressibility
- Filtering water to prevent soil migration (cavities and piping)
- Filtering water to prevent "contamination" (construction catchments, silt fences)
- Improving workability or hauling characteristics of source, disposal, or contaminated materials

To accomplish such a wide range of objectives, an array of improvement methods, approaches, and techniques may be employed.

7.1.1 Adverse Effects of Dewatering

While applications of dewatering provide many solutions for both temporary construction and permanent geotechnical improvements, there may be, on occasion, some undesirable side effects. While one of the ground improvement objectives is to cause strengthening and decreased compressibility by intentionally causing settlement, if not controlled, undesirable settlements and associated damage may be caused to adjacent structures or infrastructure. Other side effects of dewatering may include:

- Reduction in yield of neighboring water supply wells. There are certain remedies for this problem, including installation of cutoffs and/or installing recharge wells to minimize drawdown away from the project work area.
- Salt water intrusion if near a fresh–salt water boundary. This has been a major concern in areas such as Florida and Hawaii, where the fresh water supply aquifer naturally forms a pressurized lens, thus preventing long-term contamination by intrusion of salt water. When water is withdrawn, the fresh water lens recedes.
- Deterioration of previously submerged timber structures (i.e., piles). If untreated timber is exposed to oxygen due to dewatering, then aerobic organisms may attack the timber. This can be partially alleviated by injecting water near the timber substructure as has been done for historic structures in Boston, MA (Powers et al., 2007).

7.1.2 Common Drainage Applications

The fundamental approaches and treatments of hydraulic modification will vary greatly depending on whether the objective is to retain water (such as by a dam, levee, or reservoir), provide temporary dewatering (such as to provide for "dry" construction near or below naturally occurring groundwater and seepage), and/or for permanent dewatering (such as for slope stabilization, increased performance and stability of retaining walls to prevent future flooding of subsurface structures, to mitigate liquefaction potential). Other objectives are to provide "dynamic" drainage to alleviate buildup of dynamic pore water pressures (as a tool to mitigate liquefaction), or to modify the flow characteristics of the ground by altering soil permeability or seepage forces.

The following descriptions provide brief overviews of some of the most common applications of hydraulic ground modification for drainage.

7.1.2.1 Construction Dewatering

For projects where deep excavations are planned and/or the water table is known to be relatively shallow, dewatering of the project area may play a significant role in the planning and design of the construction process. *Construction dewatering*, sometimes referred to as "unwatering," may also be included as part of the permanent design plan to prevent future infiltration and/or flooding of subsurface components of a project. For many projects, construction dewatering primarily provides a temporary dry working area where the project site is initially saturated, flooded, or submerged. Good examples are drainage of swampy areas for equipment accessibility, excavations for construction below the water table, cofferdams for "underwater" construction (e.g., bridge foundations in rivers), and redirection of flow (e.g., river bypass for dam construction/repair).

The elimination or reduction of groundwater around and below open deep excavations has a number of positive attributes in addition to providing a dry workspace. The pressures exerted by water in the ground add significant load to the lateral earth pressures along the sides of the excavation. In addition, the water pressure at the base of an excavation can provide an upward force that may be enough to surpass the weight (and/or strength) of the soil in the bottom, resulting in heave, or worst case, a failure mode called "blowout," which would potentially result in a catastrophic failure and flooding of the excavation. Contractors should consider these loads and possible failure mechanisms, and design accordingly. Some structures with components below the normal water table will require dewatering during construction as well as long-term control after the project is completed and put into operation. For cases where analyses show that there would be a continual large inflow of water that needs to be evacuated, a cutoff wall may be appropriate and economical. There are many varieties of cutoff walls for both temporary and permanent applications that utilize different structural and nonstructural components, including slurries, grouts, soil admixtures, sheet piles, and steel beams. Cutoff walls are essentially hydraulic barriers (discussed below), but also may be designed to perform one or more structural functions, like serving as foundations or walls.

Construction dewatering is typically implemented prior to excavations or any actual construction where interception of a water table is expected. For some deep excavations, dewatering and excavating proceed in alternating steps, allowing the use of shallow well pumps or multistage well point systems rather than more expensive deep wells (Figure 7.1).

7.1.2.2 Permanent Drainage

Permanent/long-term drainage is implemented where persistent "dry" (or drained) conditions are desired. Examples include athletic fields, green/park spaces, green roofs (where drainage may be collected for recycling), and

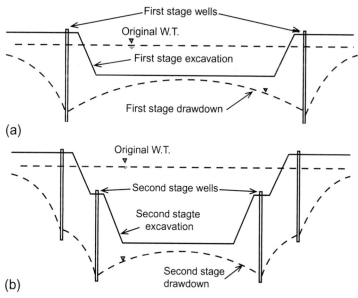

Figure 7.1 Multistage well system for excavation dewatering: (a) first stage and (b) second stage.

engineered "green" space suitable for parking or emergency vehicle use. Permanent drainage also may be included in design adjacent to foundations, and new embankments and roadways, for example. At another level, permanent drainage may be the key in designing for reclaimed land and useable land below sea (or river) level (such as large agricultural tracts in the Netherlands, Sacramento Delta, residential regions of Greater New Orleans, municipal Sacramento, etc.).

7.1.2.3 Stabilization of Slopes, Retaining Walls, and Excavations

Drainage for *slope stabilization* is discussed in detail in Section 7.5 as a relatively (theoretically) simple means to stabilize many geotechnical aspects of projects where the effects of water pressures, water weight (or added weight to soil), and/or water forces (e.g., seepage forces or gradients) can act as destabilizing components. Methods described primarily include types of drainage and filters, and typical applications where drainage is used to improve ground conditions or stability. When the goal is to increase stability of retaining walls and excavations, many of the same principles may be applied. Dewatering and draining adjacent soils can provide acceptable safety factors by improving the parameters used in respective stability analyses. Fundamental analyses of lateral earth pressures on walls and potential heave of excavation bases clearly show that the elimination of water pressures can greatly increase stability, often by as much as two times!

Drainage behind retaining walls is a critical part of the design of these structures. Simply looking at the components of lateral earth pressures shows that water (hydrostatic water pressure) can nearly double the lateral force acting on an unsupported wall. But also as part of the calculation of lateral earth pressures is the weight of the backfill material. If the soil in the presumed failure zone is kept properly drained, then not only will no water forces exist, but the weight of the material may be significantly reduced, thus further lowering the overturning forces on the wall. Similarly, stabilization of excavation walls relies on the same fundamental design parameters as used for retaining walls. Stability can be greatly improved if drainage/dewatering can eliminate (or reduce) hydrostatic forces on the sides of the excavation. In cases where there is distress to a retaining or excavation wall, a remedial measure may be to intercept or redirect any water that may be entering the soil mass behind the wall.

7.1.2.4 Forced Consolidation

Naturally occurring water in soils can be problematic for different soils under different conditions. Saturated fine-grained soils are susceptible to significant deformation and settlement through consolidation when a net load is applied, as explained in Chapter 3. Dewatering these soils through *forced and/or assisted consolidation* will reduce compressibility (and future settlements) as well as add significant strength. Preconsolidation and assisted consolidation will be addressed separately in Chapter 9. Another alternative technology to force consolidation of clays is electroosmosis, to be described in Chapter 10. This method has been effective for a number of ground improvement solutions, but has not yet gained wide acceptance or been widely applied.

7.1.2.5 Liquefaction Mitigation

Saturated loose cohesionless soils may be susceptible to liquefaction if subjected to rapid (e.g., dynamic) loading without the ability to drain excess pore water pressures. Dewatering or providing ample drainage for these soils is an improvement technique that has been used to mitigate the potential for liquefaction, and has been shown to be effective for improved sites during recent earthquake loading. The use of gravel columns acting in part as liquefaction mitigation drains was mentioned briefly in Section 6.1.2. Another option that has become more popular is installation of pressure relief wells (or drains) that provide rapid dissipation of excess water pressure buildup. A number of case studies show the success of these types of drains using either gravel columns or relatively large diameter vertical geocomposite drains (EQ drains).

7.1.2.6 Controlling Seepage and Exit Gradients

Where there is a significant flow through and/or exiting the ground, attention must be paid to the seepage forces generated by such flow in conjunction with the in situ stresses and parameters of the soil through which the fluid is passing. The *gradient*, which is a measure or calculation of the *head loss* with respect to travel distance, provides an indicator of the internal seepage forces that may be destructive. It is imperative that the gradient be designed so as not to exceed a maximum value beyond which soil erosion may be initiated. Gradients can be particularly dangerous where water exits from the ground. Examples of this would be groundwater flow emerging from a soil slope and seepage water exiting from within or beneath a hydraulic structure such as a dam or levee.

7.1.2.7 Filtering

When water flows through the ground, there are seepage forces exerted that have a tendency to carry away particles with the flow. If this type of *internal erosion* is not prevented, it can lead to severe consequences and even catastrophic failure (see discussion in Section 3.1.2). If the groundwater flow is properly filtered, migration of soil particles will be prevented while still allowing water to flow. Filtering can be accomplished with either subsequently finer soil gradations or with geosynthetic (geotextile) filters using standard criteria relating to soil grain sizes, or in the case of geotextiles, opening sizes of the fabric. Filtering and seepage control will be described further in Section 7.6.

7.1.2.8 Roadways and Pavements

Drainage is a critical component of roadway and pavement design. Related facilities include airfields, parking lots, racetracks, railway beds, and so forth. These all share a common problem in that they are exposed to substantial water inflows, but, due to their relatively flat geometries, may have difficulty draining that inflow away. This can often result in damage or increased maintenance requirements. Historically, pavements were designed to be "strong" without too much attention to drainage (Cedergren, 1989). Research conducted throughout the 1970s-1990s by the U.S. Army Corps of Engineers and the Federal Highway Administration (FHWA) clearly showed that maintaining drained components of a pavement system enhances performance and reduces maintenance; in fact, well-drained pavements will outlast undrained ones by three to four times (Cedergren, 1989). The problem becomes compounded because well-compacted soils will tend to have lower permeability and reduced drainage potential. Proper geometric design, gradation of base layers, and functional edge drains must include consideration of soil permeability, filtering, and discharge flow capacity. Guidelines and specifications of base and subbase materials are well ingrained in the design parameters of most municipalities and highway agencies. Geotextiles are now commonly installed in pavements in part to provide a filtering function. Today's modern construction techniques now make it possible to rapidly install prefabricated geocomposite edge drains with high-capacity plastic cores wrapped with geotextile, filter fabric. More recently, the use of modified geonet (geosynthetic) drains is finding its way into the design of internal pavement layers. Filtering and drainage with geosynthetics are specifically addressed in Chapter 8.

7.1.3 Common Retention Applications

As opposed to drainage applications where the object is to improve conditions by eliminating or redirecting water, there is another category of hydraulic modification with a very different goal. For structures designed to retain or convey water, or otherwise provide a permanent barrier to fluid flow, there are a number of methods to reduce the permeability or flow of water in the ground. Permanent fluid barriers may be provided by altering the inherent soil properties, deep soil mixing, grouting, constructing cutoff (slurry) walls, or by introducing a geosynthetic membrane (Section 8.3). Altering soil permeability for retaining water or creating a fluid barrier often involves physical and/or chemical techniques. Common applications include water storage or conveyance structures (i.e., dams, reservoirs, levees, culverts, canals, ditches, etc.), landfill liners and covers, containment of contaminated soil, and impoundment of mine tailings. Applicable improvement methods include a range of approaches, from mechanical densification (Chapter 5), to admixture stabilization (Chapter 11), to grouting (Chapter 12). The use of geosynthetic membranes (Chapter 8) has also become an important addition to applicable means of creating fluid "barriers" for the types of structures mentioned above and to prevent water migration that might otherwise lead to volume change or distress. For temporary fluid barriers, ground freezing (Chapter 13) is becoming a popular method.

Cutoff walls and diaphragm walls may involve a number of different methods, materials, and applications. Steel or plastic interlocking sheet piles have been used for many decades as both a temporary and permanent tool for intercepting flow and reducing seepage for building excavations, reservoir dams, and so on. Slurry trenches or diaphragm walls can be a viable (but potentially expensive) solution (Figure 7.2). They may be constructed with complete replacement or with various mixtures of native soil, bentonite, and cement. The difference in mixtures used is primarily dependent on how strong and/or rigid a wall is needed for long-term performance. Walls

Figure 7.2 Installation of a 60 m (200 ft) deep slurry cutoff wall at Clearwater Dam, Piedmont, MO. *Courtesy of Layne Christensen.*

constructed by deep mixing methods generally consist of overlapping secant piles or by specialized equipment such as *cutter soil mixing* machines (Chapter 11). Grouting methods have long been used for reducing seepage/leakage from reservoir dams, and now jet grouting has become very common for design of temporary and permanent dewatering combined with soil stabilization and structural support for many projects.

Bruce et al. (2008) describe the developing practice of "composite" cutoff walls, where a concrete diaphragm wall is constructed between two grouted rows. The drilling and grouting program provide details of the subsurface conditions so that the more costly diaphragm wall can be more efficiently and effectively constructed.

7.2 DEWATERING METHODS

The type of dewatering method(s) used for any project or to solve one or more hydraulic improvement objective(s), will largely depend on elevation difference(s) between source and disposal, as well as permeability (hydraulic conductivity) or flow capacity within the ground. The simplest mode of dewatering will consist of trenches or gravity (or siphon) wells/drains where disposal is at an elevation below the source. In these cases, minimal (if any) pumps are needed to collect, transmit, and discharge the excess (unwanted) water. Where pumps are needed to lift to significant heights and/or where small grain size limits the effectiveness of gravity drainage, more complex dewatering systems must be deployed. Figure 7.3 depicts the general applicability of some categories of dewatering systems as a function of soil grain size.

7.2.1 Types of Dewatering Systems

Given the wide variation in demands and requirements for dewatering systems, there is an equally broad variation in the types of equipment that will meet those needs. The following overview discusses several dewatering systems that can provide a wide range of ground improvement solutions. Each has certain limitations and restrictions, but all can practically, safely, and relatively economically allow construction and/or remediation in difficult situations involving groundwater.

7.2.2 Horizontal Drainage and Gravity Drains

Installation of "horizontal" drains (actually most are subhorizontal) is often cost-efficient, as it may not require pumps. This type of system may be used to permanently lower the water table by allowing gravity drainage, or it can be

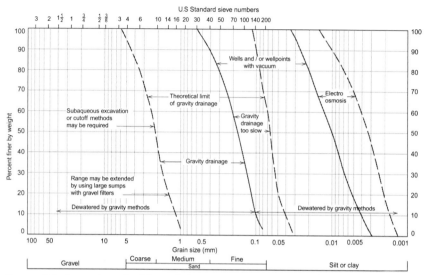

Figure 7.3 General applicability of dewatering systems as a function of grain size. *Courtesy of Moretrench.*

used for construction dewatering by temporarily pumping prior to and during construction. Dewatering depths of up to 6 m (20 ft) and horizontal drain lengths of up to 50 m (165 ft) is common. Horizontal drains are also a common solution for slope stabilization, as will be discussed in Section 7.5. Horizontal drains may be constructed as "trench" drains or "French" drains, where a free-draining material (i.e., uniform gravel) is placed in a shallow trench, typically separated by a geotextile filter. Gravity drainage is effective in sands, gravels, and some silty sands, but becomes less so in silts and clayey materials.

Vertical drains have also been effective, particularly when used in conjunction with other dewatering wells as a means to supplement drainage of stratified soils and where pore pressure relief is desired. In these cases, vertical drainage may be either up or down depending on the available drainage outlet(s) and any internal pressures and gradients in the ground or between stratified layers.

7.2.3 Shallow Well, Sump Pumping, and Wellpoints

Where the pumping depths are relatively shallow (<5 m), small, relatively inexpensive equipment may be appropriate to relieve water pressures, prevent flooding, or to maintain dry working areas. Where the surrounding earth materials are stable enough to stand up without sloughing, slumping, or significantly eroding, groundwater inflows may be allowed to flow freely into ditches or "sumps" at selected locations, at which point the water can be removed by low-cost pumps. Open, pumping from sumps and ditches can be the least expensive dewatering method if conditions are favorable (Powers et al., 2007). Caution should be exercised and other methods considered if there are concerns regarding the seepage and stability of surrounding soils (i.e., loose, high-permeability soils), adverse effects on surrounding or adjacent properties, or other foreseeable problems that may be associated with free water flow and direct pumping.

Wellpoints are typically designed as a system of relatively shallow wells that work in concert with one another because multiple wellpoints will share a common pump and piping system (Figure 7.4). The technique has been used for construction dewatering for over 80 years and is one of the most versatile "predrainage" methods because it can be effective in most types of soils and has a wide range of pumping volumes (Powers et al., 2007; www.moretrench.com). Wellpoints consist of a vacuum-type system in which several wellpoints are connected to a shared wellpoint pump and header pipe. Wellpoint systems are practical when a large number of closely spaced wells are needed. The systems are generally most suitable when water table levels only need to be lowered by no more than 5 m (16 ft) because their effectiveness is limited by available suction lift. If a greater amount of dewatering is required, then multistage pumping or deep wells are usually employed. Figure 7.5 shows an application of wellpoints to lower groundwater levels to below subgrade excavation for a 4800 m^2 (51,000 sq ft) library addition at Colgate University, Hamilton, NY.

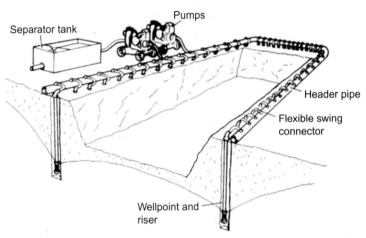

Figure 7.4 Schematic of a typical wellpoint system. *Courtesy of Pump Hire Ltd.*

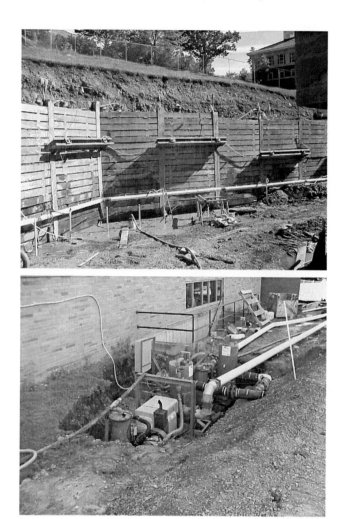

Figure 7.5 Wellpoint application and pump for a building excavation, Colgate University, NY. *Courtesy of Moretrench.*

7.2.4 Deep Wells

Where a significant drawdown depth and a large lift is required, it is most common (and efficient) to place individual submersible pumps down pre-drilled (often cased) borings with slotted or perforated pipe at the depth(s) where water is to be withdrawn. With individual pumps and the significant drilling involved, the overall costs may be higher than other dewatering methods. But improvements in well design, installation, and pump technology have made this option more cost-effective than in the past. The effectiveness of deep wells to dewater the ground is a function of pump capacity and project requirements, as well as soil and groundwater flow characteristics. These types of wells work best for soils with permeability between 10^{-1} and 10^{-3} cm/s (www.wikipedia.com).

Deep wells can vary in size from 7.5 to over 60 cm (3-24 in) in diameter and have been placed to over 60 m (nearly 200 ft) in depth (www.moretrench.com). These types of wells are typically not spaced closer than 15 m (50 ft). An example of a large, deep well dewatering case is the Beacon Hill Station in Seattle, WA, where dewatering was conducted to depths of over 62 m (190 ft) to drain and stabilize granular deposits in advance of a tunneling operations (www.moretrench) (Figure 7.6).

7.2.5 Ejector Systems

Ejector systems tie together multiple wells with a single pumping station and are typically used where the groundwater must be lowered more than 5 m, often in soil with low permeability. For these situations, they can be used at close spacing. The ejector (sometimes called *eductor*) system lifts the well water with a nozzle and venturi fed by water under high pressure, and so is not limited to suction lift. This enables the use of common pumps and close spacing of wellpoints, together with the ability to dewater greater depths, up to 30 m (100 ft) or more with a single stage (www.griffindewatering.com). In a recent project for the South Ferry Terminal, Manhattan, NY, ejector wells were used to draw a high groundwater table to a depth of 26 m (85 ft) below grade to enable deep excavation (www.moretrench.com) (Figure 7.7). High vacuum added to ejector systems can improve soil drainage in fine-grained soils.

7.2.6 Vacuum Wells

Where the permeability of a soil is very low, dewatering methods that rely on gravity to draw the water toward the wellpoint or collection location may be ineffective. Terzaghi and Peck (1967) suggested that if the average effective grain size of a soil, given by D_{10}, is less than about 0.05 mm, then capillary tension (suction) prevents the release of water. Others have suggested various criteria as limits to effective gravity drainage (e.g., 60% "fines," $k = 10^{-3}\text{-}10^{-5}$ cm/s) (Cedergren, 1989; Hausmann, 1990). Figure 7.3 shows the general range of soils applicable for vacuum dewatering in fine sands and silts.

Figure 7.6 Application of deep wells to dewater for a deep shaft and tunnel excavation at Beacon Hill, WA. *Courtesy of Moretrench.*

Figure 7.7 Application of ejector wells to dewater for a deep excavation at the South Ferry Terminal, NYC. *Courtesy of Moretrench.*

Vacuum pumps increase the effective pressure in the soil surrounding wellpoints to overcome capillary tension and greatly increase the flows to the pumps. Many successful applications using the vacuum method have been reported for stabilization of steep cut slopes since the 1940s (Terzaghi and Peck, 1967). More recently, vacuum systems have been developed as an aid to forced consolidation, both with and without additional surcharge, and to dewater dredged organic silts, which are often contaminated with heavy metals (i.e., cadmium, lead, and mercury). Vacuum-assisted consolidation will be discussed in Section 9.3. Adequate care must be taken to ensure that connections are tight throughout vacuum systems and that each system is sealed where the wells or drains are exposed near the surface and atmospheric pressures. A review of successful case studies and design literature suggests that vacuum wells must have relatively close spacing (approximately 1-2.5 m) to be effective, given the dissipation of pressure differential over relatively short distances within a soil mass.

Vacuum pumps may also be used in conjunction with other pumping and dewatering systems. Examples include vacuum pumps used in conjunction with traditional, shallow-depth centrifugal pumps to increase the pressure differential and efficiency of wells in low permeability sands and silts, and with submersible pumps to increase effectiveness of deep wells (Hausmann, 1990). Vacuum pumps have also been used to improve efficiency with horizontal drains used for slope stabilization.

7.2.7 Electroosmosis

Utilizing the concepts of electrical charge in fine-grained soil particles (principally clays) and the electrical "bonds" holding ions and the double layer of water dipoles to the particles, it has long been recognized that liquids could be moved within a porous media under the influence of an externally applied electric potential. This phenomenon was discovered more than 200 years ago by Reuss (Cedergren, 1989), and was described in terms of "modern" soil improvement by Casagrande (Turner and Schuster, 1996). The principles of *electroosmosis* (or electrokinetic dewatering) and some ground improvement applications are described in Chapter 10. The fundamental process is that the application of an electric potential (DC current) will draw the dipolar water molecules toward the negative terminal (cathode), where it can be removed from the ground. This results in dewatering by drawing water out of the soil and reducing moisture content. Electroosmosis has been demonstrated to be effective in stabilizing soft silts and clays, removing contaminants, and forced consolidation, among other applications. Electroosmosis has been used for slope stabilization applications, as will be described in Section 7.5.2.

7.3 WELL HYDRAULICS AND DEWATERING DESIGN

In order to better understand the requirements and capabilities of a dewatering system, one needs to first understand the basics of well hydraulics and soil water storage/discharge capacity. In order to design any dewatering pumping system, one must be able to estimate the total amount of water that needs to be pumped in order to achieve the desired project goals, and the amount of water that can reasonably be expected to be drawn from an individual well. The combination of these estimates will play a vital role in designing an adequate dewatering system, including placement and spacing of wells or wellpoints. Depending on the complexity of the aquifer or soil mass from which water is to be drawn, dewatering design can use simplified analytical models, which inherently assume pumping is in equilibrium or a steady state of flow. For more complex situations, numerical models are often used for better accuracy. For the most basic analytical analyses of water drainage to wells, a number of simplifying assumptions are usually made. First, the water bearing strata, or *aquifer*, is assumed to be horizontal. Second, Darcy's Law is assumed to be valid such that all flows are considered to be laminar, and that the flow rate is directly proportional to soil permeability and hydraulic gradient. A simple method for estimating the hydraulic gradient can be made using the Dupuit-Theim approximation. According to this method, the hydraulic gradient is equal to the slope of the drawdown curve at any point.

Three critical parameters of interest are the location of the drawdown curve (identifying the zone where dewatering will take place), the lateral extent of drawdown influence, and the discharge flow rate. In the theoretical or idealized case for a well in a water table aquifer, the discharge rate can be approximated using a simple equation (Equation 7.1) and geometries as depicted in Figure 7.8.

$$Q = \frac{\pi k \left(H^2 - h_w^2 \right)}{\ln R_o / r_w} \tag{7.1}$$

where Q is the flow rate, k the coefficient of permeability, H the height of original water table (saturated thickness), h_w the height of water table at the well, R_o the radial distance of the zone of drawdown influence, r_w the radius of the well.

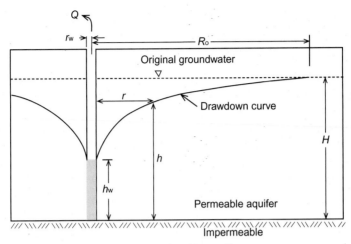

Figure 7.8 Idealized drawdown in a water table aquifer.

The height (h) of the phreatic surface (drawdown curve) at any distance (r) from the well can be estimated as

$$h = \sqrt{H^2 - \frac{Q}{\pi K} \ln \frac{R_o}{r}} \qquad (7.2)$$

It should be noted that Equation (7.2) will only give adequate results for h when r is greater than H.

In many cases, dewatering problems and designs can be adequately solved with simple analytical models. For other hydraulic situations, such as confined or mixed aquifers, complex geometries, or other more complex flow regimes, one should consult a more detailed reference on dewatering, such as Powers et al. (2007) or Cedergren (1989). The use of finite difference and finite element numerical models has increased the ability to accurately model groundwater to accommodate heterogeneous aquifers, anisotropic properties, transient flows, and even three-dimensional flows.

7.4 DRAINAGE CAPACITY, PERMEABILITY, AND TESTS

7.4.1 Groundwater Flow Terminology

Before one can understand the hydraulics and analyses of dewatering, some fundamental terminology and basics of flow in the ground must be defined. *Porosity* rather than void ratio is typically used for most dewatering applications. Porosity is helpful for understanding permeability (hydraulic conductivity) and storage capacity of a soil mass. Void ratios are more appropriately used by geotechnical engineers for solving settlement and density problems. The two terms are closely related and can be calculated from one another through phase relationships.

Porosity (n) is defined as the percentage of the total soil volume that is void space:

$$n = \frac{V_v}{V_t} \tag{7.3}$$

Void ratio (e) is defined as the ratio of void space in relation to the volume occupied by solid soil particles:

$$e = \frac{V_v}{V_s} \tag{7.4}$$

One must be careful to consider that the *effective porosity*, that which allows flow through the ground, may be considerably lower than that calculated due to interconnectivity between pore spaces and to the fact that static water adheres to soil grains through capillary surface tension and electrical bonding (Powers et al., 2007). Porosity is dependent more on the range (or gradation) of particle sizes than on actual particle sizes. It will also depend on density of grain packing, often described in terms of *relative density* (D_r), which is defined as the actual density condition of a soil between its minimum and maximum "possible density." D_r is given as a percentage between 0% (loosest possible state) and 100% (densest possible state). Due to sampling disturbance effects, the relative density of coarse-grained soils is most commonly determined by correlations to field investigation tests, such as the SPT and CPT penetration tests described earlier in Section 3.2.3.

Obviously, the capacity of a soil to transmit and convey water is a fundamental parameter in the design and functionality of any drainage system where the water travels through a soil mass. In that regard, evaluation or estimation of the ability for water to flow through the soil is paramount. This can be done through either laboratory tests of representative samples, or with field tests, or both. The index parameter used to report the rate of fluid flow through the ground is known as the soil *coefficient of permeability* (or *hydraulic conductivity*), and is represented in terms of length/time (e.g., cm/s). This is in contrast to the volumetric flow rate (used for pumping or discharge rate) given in units of volume/time.

The measurement or estimation of permeability (hydraulic conductivity) is not a trivial task for several reasons. First, soil permeability varies more widely than any other soil parameter or property. Measured values span more than 10 orders of magnitude. Because of this, the accuracy and reporting of soil hydraulic conductivity is generally given in terms of its order of magnitude (e.g., 2.4×10^{-5} cm/s). Second, under typical field conditions, even for relatively homogeneous materials, hydraulic conductivity may be highly anisotropic, especially for fine-grained soils or other geologic media that have natural anisotropy, such as may be due to depositional or weathering characteristics. There are also complicated two- and three-dimensional effects, some of which may be modeled by flow nets, but often are not well represented by idealized flow patterns. In fact, if one considers the actual flow path of individual water droplets, it is actually far from laminar flow lines as water follows the tortuous path of voids within a soil (or ground) mass. In addition, hydraulic conductivity will vary with changes in pressure, temperature, fluid viscosity, and degree of saturation. Nevertheless, rather rash assumptions (i.e., steady-state, laminar flow, etc.) are often made for the flow regime of water within the ground—which may not be as poor a compromise as it may outwardly appear, given the other unknowns and variability.

As mentioned, the coefficient of permeability (hydraulic conductivity) for a soil can be measured in either the laboratory or in the field. Each approach has its advantages and drawbacks. Laboratory tests can be carefully controlled and monitored, but require a sample that is representative of the field conditions in terms of soil structure, anisotropy, and flow direction. Furthermore, laboratory tests are generally performed with a uniform "one-dimensional" flow, whereas in the field groundwater flow is actually two- or three-dimensional. Common laboratory procedures include the constant head test and falling head test, performed by gravity feed under atmospheric conditions, as well as tests performed at controlled pressures or pressure gradients, usually under triaxial stress conditions. There is also a procedure to determine hydraulic conductivity from time rate of consolidation testing. But all of these tests are limited to the confines of the laboratory and small sample sizes. Field tests, on the other hand, are not hindered by sample disturbance or problems with "representative" material properties, as they test the actual in situ ground. However, field testing has some disadvantages; for example, the actual flow paths traveled are not always clear. Field permeability tests typically include pumping tests and borehole tests, where the rate of rise or fall of the piezometric head is monitored, or pumping rates measured.

Hydraulic conductivity of some granular soils can also be roughly approximated by empirical methods using grain size analysis curves. The Hazen formula relates hydraulic conductivity to the D_{10} grain size (mm):

$$k = (0.1 \times D_{10})^2 \, \text{m/s} \tag{7.5}$$

The Hazen formula can provide reasonable approximations for relatively loose, uniform granular soils, but it does not take into consideration soil density or nonuniform gradations. Prugh developed empirical correlations for hydraulic conductivity that combine laboratory and field investigations based on median grain size (D_{50}), gradation uniformity, and soil density (Powers et al., 2007). Estimates of hydraulic conductivity using Prugh's interpretation are shown in Figure 7.9.

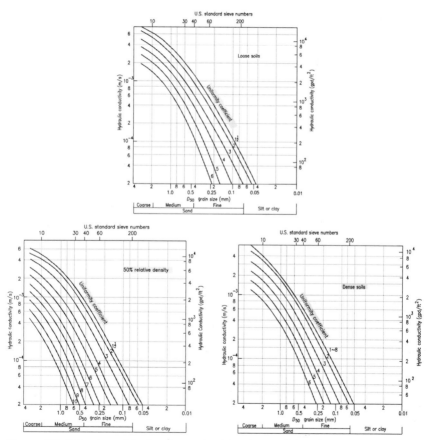

Figure 7.9 Estimates of hydraulic conductivity (permeability) based on grain size distributions and density. *Courtesy of Moretrench.*

Another complication with estimates of hydraulic conductivities from laboratory tests or empirical estimates based on grain size analyses is that the samples tested may not accurately represent stratified soil layers that commonly occur in nature due to geologic deposition. For field dewatering or drainage applications, evaluation of the actual permeability coefficients may not be essential, as the use of test and observation wells may be sufficient to evaluate flow (pumping) rates and subsequent drawdown parameters.

7.4.2 Drainage Capacity

The volume of water that will drain by gravity from a given volume of soil is called the *specific yield*. The volume of water that will be released per unit volume of water per unit change in head (or pressure) is called the *specific storage*. It is important that the drainage system is equipped to handle the maximum likely discharge. But note that not all the water stored in a soil will drain by gravity. The amount of water that remains in the soil by capillarity and surface tension is called *specific retention*. The specific yield will decrease with finer-grained soils, as these soils will exhibit higher capillarity. In fact, for clays the specific yield may approach zero. The implications of specific yield on dewatering design will be significant, and situations with low specific yield will require consideration of pumping and dewatering methods other than those driven by gravity alone.

7.5 SLOPE STABILIZATION BY DEWATERING/DRAINAGE

While there are many approaches and methodologies to slope stabilization, including a number of ground improvement techniques, dewatering a slope is a fundamentally simple concept for improving stability. Because of its ability to be highly efficient in terms of design and construction costs, and success at stabilizing a wide variety of slope stability cases, drainage of both surface and subsurface water is the most widely used slope stabilization method (Turner and Schuster, 1996). From basic slope stability theory and analyses, it is known that the existence of water within a potential slide mass generally always reduces stability. There are several reasons for this. First, as potential instabilities are fundamentally gravity driven, the added weight provided by water in the soil mass adds to the driving forces of the potential sliding mass. Second, any positive pore water pressures generated at the potential slide surface will reduce effective stresses, thereby reducing resisting frictional strength. Third, any seepage forces produced by water flow in the direction of the potential slide will reduce slope stability by adding to driving forces. Drainage of water will also reduce the risk of surface and internal erosion (piping). Therefore, by dewatering and/or redirecting water so that it does not reach the potential slide surface or slide mass, pore water pressures and seepage forces are reduced (if not eliminated), and added water weight may also be reduced. This is why drainage is one of the most important of all stabilization methods considered for remedial correction of active or incipient landslides, or for prevention/mitigation of hazard by increasing slope stability. There are a number of drainage techniques that are applicable to slope stabilization. Some of these will be discussed here.

7.5.1 Surface Drainage

Surface drainage is essential for treatment of many slopes. It requires minimal engineering while providing an effective means of aiding in slope stabilization. Proper collection and redirection of surface water will ensure that runoff will not erode the surface soils or infiltrate a slope, thereby avoiding additional seepage forces and added water weight to the slope mass. A number of techniques may be employed either as remedial work or as preventative design. Surface drainage systems are often simply concrete-lined channels (ditches) or corrugated steel pipes strategically placed at the head of slopes or at berms to divert water that has a tendency to collect. Another option is the installation of one or more shallow interceptor drains placed at strategic locations on a slope to catch and discharge both surface and near-surface water, so as to maintain any groundwater at a controlled depth within the slope (Figure 7.10). Designs should consider maximum inflow volumes, which can be calculated using parameters such as:

• Area and shape of catchment basin

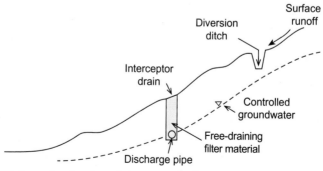

Figure 7.10 Example of surface drainage methods.

- Rainfall intensity
- Duration of inflows
- Infiltration coefficient of the surface and subsurface soils (based on ground cover and vegetation)

Adequate drainage (discharge) capacity must also be incorporated into designs to dispose of the collected water so as not to disrupt any facilities downstream of the slope. Culvert and drain design can be calculated by relatively simple hydrology equations given the input parameters mentioned above.

While ideally implemented in slope design or development, catchment and redirection of surface flows are often added or upgraded after an instability or drainage problem is recognized. Temporary measures in response to rainfall-induced sliding or other failure triggered by added water (e.g., water/drain/sewer break) may include sandbagging, ditches, redirection with plastic pipe, and even ground freezing (Chapter 13).

Additionally, there are a number of other measures that may be taken to promote rapid runoff by preventing infiltration to improve slope stability. These include seeding/mulching, using shotcrete, thin masonry, riprap/rockfill, or paving (Turner and Schuster, 1996).

7.5.2 Subsurface Drainage

Slope stability calculations clearly show that the stability (usually expressed as *factor of safety*) against sliding decreases when the potential slide mass includes a phreatic surface (i.e., water table). The higher the groundwater level is above the potential slide surface, the greater the reduction in stability. The fundamental reasoning for this was explained in the introduction to Section 7.5. Therefore, to maximize the stabilization effects by drainage, groundwater should be kept from entering the potential slide mass or slope as much as possible. Some common subsurface drainage methods employed for slope stabilization include:

- Drainage blankets
- Trench drains or cutoff drains
- Horizontal drains
- Relief drains
- Drainage galleries and tunnels
- Vacuum dewatering, siphoning, and electroosmosis

Wherever possible, these systems will drain by gravity. However, for some cases pumps may occasionally be employed to assist in removing groundwater. It should be easy to understand that incorporating these types of drainage methods into initial design and construction rather than installing as remedial work, is clearly advantageous and likely more cost-effective, especially for "developed" slopes. Historically, drainage of slopes for stabilization has been employed mostly as a remedial measure and continues to be one of the most common solutions for stabilizing landslides. But installation of drains for increased slope stability is now more commonly used in preventative design (Turner and Schuster, 1996). Most slopes and embankments will possess unique characteristics that will require individual designs. A number of different factors need to be considered in designing an effective slope drainage system. These include characterization of the groundwater regime, groundwater recharge and response to rainfall, type of drain(s), applicability of construction method(s), and necessary maintenance. Many drainage applications will ultimately include combinations of drain types and/or methodologies. Where natural slopes are encountered, drainage design may be more complex and will be a function of the subsurface structure of the ground and flow regime upslope of, or behind, the potential slide mass.

7.5.2.1 Drainage Blankets
Drainage blankets are used beneath or behind constructed slopes or embankments to intercept groundwater, thus ensuring that it will not enter the engineered soil mass. When an embankment or slope is to be constructed over a relatively shallow deposit of "poor" material underlain by stable material, the most reasonable and economical solution is often to excavate the poor soil. If future seepage into the engineered slope or embankment mass is a concern for these situations, a *blanket drain* constructed of free-draining granular material with an adequate discharge system (possibly a drainage well or perforated pipe drain) may be desirable beneath the constructed earthfill. Drainage blankets are sometimes used in the downstream portion of embankment dams to prevent water from accumulating, thus increasing the stability of the downstream slopes of these embankments. An example of the use of a drainage blanket beneath a constructed embankment is shown in Figure 7.11.

7.5.2.2 Trench Drains
If the extent of poor soil is large or too deep to be economically removed and replaced, then deep trench drains may be more appropriate to intercept and draw away unwanted groundwater. Trenches are typically excavated by backhoe or clamshell excavators perpendicular to the direction of the slope and backfilled with free-draining material. Adequate discharge must also be

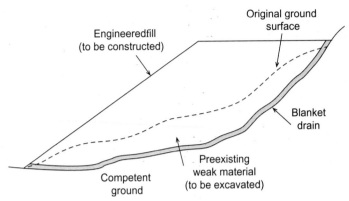

Figure 7.11 Example of a drainage blanket for slope stabilization.

provided for the maximum design flow. Installation of trench drains also may have the added benefit of providing increased resistance to sliding, as the compacted granular fill will act as a "key" into the weaker material beneath.

7.5.2.3 Cutoff Drains

Where groundwater is relatively shallow, *cutoff drains* can be used to intercept the flow and redirect it away from near-surface soils or away from potential slide surfaces. The main difference between trenches and cutoff drains is that trench drains tend to be relatively wide areas of free-draining fill, while cutoff drains are typically constructed using a perforated pipe embedded in a narrow trench of free-draining material, with a membrane or low permeability barrier downstream to cut off any near-surface flow downstream of the drain. Impermeable material is usually compacted in the top of the trench to prevent surface infiltration (Figure 7.12). As always, proper filtering and drainage criteria need to be observed, including properly matching pipe perforation sizes and surrounding free-draining material.

7.5.2.4 Relief Wells and Drainage Wells

Where the required depth of drainage is relatively deep, making conventional excavation to the required depth unfeasible and/or uneconomical, vertically drilled wells may be the best alternative. Ideally, they will drain by gravity into a free-draining stratum or subhorizontal discharge drain. But in most cases, these types of deep wells must be equipped with pumps to discharge the collected water in order to maintain working drainage conditions.

As the name implies, *relief wells* (or pressure relief wells) are intended to relieve subsurface water pressures and drain surrounding stratum by

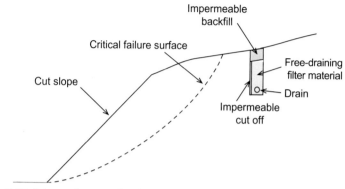

Figure 7.12 Schematic example of a cutoff drain.

providing a permanent drawdown or depression of the water table, or at very least provide relief for any excess water pressures that may be generated (as in the case of liquefaction mitigation drains discussed earlier in Section 7.1.1). Relief wells are typically drilled with 0.4-0.6 m (1.3-2 ft) diameters fitted with a 10-20 cm (4-8 in) partially slotted or perforated pipe, and backfilled with a free-draining material that will act as a filter for the drainpipe. In some cases, relief wells up to 2 m in diameter have been installed. Relief wells have been routinely installed up to 50 m (160 ft) depths, with typical spacing of 5-12 m (15-40 ft) (Abramson et al., 2002). Relief wells have been instrumental in stabilizing excavation walls, embankment dams, levees, and slopes where there is a tendency for buildup of potentially destabilizing excess water pressures. They have also been used to prevent "blowout" where water pressures and vertical exit gradients may become high, and have been incorporated in design of flood control levees.

To increase the effectiveness and efficiency of relief wells, a system called RODREN was introduced in the 1990s in Italy (Bruce, 1992). The principle of the RODREN system is that large diameter, vertical drains are interconnected by small-diameter (3-4 in) drains at their bases, which are then connected to a gravity discharge horizontal drain to the slope face or toe. This not only alleviates the need for pumping, but also makes for more effective drainage by maximizing the collective drawdown of all the wells in the system.

7.5.2.5 Horizontal Drains

Horizontal drains are commonly installed to draw down water levels in slopes to add stability. They do this by unweighting the slope material and reducing

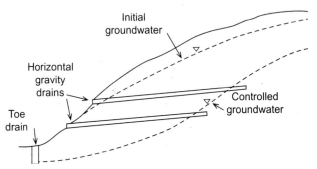

Figure 7.13 Example of subsurface drainage with horizontal and toe drains.

unwanted excess pore pressures as described in Section 7.1.1 (Figure 7.13). Horizontal drains are generally the best alternative when the depth (or distance into a slope) to desired dewatering is great and where the costs of excavation and/or placement of a trench or blanket drain are deemed to be uneconomical or unrealistic. Due to their relatively rapid installation speed, horizontal drains are often a good choice for stabilizing active or incipient slides. Many excellent case studies have shown the economical and practical solution to (often persistent) slope stability problems (Black et al., 2009; Machan and Black, 2012; Rodriguez et al., 1988; Roth et al., 1992). Horizontal drains are conveniently used where the drainage water can be discharged at the face of a slope or bench, or collected at the toe of a slope into a suitable discharge drain or other outlet (e.g., natural streambed, waterway, or culvert). While termed "horizontal" drains, these types of drains are usually installed at small inclines of 2-5° so they can drain by gravity. Drains are typically installed as small-diameter perforated pipes, typically 5-6 cm (2-2.5 in) in diameter, wrapped with geosynthetic filters in slightly larger diameter "horizontally" drilled holes (Abramson et al., 2002).

Installation of horizontal drains may be difficult in ground that may collapse the drain hole after drilling but prior to placement of the drain. Examples include sandy soils, ground with cavities, and fractured ground. Some efforts have been made to advance casing while drilling, into which a drain can be placed and then casing withdrawn. But often, casing may not be economically feasible. Other difficulties with installation have been encountered with soil containing boulders, cobbles, or other rock fragments.

Horizontal drain systems can be effective at lowering the water table and relieving groundwater pressures in a wide range of soil types, including relatively low permeability clays. In some cases, drainage of low permeability soils has been enhanced with the use of vacuum systems, as these will greatly increase the pressure differential toward the drains. Vacuum-assisted horizontal drains can also aid in stabilizing slopes by redirecting seepage forces toward the drains and away from the slope surface, and actually increasing normal stresses (and subsequently shear strength) on the potential failure surface(s). Spacing of horizontal drains will depend on the permeability and the volume (quantity) of the anticipated flow. Drains may be spaced apart by approximately 2-60 m (6-200 ft) or more, depending on the particular case. Closer spacing is typically used for lower permeability and more critical situations.

Horizontal drains have been installed as long as about 270 m (890 ft), while more commonly limited to about 60 m (200 ft). The length designed should ensure that water is drawn down and away from any potential (or active) slide mass. Such designs should include multiple cross-sections of critical failure surfaces, as well as geologic and hydrologic profiles where possible. It is usually recommended that the first 2 m (6 ft) or so of the drainpipe nearest the slope face not be perforated to avoid intrusion of roots or looser material near the surface. Drains may also be installed at multiple elevations, with the lower elevations (and longer lengths) typically providing the greatest drop in water table elevation.

Black et al. (2009) describe a test program where horizontal drains were installed (with moderate difficulty) to stabilize an approximately 30 m (100 ft) deep clay landslide in western New York State. The landslide had been activated when fill was placed near the head of the slope as part of highway construction. Stability analyses determined that excess pore water pressures were a significant factor contributing to the slide movement. In addition, observation wells showed a relatively high water table within 2 m (6 ft) of the ground surface. Nineteen horizontal test drains were installed with lengths from 134 to 230 m (440-750 ft) near the basal shear zone of the slide to investigate their effectiveness. It was reported that more than 90% of the installed drains produced discharge water and were still discharging water a year after installation. Observation piezometers showed a gradual drawdown of up to 3.8 m (12 ft) during the first 6 months, coincident with a significant decline in slide movement. Success of the test program resulted in a decision to install a broad system of horizontal drains across the approximately 100 m (330 ft) wide, 400 m (1300 ft) long landslide to mitigate the slope movements. A total of 173 drains were installed with lengths ranging between 125 m (410 ft) and 271 m (890 ft), resulting in water table drawdowns of 2-4 m (5-15 ft). These successes have led to New York's Department of Transportation embarking on another nearby highway project to install around 10 km of horizontal drains to reduce/eliminate movement in a slide that has persisted since 1948 (Poelma, M., 2013. Personal communications). Between this project and another in Wyoming (described by Machan and Black, 2012), over 81,000 lineal meters (2,66,000 ft) of drains were installed.

Santi et al. (2003) described installing horizontal prefabricated "wick" drains to assess their effect on slope stabilization at five sites. They concluded that the driven drains provided an effective and economical method of stabilizing landslides.

More recently, *launched horizontal drains* have been introduced for slope stabilization (www.geostabilization.com). These drains are installed as a perforated, hollow soil nail by a high-powered air cannon used for soil nailing (to be described in Chapter 15). These drains are made of perforated steel or fiberglass pipes that also provide tensile and shear contributions offered by soil nails. Launched horizontal drains can be installed rapidly (up to 200 drains per shift) and can be spaced as close as 1 m on center, greatly increasing the coverage area and shortening drainage paths (Figure 7.14).

7.5.2.6 Drainage Tunnels (Galleries)

Drainage tunnels are sometimes used when the volume of potential slide mass, along with depth and length of required drainage, is very large. They can also provide a feasible solution when access or topography makes installation of horizontal drains impractical. Drainage tunnels (also known as galleries) are relatively large holes (commonly 2-3 m) drilled by conventional mining

Figure 7.14 Drainage of a slope with closely spaced launched horizontal drains. *Courtesy of GeoStabilization International.*

methods; they may be used in conjunction with adjacent smaller vertical and subhorizontal drains drilled from within the tunnels to drain water and relieve water pressures within large soil and rock masses. The tunnels may run parallel to the direction of the slope or perpendicular to the slope (cross slope), acting as gravity collectors and discharge drains for many smaller drains. While relatively expensive with respect to other drainage techniques already discussed, for very large and "critical" projects, they may be appropriate. An example of effective slope stabilization using drainage tunnels was described by Rodriguez et al. (1988), where approximately 200 m (650 ft) tunnels were drilled at an average depth of about 15 m (50 ft) beneath a slope to support a new highway embankment.

Another good example of effective drainage tunnels was described by Millet et al. (1992). In this case, drainage was provided beneath a landslide mass adjacent to the reservoir above the 72 m (236 ft) Tablachaca Dam in Peru. In this case, a 300 m (1000 ft) wide, 350 m (1150 ft) high landslide was stabilized with a combination of techniques, including a gravity earth buttress constructed at the toe of the slide mass, rows of soil anchors, and a system of drains connected to drainage tunnels. The length of tunnels and galleries totaled more than 1500 m (5000 ft). Radial drains drilled from within the tunnels totaled an additional approximately 3300 m (11,000 ft). Discharge from the tunnel system was estimated at about 75-150 l/min (20-40 gpm).

Additionally, drainage tunnels allow for inspection and detailed evaluation of the subsurface conditions beneath landslide masses and/or the hydrologic conditions that may be contributing to sliding. The findings from installation drilling of the tunnels may be instrumental in design of the final drainage system.

7.5.2.7 Vacuum Dewatering, Siphoning, and Electroosmosis

Although much less common, a few other techniques for dewatering for slope stabilization should also be mentioned. While not new ideologies, these methods have not yet gained widespread usage due to underdeveloped technologies, costs, or other uncertainties. The use of vacuum in drilled wells has been employed as a temporary emergency measure for arresting active landslides in fine-grained slopes. Application of a vacuum increases soil suction (or effective stress), thereby increasing slope stability. In addition to the transient strength increase while vacuum is applied, prolonged duration will promote consolidation of clay soils. Depths of 30-35 m were reportedly successfully stabilized after vacuum was applied for a period of 2-4 weeks (Turner and Schuster, 1996).

Siphon drains were reportedly used to successfully dewater and stabilize unstable slopes in France (Gress, 1992). An advantage of siphoning is that water can be vertically extracted without powered pumps. In this case, siphoning was accomplished with sealed PVC piping.

The principles of electroosmosis were introduced as early as the 1940s by Casagrande, and have been applied sporadically for slope stabilization since at least the 1960s (Turner and Schuster, 1996). Electroosmosis, when used as a dewatering technique, can effectively consolidate fine-grained soil and increase shear strength. A number of successful applications have shown the merit of this process of dewatering (Turner and Schuster, 1996), and methods have been developed that allow free water flow without pumping. But costs still remain high, and acceptance of this approach has yet to gain mainstream popularity. Electroosmosis will be discussed in more detail in Chapter 10, along with other applications of this stabilization method.

7.6 FILTERING AND SEEPAGE CONTROL

One aspect of hydraulic modification involves improvements to ensure that the water flowing through the ground does not adversely affect the soil mass through which it travels. This is designed or corrected by providing proper filtering of the water passing through a soil. Filtering of water in geotechnical applications provides a number of important functions. When water passes through the ground from a finer-grained soil to a coarser-grained soil, there is a potential for some of the finer-grained material to erode into the void spaces of the coarser-grained soil. This is referred to as *internal or piping erosion*. The result may be to reduce the flow through the coarser material by clogging the pores (voids), and/or create cavities in the finer-grained soil mass. Voids or cavities created in this way may have the potential to collapse due to overburden stresses, which may, in turn, propagate to cause additional deformation.

The worst-case scenario is that the eroded cavities progress to the point of failure of the geotechnical structure. An example of this type of failure is the "pumping" of fines from the subgrade or subbase beneath a roadway due to the transient live vehicle loads creating voids beneath the pavement. This type of loading can lead to collapse of the pavement and/or the generation of "potholes." While not necessarily catastrophic, this type of failure may be a safety hazard and could lead to costly repairs and potential damage to vehicles. This type of internal erosion, often gone unnoticed until too late, can quickly propagate through a hydraulic structure (i.e., dam or levee), creating an internal "pipe" that may result in severe damage or collapse (piping failure) of the structure. In the extreme case, progressive piping has been the cause of catastrophic failures of earthen dams and levees, such as the collapse of the Teton Dam in 1976 mentioned in Chapter 3, or in the creation of fatal "home-swallowing" sinkholes.

For any drain and/or pumping design, it is important that the drain not become excessively clogged with soil particles from the mass being drained. Excessive material should not be allowed to flow into the drain. Historically, this had been accomplished primarily through the use of graded soil filters. By designing and controlling the proper gradation of adjacent soils through which water is flowing, proper filtering of the water can take place to mitigate internal erosion, high seepage pressures, or clogging problems. This type of design has been understood and implemented for many years and is generally known as *soil filter criteria*. Initially developed by Terzaghi and Peck in 1948 (1967) after experimentation on numerous soil filters, two basic criteria were derived based on the soil gradations:

(1)   The voids in the filter material (downstream soil) must be small enough to retain and prevent migration of the upstream material being filtered.

In order to satisfy this criterion, it was found that if the larger sized particles of the soil being filtered were retained, then the finer portion would also be protected. The effective void size of the filter is a function of the finer grain sizes and is taken to be about $1/5$ D_{15} of the filter gradation. This resulted in a simple equation relating D_{15} of the filter ($D_{15(\text{filter})}$ = screen diameter through which 15% of the filter material will pass), to D_{85} of the soil being filtered ($D_{85(\text{soil})}$ = screen diameter through which 85% of the soil being filtered will pass).

$$\frac{D_{15(\text{ filter})}}{D_{85(\text{soil})}} < 5 \qquad (7.6)$$

(2)   A soil filter must be sufficiently more permeable than the soil it is filtering.

In other words, there must be adequate flow so no buildup of hydrostatic pressures or seepage forces occurs. This generally requires that the gradation of the finer portion of the downstream filter material is significantly greater than the finer portion of the material being filtered.

$$\frac{D_{15(\text{ filter})}}{D_{15(\text{soil})}} > 4 \qquad (7.7)$$

Some granular filter designs address the assurance of adequate flow by simply requiring that the filter material have a permeability that is a multiple of the permeability of the soil being filtered. One criterion used by the FHWA requires a granular filter to be at least 10 times more permeable that the soil being filtered (www.fhwa.dot.gov). The US Navy (www.wbdg.org; US Department of Defense, 2005) added a few requirements to the Terzaghi and Peck criteria to maintain compatibility between filter and soil, resulting in the following:

$$\frac{D_{50(\text{filter})}}{D_{50(\text{soil})}} < 25 \qquad (7.8)$$

$$\frac{D_{15(\text{filter})}}{D_{15(\text{soil})}} < 20 \qquad (7.9)$$

The US Navy also added a requirement that a filter soil should have no more than 5% passing a #200 sieve to add stability to soil filters. And when soil filters are used around perforated drainpipes, the following requirements are made to prevent filter material from passing into the perforation slots or holes of the drains:

$$\frac{D_{85(\text{filter})}}{\text{Slot width}} < 1.2 - 1.4 \qquad (7.10)$$

$$\frac{D_{85(\text{filter})}}{\text{Hole diameter}} < 1.0 - 1.2 \qquad (7.11)$$

Filtering can also provide a means to control transport of soil materials in surface runoff. Examples include filtering of "dirty" water runoff from construction sites or other overland flows, and surface erosion control. This type of filtering is usually done with geotextile filters or geotextile-wrapped granular soils. Geotextile filters have taken the place of soil filters in many designs and applications due to ease of installation, reliability and uniformity, and, often, economic savings. A discussion of geotextiles used as filters is included in Chapter 8.

Seepage from water retention or conveyance structures can be an issue if the amount of seepage adversely affects a component of a project. Simply put, if a water storage (reservoir) or water retention facility cannot meet its intended performance goals due to seepage of water, then it may be advantageous to use mitigation measures to reduce the losses. This may involve lining the upstream area suspected of allowing excessive seepage with a low permeability cap (or "impermeable" membrane), or installing a barrier through the suspected seepage zone, such as with sheet piling, a slurry wall, a soil-mixed wall, or a grout curtain, as described in Section 7.1.2. In instances where the leakage does not present a danger or hazard, the economics need to be carefully weighed to determine the cost-effectiveness of such a repair. Remember that excessive seepage and associated hydraulic gradients can, unfortunately, promote more serious consequences (as described previously). Another solution for control of potentially damaging seepage from within embankments and slopes is to install properly filtered internal drains to collect and discharge seepage water, as described for slope stabilization in Section 7.5.

7.7 MEMBRANE ENCAPSULATION

Membrane encapsulation creates/installs a membrane to prevent migration of water. The approach has been primarily used to control volume change in expansive soils or soils susceptible to frost heaving. Essentially, this provides a means to maintain consistent water content, thereby reducing the potential for volume change. Shrink/swell in soils is primarily a result of change in moisture levels. As long as the moisture is consistently maintained, a soil will neither shrink, nor swell. One approach is to induce high moisture by injecting or flooding with water, and then maintain the elevated moisture by "sealing" the moist soil with a membrane. Sometimes, a chemical additive, such as potassium chloride-based materials, will be added to attract and hold moisture.

The membrane, or seal, may be achieved in a few different ways. Geosynthetic membranes may be used, but obviously may be applicable only to new earthwork construction. Geomembrane liners, typically high density polyethylene, can be placed so that earth materials are located on top of the liner (or within a geomembrane-lined trench). Then the liner is folded over the top of the soil with enough overlap to "seal" the enclosed soil. If successful performance relies on a secure seal, then care must be taken to assure no leaks, both immediately following construction, as well as for the design lifetime of the project. Commercially produced nitrile *"superbags"* are available to take out much of the uncertainty of a manually constructed "good seal." A discussion of using geomembranes as hydraulic barriers is included in Section 8.2.

Alternatively, membrane seals may be constructed in place by injecting grout barriers of urethane or other chemical mixtures into the ground, either by hydrofracturing the ground along preferred planes and filling the fractures with grout materials, or by permeation grouting (forming treated zones of "impermeable" material). One must exercise caution when it comes to the reliability of these types of seals, as it is difficult, if not impossible, to ensure complete interlocking of the grouted planes that cannot be observed. More recently, jet grouting has been employed to create high-quality, soil-mixed barriers with reasonable success. These grouting techniques will be discussed in Chapter 12.

Ground freezing has also been successfully used for temporary sealing of a soil mass during construction and also for emergency containment of contaminant spills or leaks. Ground freezing will be described in Chapter 13.

7.8 ALTERING SOIL/GROUND HYDRAULIC PROPERTIES

Another approach to modifying the hydraulic properties of a soil mass is to physically and/or chemically modify the soil or ground. Densification was discussed earlier in Chapters 4–6, and can result in significant decrease in the hydraulic conductivity (permeability) of a soil. Control of water content and method/equipment used in compaction has also been shown to play a big role in achieving good water retention and reducing seepage for hydraulic structures (dams, canals, reservoirs, etc.) and landfills.

Gradation control is a method whereby certain grain sizes are retained or discarded to achieve preferred hydraulic properties. This will be discussed in Section 11.2.1. Well-graded granular soils will inherently have a lower hydraulic conductivity than a more uniform material, or one without appreciable fines. An "open-graded" granular soil usually refers to a fairly uniform coarse sand or gravelly material that is often used to keep an area well drained, such as adjacent to foundations, behind retaining walls, surrounding buried utility lines, and so on. Open-graded gravels are also used as drains themselves, as in the case of vertical gravel drains, or "French" drains.

Mixing other materials with engineered soil or in place in the ground can serve as a means of altering hydraulic properties. Admixtures of lime, cement, bitumen, urethanes, polymeric grouts, bentonite, and others are commonly mixed with natural soils to create zones of very low seepage or "impermeable" barriers. Discussion of these applications will follow in Chapter 11. And finally, various grouting methods, some already mentioned, are often used primarily for seepage/leakage control. Grouting methods and applications will be addressed in more detail in Chapter 12.

REFERENCES

Abramson, L.W., Lee, T.H., Sharma, S., Boyce, G.M., 2002. Slope Stability and Stabilization Methods, second ed. John Wiley & Sons, Inc, 717 pp.

Black, B.A., Machan, G., Peolma, M., 2009. Horizontal drains in a clay-landslide stabilization test program. Transport. Res. Rec. 2116, 35–40.

Bruce, D.A., 1992. Two New Specialty Geotechnical Processes for Slope Stabilization. Geotechnical Special Publication No. 31, ASCE, pp. 1505–1519.

Bruce, D.A., Dreese, T.L., Heenan, D.M., 2008. Concrete walls and grout curtains in the twenty-first century: the concept of composite cut-offs for seepage control. In: USSD 2008 Conference, Portland, OR, 35 pp.

Cedergren, H.R., 1989. Seepage, Drainage, and Flow Nets, third ed. John Wiley & Sons, Inc, 465 pp.

Gress, J.C., 1992. Siphon drain: a technique for slope stabilization. In: Proceedings, Sixth International Symposium on Landslides, vol. 1. A. A. Balkema, pp. 729–734.

Hausmann, M.R., 1990. Engineering Principles of Ground Modification. McGraw-Hill, Inc, 632 pp.

Machan, G., Black, B.A., 2012. Horizontal drains in landslides: recent advances and experiences. In: Proceedings of the 11th International Symposium on Landslides, Banff, Alberta, Canada.

Millet, R.A., Lawton, G.M., Repetto, P.C., Garga, V.K., 1992. Stabilization of Tablachaca Dam landslide. Geotechnical Special Publication No. 31, ASCE, pp. 1365–1381.

Powers, J.P., Corwin, A.B., Schmall, P.C., Kaeck, W.E., 2007. Construction Dewatering and Groundwater Control: New Methods and Applications, third ed. John Wiley & Sons, Inc, 636 pp.

Rodriguez, A.R., Castillo, H.D., Sowers, G.F., 1988. Soil Mechanics in Highway Engineering. Trans Tech Publications, Ltd., London, 843 pp.

Roth, W.H., Rice, R.H., Liu, D.T., Cobarrubias, J., 1992. Hydraugers at the via De Las Olas Landslide. Geotechnical Special Publication No. 31, ASCE, pp. 1349–1364.

Santi, P.M., Crenshaw, B.A., Elifrits, C.D., 2003. Demonstration projects using wick drains to stabilize landslides. Environ. Eng. Geosci. IX (4), 339–350.

Terzaghi, K., Peck, R.B., 1967. Soil Mechanics in Engineering Practice, second ed. Wiley, New York, 729 pp.

Turner, A.K., Schuster, R.L., 1996. Landslides: investigation and mitigation. In: Transportation Research Board Special Report 247. National Academy Press, Washington, DC, 673 pp.

U.S. Department of Defense, 2005. Unified Facilities Criteria, Soil Mechanics, UFC 3-220-10N, 394 pp.

https://www.fhwa.dot.gov/engineering/hydraulics/pubs/09112/page16.cfm (accessed 09.14.13.).

http://www.geostabilization.com (accessed 01.15.14.).

http://www.griffindewatering.com/dewatering/eductor_system (accessed 02.16.14.).

http://www.google.com/patents/US7454847 (accessed 08.29.13.).

http://www.hpwickdrains.com (accessed 08.30.13.).

www.mirafi.com (accessed 08.21.13.).

http://www.moretrench.com (accessed 09.15.13.).

http://www.moretrench.com/cmsAdmin/uploads/Beacon_Hill_Station_Seattle_Washington.pdf (accessed 09.23.13.).

http://www.wbdg.org/ccb/DOD/UFC/ufc_3_220_10n.pdf (accessed 09.26.13.).

http://en.wikipedia.org/wiki/Dewatering (accessed 08.15.13.).

CHAPTER 8

Geosynthetics for Filtration Drainage, and Seepage Control

While the use of geosynthetics has proliferated throughout a wide range of geotechnical applications, special attention to hydraulic improvements for filtration, drainage, and seepage control seems appropriate for the discussion of hydraulic modification. In fact, the use of geosynthetics for drainage and filtration has often taken the place of conventional applications that had historically been engineered with carefully graded earth materials. This not only has led to economic savings and ease of construction, but also has been attributed to providing a more uniform and safer solution by averting natural variability in materials and workmanship, less potential for segregation of materials under hydraulic gradients, and reduced exposure to piping. For other applications where the result is to improve the ground by reducing or eliminating seepage, a few types of geosynthetics can provide "hydraulic barriers" to prevent any significant flow, where naturally occurring soils may be insufficient or impractical. A third category of applications with geosynthetics combines different types of materials and serves multiple functions. These are referred to as *geocomposites*. While geosynthetics have provided significant economic, construction, and performance advantages, there can be intrinsic problems, including the loss of ability to "self-heal" after rupture, chemical and/or biological degradation of the geosynthetic materials, and long-term flow compatibility. Other primary functions, including separation and soil reinforcement, are addressed in Chapters 14, 16, and 17.

A wealth of information and resources on geosynthetics can be found through organizations such as: Industrial Fabrics Association International (IFAI; www.ifai.com), the North American Geosynthetics Society (NAGS; www.geosyntheticssociety.org), the Geosynthetics Institute (GSI; www.geosynthetic-institute.org), or from comprehensive texts such as *Designing With Geosynthetics* by Koerner (2005, 2012) and *Geosynthetic Engineering* by Holtz et al. (1997).

Soil Improvement and Ground Modification Methods

8.1 GEOTEXTILES FOR FILTRATION AND DRAINAGE

Geotextiles are continuous sheets of woven, nonwoven, knitted, or stitched fibers (Figure 8.1). They may be made of a variety of types of fibers, including natural yarns or polymeric fibers. Geotextiles are the principle geosynthetic materials used for *separation* (between dissimilar soil types/gradations) and *filtration* of flowing water (into, through, or exiting a soil mass). Geotextiles may also provide significant *reinforcement* by providing tensile strength and shear resistance, and certain types are used for erosion control. The reinforcing function of geotextiles will be discussed later in Chapter 14. In some applications, they may also provide a limited *drainage* function as described in Section 8.2.

Some newer geotextiles not only provide separation and filtering, but also claim to provide a substantial in-plane drainage capacity based on the in-plane permeability of the geotextile fabric (see Section 8.2).

8.1.1 Filtering and Geosynthetic Filtering Criteria

Historically, traditional filters were designed by placing carefully graded soils in "zones" so that soils would be successively coarser in the direction of flow. If engineered correctly in this manner, each successive soil filter would satisfy the basic filter criteria: (1) the filter should have small enough openings

Figure 8.1 Photograph of a range of typical geotextiles used in geotechnical applications. *Courtesy of Geosynthetic Institute.*

(voids or pore spaces) to prevent migration of the soil being filtered (i.e., the "upstream" soil from which the flow is entering the filter), and (2) the filter should be coarse enough (e.g., have large enough openings/voids) to adequately allow the flow to pass through (without generating excess pore water pressures or seepage forces). Simple criteria were initially developed by Terzaghi and Peck (Terzaghi et al., 1996) after experimentation on numerous soil filters based on the soil gradations, and subsequently modified as described in Chapter 7.

With the same basic philosophy and concepts used for soil filters, design criteria have been developed for geotextile filters. Fundamentally similar to those for soil filters, criteria for geotextile filters include (1) soil retention, (2) adequate flow, and (3) long-term flow compatibility (clogging resistance).

Geotextiles are permeable fabrics, typically made from polypropylene or polyester. They may be woven or bonded by other means (i.e., needle-punched, heat bonded). The fundamental parameters important to the success, performance, and functionality of geotextiles for filtration closely mimic the criteria for soil filters. These generally include the ability to allow adequate flow from a soil across the plane of the geotextile with limited soil loss over a design service life of the soil-geotextile system. Specific filter criteria are discussed in the following section.

In the case of geotextiles, soil retention is addressed by making the geotextile voids small enough to initially only retain the coarser soil fraction. While this might at first seem counterintuitive, the coarser fraction is targeted in these designs as research has shown that a process called "bridging" causes the buildup of coarser-sized particles to eventually block the finer-sized grains. On the other hand, the retention criteria for soil filter design described in Chapter 7 has similar characteristics, as it targets the D_{85} of the soil being filtered. The design formulas for geotextiles typically use soil particle size characteristics and compare them to the size(s) of the openings in the geotextile fabric. The most common geotextile opening size used, defined as O_{95}, refers to the opening size that will retain 95% of uniform-sized spheres by dry sieving. In the United States, this is called the *apparent opening size* (AOS) obtained by dry sieving different sized uniform spheres. ASTM D4751 provides guidance for obtaining AOS. In Europe and Canada, testing is done by wet or hydrodynamic sieving to derive the filtration opening size. Similar to the principles of wet sieving for soil gradation, these processes may be preferable (Koerner, 2005), as they are believed to provide more accurate results and more closely represent the actual flow and filtering conditions.

The simplest soil retention design procedures are based on the percentage of soil finer than the #200 sieve (0.074 mm). For example, an AASHTO guideline recommends (Koerner, 2005):

- $O_{95} < 0.60$ mm, (AOS $\geq$ #30 sieve) for soil with $\leq 50\%$ passing #200 sieve
- $O_{95} < 0.30$ mm, (AOS $\geq$ #30 sieve) for soil with $> 50\%$ passing #200 sieve

Several other researchers have made recommendations based on more complete grain size distributions. A simple, but widely used relationship was outlined by Carroll (1983):

- $O_{95} < (2\text{-}3)\ D_{85}$ (where D_{85} is the grain size for which 85% of the soil is finer)

More detailed design includes information on soil hydraulic conductivity, plasticity (PI), percent clay, and undrained shear strength, such as described in NCHRP Report 593 (NCHRP, 2007).

To address the criteria for adequate flow, many designs simply require a geotextile to have a permeability that exceeds a certain multiple of the soil permeability. These multiples typically range from about 4 up to 10 times (or more), depending on the critical nature and severity of the application. Because of this, the geotextile permeability (hydraulic conductivity) is needed. When flow is perpendicular to the plane of the fabric ("cross-plane"), the geotextile permeability is typically computed by the term *permittivity*, often used by the geotextile industry, which includes the thickness of the geotextile. This mitigates any discrepancies with variability common to relatively thick and compressible fabrics. Permittivity is defined as

$$\psi = \frac{k_n}{t} \tag{8.1}$$

where ψ is the permittivity (s^{-1}), k_n the cross-plane permeability (n for flow normal to the plane, cm/s), t the fabric thickness at a specified normal pressure (cm).

Permittivity is often used when comparing geotextiles of different thicknesses. The testing procedure for measuring permittivity is fundamentally the same as for measuring soil permeability. Geotextile permeability can be obtained by simply multiplying the measured permittivity by the fabric thickness. The geotextile manufacturer typically provides these values.

Long-term flow compatibility (anticlogging criteria) may be based on the percent open area (POA) for woven geotextiles, and on porosity for non-woven fabrics (www.fhwa.dot.gov). POA is a comparison of the total open area to the total area of the geotextile. Porosity is the relationship between the volume of voids and the total volume of the geotextile, reported as a percentage by volume. NCHRP Report 593 recommends a POA >4% for woven geotextiles or a porosity >30% for nonwoven fabrics. When geotextiles used as filters fail, the most likely reason is clogging. Koerner (2005) describes a number of scenarios where excessive clogging has been observed from experience. One approach is to allow a certain amount of fine sediment to pass through a more open geotextile with POA >10% for woven fabrics or porosity >50% for nonwoven fabrics. But in order to use these criteria, one must feel confident that neither the loss in retention, nor the passing of material into the drainage system, would promote a significant problem to the application for which it is used.

When designing geosynthetic filters, it is important to consider both survivability (ensuring resistance to installation damage) and durability (resistance to chemical, biological, and ultraviolet light exposure). AASHTO M288-06 provides specifications for allowable strength and elongation values. ASTM 5819 provides a guide to choosing appropriate durability test methods. Additional ASTM standards listed at the end of this chapter are available for specific strength and survivability tests.

8.1.1.1 Geotextile Filter Applications

The use of geotextile filters have become commonplace for a wide array of applications. If properly designed, a geotextile filter may act as the sole filtering medium between an appropriate soil and drain or well. Where plastic pipe is installed for drainage or a well, a fabric "sock" is often placed around the (sometimes open) end of the pipe, covering the perforations to prevent material from entering the drain. For many applications where granular soil filters had traditionally been used, geotextiles have taken their place in new designs and construction. Common applications include filtering for gravel drains, filters for zoned earth dams, and filtering within engineered roadway layers. Commonly, geotextile filters used for highway applications involves geocomposites to provide filtering and drainage as described in Section 8.2.

Geotextile filters also have become an integral part of drainage design behind both rigid and flexible retaining walls. As described in Chapter 7, the need to provide long-term drainage so as to prevent buildup of hydrostatic pressures behind retaining walls is paramount to their survivability. Geotextiles have become the norm for filtering drainage water from the backfill soil to assure that drainage will continue unobstructed. In some cases where the volume of drainage water is very small, the geotextiles may provide for both functions of filtering and drainage. For flexible wall systems, such as might be constructed with stone-filled, wire baskets (gabions) or other free-draining wall systems, geotextiles are now used almost exclusively to provide the necessary filtering function. When used in this fashion, they are actually also functioning as a separator between different materials. In another type of application, geotextiles have been used as filters beneath coastal erosion control structures made of placed rock riprap, articulated concrete blocks, or concrete block mattresses. These types of erosion control, or *armoring*, can be used to face the upstream slopes of earth dams subjected to wave action. In these applications, the filters may need to be designed to handle flow in both directions, as tides and waves can push water through the protective stone/blocks, building up water pressures in the soil beneath, which must then be dissipated back through the erosion protection without loss of the soil material beneath. In another type of erosion protection application, geotextiles are often secured directly to the ground surface of steep slopes subjected to heavy rainfall and/or prone to surface sloughing.

A surface application of geotextiles as filters is for control of "dirty" construction runoff by constructing silt fences or fabric-wrapped, granular material to trap the suspended, fine-grained material and allow relatively clean water to flow away. While often not designed specifically for each application, there are readily available products available for construction runoff filtering. Where large amounts of runoff are expected, *silt fences* may be constructed to capture transported sediments. The design of silt fences is based on the amount of flow expected and relies on a certain amount of intended clogging of the fabric in order to form a sediment "trap." While not really a soil improvement application, it is related to similar flow and filtering functions and criteria. Silt fence design is covered by geosynthetic references such as Koerner (2005) and Holtz et al. (1997).

Another filtering application is in the stabilization of dredged materials and other high water content "sludges" by placing these materials into fabric containment "bags." This allows fluid to drain out of the material, making it easier to transport and/or dispose of. Commonly known as *Geotubes*®, they have also been used extensively for shoreline protection, breakwaters, levees, beach rebuilding, and as a component for reclaimed land. These confinement applications will be discussed in Chapter 16. Figure 8.2 depicts an example of dewatering of marine spoils with Geotubes.

Figure 8.2 Dewatering of dredged sediment with Geotubes. *Courtesy of Infrastructure Alternatives.*

8.1.2 Geotextile Drains

As mentioned above, the use of geotextiles for hydraulic applications is primarily for filtering functions, but geotextiles can provide drainage under certain conditions. When geotextiles are placed in an application where fluid flow occurs within the plane of the fabric, they can provide a limited amount of drainage. Except for the consideration of flow direction, the criteria for soil retention and long-term flow compatibility described in the previous section on filtration are virtually the same. This leaves only the discussion of adequate flow and in-plane permeability. Just as the variable thickness due to compressibility was addressed for measurement of cross-plane permeability for the filtering function, it is handled similarly for in-plane drainage. For this, the geotextile industry uses a term called *transmissivity* to describe the flow rate within the fabric. Transmissivity is defined as

$$\theta = k_p \times t \qquad (8.2)$$

where θ is the transmissivity (cm^2/s), k_p the in-plane permeability (cm/s), t the thickness (cm).

A recent innovation integrating hydrophilic and hygroscopic yarns into a high-strength woven geotextile incorporates a true wicking component that draws water from the ground and is able to transport it away from critical components of projects located in high moisture environments, or where moisture introduced by rain or snow can be drawn out of the subsoil (www.tencate.com). Figure 8.3 shows a graphic of how these types of geo-textiles are able to draw water out of a soil without pumping, vacuum, or relying on induced pore pressures. Figure 8.4 shows a close-up view of the wicking geotextile fabric. Case studies have shown very good results in reducing standing water and frost heaving for difficult roadway applications. This type of geotextile has been successful in reducing perpetual frost heave problems in Alaska highways, as well as reducing pumping and flooding problems in other environments with high moisture soils.

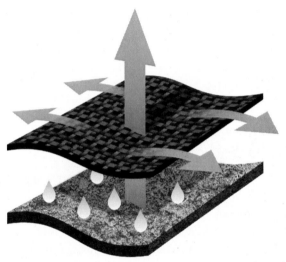

Figure 8.3 Mirafi's H₂Ri woven geotextile capable of wicking moisture from subgrade soils. *Courtesy of Tencate-Mirafi.*

Figure 8.4 Close-up view of wicking geotextile fabric. *Courtesy of Tencate-Mirafi.*

Just as with filters, geosynthetics have now become commonplace as substitutes for natural soil materials for drainage applications. In some cases, thick, nonwoven geotextiles can provide some amount of drainage function. While this may be adequate for low-volume flows, the drainage capacity is highly dependent on stresses that will compress the fabrics and reduce flow area.

8.2 GEONETS, GEOCOMPOSITES, AND MICRO SIPHON DRAINS

In-plane drainage using geosynthetics is usually designed with either *geonets* (usually in combination with a geotextile) or with a *geocomposite*, a class of geosynthetics often designed principally for in-plane drainage. These hybrid geosynthetics are made by combining different types of geosynthetic components, and serve the purpose of providing both filtration and drainage. Geocomposites are typically combinations of a drainage (and sometimes barrier) material with a geotextile filter to prevent soil migration into the drainage system. Geosynthetics used for drainage include perforated plastic pipes (or "geopipes"), geonets (ribbed materials intended to convey in-plane flow), and corrugated geomembranes (which can provide substantial in-plane flow capacity as well as a hydraulic barrier). Geosynthetic hydraulic barriers are discussed in Section 8.3.

Geonets are typically formed by two biplanar sets of relatively thick, parallel, polymeric (usually polyethylene) ribs bonded in such a way that the two planes of strands intersect at a constant acute angle, forming a diamond-shaped pattern (Figure 8.5). The configuration of the nets form

Figure 8.5 Biaxial geonet.

a network with large porosity that enables relatively large in-plane fluid (and/or gas) flows. While they have considerable tensile strength, they are used exclusively for drainage applications. Their initial use was almost exclusively for environmental applications, such as hazardous, liquid, waste impoundment, or landfills to collect and drain leachate fluids, and for leak detection. Geonets have also been shown to provide effective capillary breaks where moisture intrusion due to capillary rise is a concern. They have now become more widely used for drainage behind retaining walls, in slopes, in hydraulic structures (e.g., dams and canals), in large horizontal areas (e.g., golf courses, athletic fields, and plaza decks), and as drainage blankets beneath surcharge fills and embankments. In order to prevent soil intrusion into the voids, geonets are generally used in conjunction with geotextiles and/or geomembranes (Figure 8.6). While traditional biaxial geonets were never really intended to support any tensile or shear load, newer triaxial versions of geonets have been designed to provide even greater flow with added load capacity in both compression and shear (Figure 8.7). The triplanar structure provides minimal geotextile intrusion and greater flow capacity through longitudinal channels. Their higher rigidity, tensile strength, and compressive resistance make them suitable for application within roadway pavement systems, beneath highways and airfields, and beneath concrete building slabs.

Where larger drainage volumes are needed, geocomposites consist of corrugated, "waffle" type, or "dimpled" geomembrane cores with large porosity, attached to a geotextile for filtration and to prevent soil intrusion (Figure 8.8). Geocomposite drains may be configured to act as a central drain

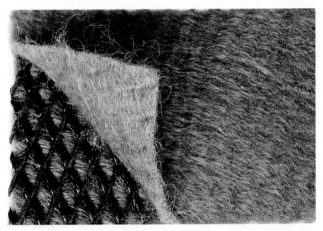

Figure 8.6 Geonet geocomposite drain.

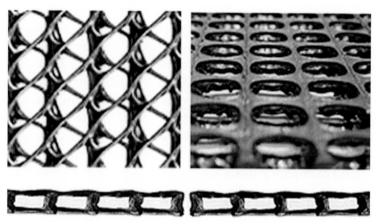

Figure 8.7 Example of triaxial geonets. *Courtesy GSE Environmental, LLC.*

Figure 8.8 Example of geocomposite "waffle" sheet drain. *Courtesy of Tencate-Mirafi.*

where a geotextile fabric fully wraps a plastic core. Thin strip drains (often called *prefabricated vertical drains*) are used as an aid to consolidation or preconsolidation (Chapter 9) and can be driven directly into soft ground with specialized equipment. Thicker versions with much higher volume cores are also now very common for installation as edge drains for roadway applications or for horizontal drainage within a soil mass or between placed fills.

Large sheets of geocomposite drains, consisting of an impermeable waffle core with geotextile filter on only one side, are now commonly used adjacent to foundations and behind retaining walls to mitigate any hydrostatic pressures beneath and behind these structures. These types of drains perform multiple functions of filtering, drainage, and hydraulic barrier, all at once.

A relatively new application of geocomposite drain, called *earthquake drains*, consisting of a perforated pipe covered with a durable filtering geotextile, can be installed for liquefaction mitigation in loose sandy soils (Figure 8.9). Liquefaction failure was introduced and described in Chapter 3. The premise is that when dynamic loading generates excess pore pressures during an earthquake, proper drainage that can quickly dissipate the excess water pressures will prevent those pressures from reaching high enough levels to initiate liquefaction. High discharge capacity drains, closely spaced, can provide such a system. One success story of this application was reported after the June 1999 earthquake in the British West Indies, which delivered estimated ground accelerations of $0.12g$. E-QUAKE® drains, developed by Geotechnics America, Inc., had been installed in lieu of significantly (about three times) more expensive stone columns in approximately half of the expected construction time for a Hyatt Regency Hotel and Casino built over a thick layer of loose saturated sands (www. geotechnics.com). Although liquefaction did occur in some areas where no mitigation had been done, the drains provided enough drainage to keep excess pore pressure ratios to less than 0.6, so that liquefaction within the treated areas was prevented. Design of earthquake drains is based on design

Figure 8.9 Earthquake drain installation. *Courtesy of HB Wick Drains.*

earthquake levels along with permeability and compressibility of the on-site soils (UCB/EERC-97/15). Other earthquake drain applications have been and continue to be done as an economical and efficient alternative to other liquefaction mitigation techniques. Earthquake drains can be installed to depths of up to 45 m (150 ft) in very loose, low-bearing capacity materials (www.geotechnics.com).

A relatively new member of the geosynthetic drain category includes *micro siphon drain* belts, such as that produced by Smart Drain (www. smartdrain.com) (Figures 8.10 and 8.11). These types of drains operate on the principle that their corrugated longitudinal openings are small enough

Figure 8.10 Photograph of a micro-siphon. *Courtesy of Smartdrain.*

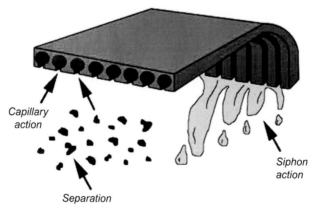

Figure 8.11 Schematic of the siphoning action of a micro-siphon drain. *Courtesy of Smartdrain.*

to create a siphoning action as a function of capillary forces. This siphoning allows the drains to be used over irregular terrain that may even include some upslope sections. A significant added benefit to these types of drain materials is that they have a very high resistance to clogging as compared to geotextile filters, which are most commonly used with perforated pipes, gravel drains, or other geosynthetic drains. Results of a recent study showed that very little reduction in flow rates was observed over time, even after two to four times the typical clogging load was applied (Sileshi et al., 2010). The capillary action and very small opening size of the micro siphons separates water and soil as opposed to conventional filtering, thus allowing application of the drain, even directly to fine-grained soil.

These types of micro siphon drains have been used to rapidly drain large horizontal "green" spaces such as golf courses, athletic fields, parks, and "green roofs." They also seem to be effective as drains adjacent to foundations and behind retaining walls, and as seepage control in embankment dams (www.smartdrain.com; Ming and Chun, 2005). Figure 8.12 shows some examples of micro siphon drain applications.

8.3 GEOSYNTHETIC HYDRAULIC BARRIERS

Two types of geosynthetic products are commonly used as hydraulic barriers in geotechnical applications: geomembranes and geosynthetic clay liners (GCLs).

8.3.1 Geomembrane Seepage Barriers

Geomembranes are most commonly made from various densities and thicknesses of polyethylene, polyvinyl chloride, and polypropylene (Figure 8.13). While no material is actually "impermeable," geomembranes provide a barrier with an effective permeability (hydraulic conductivity) on the order of 10^{-10} to 10^{-13} cm/s (Koerner, 2005). While this appears to be a "surefire" solution for any seepage or leakage problem, the barrier systems are only as good as their weakest link, which in the case of geomembranes are seams, defects, or damage. Geomembranes are commonly used in a wide variety of applications wherever a hydraulic (or vapor) barrier is needed for new construction. This will include, but is certainly not limited to, the following:

- Liners for any water retention application such as reservoirs, canals, ponds, and emergency spillways
- Buried liners (or secondary liners) for retention of waste containment leachate

Figure 8.12 Typical installations of micro-siphon drains. (a) Horizontal installation. (b) Vertical wall installation. *Courtesy of Smartdrain.*

- Seepage prevention/control (cutoffs) within embankment dams, transportation facilities, and geoenvironmental applications
- "Waterproof" facing of earth, rockfill, roller compacted concrete, concrete dams, tunnels, and pipelines
- To prevent water migration and control volume in expansive soils and frost susceptible soils

Figure 8.13 Example of geomembranes (GSI). *Courtesy of Geosynthetics Institute.*

- As a cover to prevent infiltration of rainfall into landfills, roadway base layers, etc.

A relatively new (but still not too common) application for a positive cutoff is to insert interlocking vertical geomembrane panels (or in some cases a continuous membrane) into a slurry trench (Mitchell and Rumer, 1997).

8.3.2 Geosynthetic Clay Liners

GCLs are made by bonding, needle-punching, or stitching very low permeability material (i.e., natural sodium bentonite clay) to geosynthetic materials (usually textiles) to create an economical, long-term solution where hydraulic barriers are required (Figure 8.14). Bentonite is an extremely absorbent,

Figure 8.14 Examples of typical geosynthetic clay liners (GCLs).

granular clay formed from volcanic ash. Its high net negative charge attracts water, hydrates rapidly, and swells to form a tight seal. GCLs are primarily used as substitutes for (or to complement) conventional compacted clay liners or geomembranes for surface water impoundment, secondary containment, and landfill liners and covers. They typically provide significant cost advantages, ease of installation, and increased performance as compared to traditional compacted clay liners. In addition, they have the ability to "self-repair" or "self-heal" after sustaining minor damage (e.g., small rips or holes) due to the swelling characteristics of the clay materials from which they are made. Reports of laboratory tests demonstrated that a hole up to 75 mm in diameter in a GCL will heal itself (www.epa.gov). While susceptible to transport and installation damage, this attribute is a distinct advantage over some other barrier systems.

GCLs were initially developed in the late 1980s, both in the United States with a produced called Claymax® by CETCO (Colloid Environmental Technologies Company), and at about the same time in Germany with a product called Bentofix® by Terrafix (NAUE in Europe). Claymax was produced by placing a bentonite clay mixed with an adhesive to bond the clay between two geotextiles. Bentofix was manufactured by placing a layer of powdered bentonite between two geotextiles and then needle-punching the layers together (Figure 8.15). There are now many GCLs available in addition to those mentioned above, including: GSE Environmental's Gund-Seal® (which combines a conventional GCL with a polyethylene geomembrane) and BentoLiner®; CETCO's Bentomax®, Resistex®, Akwaseal®, and InterLok®; as well as several others from additional manufacturers. More

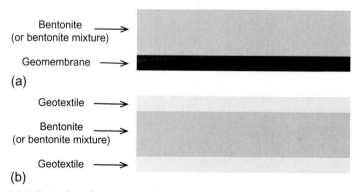

Figure 8.15 Typical configurations of GCL's. (a) Bentonite "glued" to a geomembrane. (b) Bentonite sandwiched between two geotextiles (stitched or needle-punched).

suppliers and GCL products can be found in the Geosynthetics Specifiers Guide (www.geosyntheticsmagazine.com/specifiersguide).

GCLs are now available with a number of advanced attributes, including high-strength geotextiles and roughened geomembranes for added shear resistance, and added high-viscosity polymers for even lower permeability. There are a number of ASTM specifications for installation and testing for GCLs listed at the end of this chapter. The manufacturers supply most of the parameters needed for design. Test values are also listed in the Geosynthetics Specifiers Guide.

RELEVANT ASTM STANDARDS

D4355-07 Standard Test Method for Deterioration of Geotextiles by Exposure to Light, Moisture and Heat in a Xenon Arc Type Apparatus, V4.13

D4491-09 Standard Test Methods for Crosshole Seismic Testing, V4.13

D4533-11 Standard Test Methods for Crosshole Seismic Testing, V4.13

D4632-13 Standard Test Method for Grab Breaking Load and Elongation of Geotextiles, V4.13

D4751-12 Standard Test Method for Determining Apparent Opening Size of a Geotextile, V4.13

D4759-11 Standard Practice for Determining the Specification Conformance of Geosynthetics, V4.13

D5887-09 Standard Test Method for Measurement of Index Flux Through Saturated Geosynthetic Clay Liner Specimens Using a Flexible Wall Permeameter, V4.13

D5889-11 Standard Practice for Quality Control of Geosynthetic Clay Liners, V4.13

D5819-05(2012) Standard Guide for Selecting Test Methods for Experimental Evaluation of Geosynthetic Durability, V4.13

D5890-11 Standard Test Method for Swell Index of Clay Mineral Component of Geosynthetic Clay Liners, V4.13

D5891-09 Standard Test Method for Fluid Loss of Clay Component of Geosynthetic Clay Liners, V4.13

D5993-09 Standard Test Method for Measuring Mass Per Unit of Geosynthetic Clay Liners, V4.13

D6102-12 Standard Guide for Installation of Geosynthetic Clay Liners, V4.13

D6243-13a Standard Test Method for Determining the Internal and Interface Shear Strength of Geosynthetic Clay Liner by the Direct Shear Method, V4.13
D7702-13a Standard Guide for Considerations When Evaluating Direct Shear Results Involving Geosynthetics, V4.13
Reference: ASTM Book of Standards, ASTM International, West Conshohocken, PA, www.astm.org.

REFERENCES

Carroll Jr., R.G., 1983. Geotextile filter criteria. In: Engineering Fabrics in Transportation Construction. Transportation Research RecordTransportation Research Board, Washington, DC, pp. 46–53.

Holtz, R.D., Christopher, B.R., Berg, R.R., 1997. Geosynthetic Engineering. BiTech Publishers Ltd, 451 pp.

Koerner, R.M., 2005. Designing With Geosynthetics, fifth ed. Pearson Education Inc, 796 pp.

Koerner, R.M., 2012. Designing With Geosynthetics, sixth ed. Xlibris Corp, 914 pp.

Ming, H.Y., Chun, H.M., 2005. Smart seepage solutions. International Water Power and Dam Construction, October.http://www.waterpowermagazine.com/features/featuresmart-seepage-solutions.

Mitchell, J.K., Rumer, R.R., 1997. Waste containment barriers: evaluation of the technology. In: In Situ Remediation of the Environment. ASCE, pp. 1–25, Geotechnical Special Publication 71.

National Cooperative Highway Research Program (NCHRP), 2007. Countermeasures to protect bridge piers from scour. NCHRP Report 593. Transportation Research Board, Washington, DC, 284 pp.

Sileshi, R., Pitt, R., Clark, S., 2010. Enhanced biofilter treatment of urban stormwater by optimizing the hydraulic residence time in the media. In: Proceedings ASCE/EWRI, Watershed 2010: Innovations in Watershed Management under Land Use and Climate Change, Madison, WI.

Terzaghi, K., Peck, R.B., Mesri, G., 1996. Soil Mechanics in Engineering Practice. Wiley, New York.

http://www.smartdrain.com (accessed 09.25.13.).

http://www.tencate.com (accessed 8/20/13).

CHAPTER 9

Preconsolidation

Preconsolidation was introduced in Chapter 6 as a method of deep densification, and then mentioned again in Chapter 7 as a method of hydraulic modification because it technically is a means of dewatering saturated fine-grained soils. While preconsolidation is both of these, it is for the most part a method to improve a site by reducing future settlements and increasing strength. Thus, it provides a direct benefit to improved foundation performance, allows more economical solutions to constructing projects on soft, compressible soils, and even permits economical construction where it may not otherwise be feasible. This chapter presents the current state of practice and methodologies to improving a site by preloading and draining prior to construction.

9.1 PRECONSOLIDATION CONCEPTS AND METHODOLOGIES

When a load is applied from a new structure, embankment, or fill to a site underlain by soft saturated fine-grained soils, the load will initially be taken in part by the relatively incompressible water in the soil pores, transferring that load to excess pore water pressures. With time, the excess water pressure will dissipate as the load is transferred to the soil matrix, the soil consolidates, and settlement occurs. Long-term settlement can be the most critical parameter for many types of construction over soft compressible soils. The fundamental concept of preconsolidation is to load the soil prior to construction such that the soil can be compressed, thereby strengthening the soil and greatly reducing future settlement once the project is completed. Consolidation settlement is a stress- and time-dependent process based on applied load and soil compressibility parameters, as well as geometry and drainage conditions. Therefore, variation of each parameter will play an important role in the magnitude (amount) and rate (time) at which desired consolidation may be achieved. As consolidation is often difficult to accurately predict, it is imperative to closely monitor actual field progress of deformation and pore pressure generation/dissipation, and adjust prediction analyses accordingly.

A very simple approach to preconsolidation is to apply a surcharge load approximately equal to the final design load anticipated for the completed

Soil Improvement and Ground Modification
Methods
209

project and allow the ground to naturally consolidate to the point where any predicted remaining settlement over the life expectancy of the project falls within tolerable limits. An important point that must be addressed is to ensure that the bearing capacity of the ground can safely handle the applied load. In some cases, where the bearing capacity of the foundation soil is too low, the surcharge may have to be applied in stages, allowing intermittent levels of consolidation (and associated strength gains) to be achieved prior to applying subsequent stage(s). Another practical design component is to ensure that adequate drainage is provided for discharge of the expelled water. Drainage is typically provided by placing a layer of free-draining material between the foundation soil and surcharge load. Alternatively, geocomposite drains may be utilized for this purpose. For smaller projects where a significant wait time is acceptable, this simple load application (with adequate drainage at the surface beneath the surcharge load) may be a feasible solution.

If there is sufficient bearing capacity in the foundation soil, a larger load than the final design load (excess surcharge) may be applied to expedite preconsolidation. This is possible *only* as long as the load is not so great as to cause a bearing failure or excessive deformations in the subsurface soils. As exemplified in Figure 9.1, use of a surcharge larger than the final anticipated load (excess surcharge) will result in a settlement curve that is initially steeper, causing the required preconsolidation to take place in a much shorter period. When the required amount of settlement has occurred,

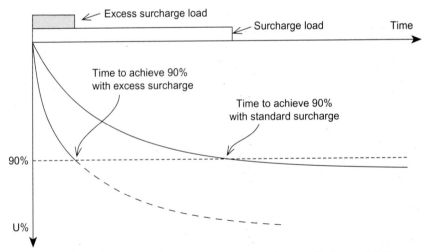

Figure 9.1 Effect of excess surcharge on time rate (and amount) of consolidation.

the preload can be removed and construction initiated. As the time rate of consolidation is exponentially faster at the beginning of the application period, then even a relatively small amount of excess surcharge will greatly expedite the time it takes to reach the target level of settlement and shorten the time needed to initiate construction.

9.2 USE OF VERTICAL DRAINS

The natural process of consolidation under a design construction load may take many years, all the while causing potential settlement problems for the constructed project. Drains can dramatically speed the time to reach a desired level of consolidation to increase strength and reduce future settlement. Vertical drains to aid in preconsolidation have been widely used for many decades. Initially constructed as predrilled sand drains, these were relatively expensive and had certain practical limitations (e.g., depth). For most preconsolidation applications, sand drains have been replaced by prefabricated (geosynthetic composite) vertical drains or *strip drains*. These are often referred to as "wick drains" in the United States, but that name is actually a misnomer, as the materials composing the drain are actually hydrophobic, and the drains do not wick water. The water is actually "pushed" into the drains by differential pressure. The water then flows up (or in some cases, down) to where it can freely discharge. With the development and use of prefabricated vertical drains (PVDs) since the 1970s, economics of vertical drainage have been greatly improved and limitations minimized (i.e., depths of up to and exceeding 65 m can often be achieved with relative ease and efficient construction). The flow capacity of such drains is typically several times greater than that of most other types of vertical drains. Today's geocomposite vertical drains usually consist of a relatively thin, rectangular, flexible polypropylene core, providing significant longitudinal flow capacity on both sides. Figure 9.2 depicts a typical PVD drain construction. The core is surrounded by a strong nonwoven (usually heat bonded polypropylene) geotextile, which acts as a filter and separator to keep surrounding soil from entering the drain core. Typical drains used for preconsolidation are approximately 10 cm (4 in) wide by 0.4 cm (0.15 in) thick, but thicker versions are available for increased flow capacity. Even with typical dimensions, the drains are capable of handling a significant discharge flow (Figure 9.3).

The use of vertical drains greatly expedites the consolidation process by shortening the drainage path length, as well as allowing horizontal drainage, which is the preferential direction of flow with highest permeability in

Figure 9.2 Example of a strip drain used for consol.

Figure 9.3 Example of discharge from a prefabricated vertical drain (PVD) drain.

naturally horizontally deposited fine-grained sediments (Figure 9.4). The use of vertical drains in forced consolidation applications can speed the time to reach acceptable levels of bearing capacity (shear strength) and reduce expected future settlements beneath loads from decades to months or less, depending on project and site specifics (Figure 9.5). The following example provides a simple calculation of this.

Example—Time rate of settlement with and without vertical drains

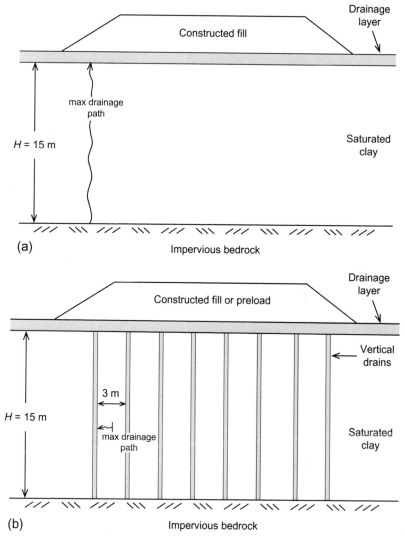

Figure 9.4 Shortening of drainage path with vertical drains.

Given a 15-m thick layer of saturated clay over an impermeable bedrock (Figure 9.4a), the maximum drainage path (single drainage) is 15 m. Then the estimated time to achieve 90% consolidation can be calculated as

$$t_{90} = \frac{T_{90} H_{dr}^2}{c_v} = \frac{0.848(15\,\text{m})^2}{c_v} = \frac{190.8\,\text{m}^2}{c_v}$$

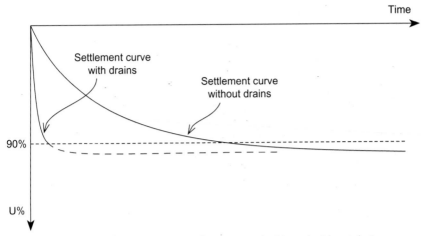

Figure 9.5 Example of time-rate curves for preconsol with and without drains.

If vertical drains are installed in a triangular grid pattern so that the distance between drains is 3 m (Figure 9.4b), then the maximum drainage path is now $3/2$ m $= 1.5$ m. Now the estimated time to achieve 90% consolidation can be calculated as

$$t_{90} = \frac{T_{90}H_{dr}^2}{c_v} = \frac{0.848(1.5\,m)^2}{c_v} = \frac{1.91\,m^2}{c_v}$$

So the calculated time to achieve 90% consolidation with vertical drains is *100 times* faster! This is impressive enough, but in reality the consolidation time may be accelerated even more because the fluid flow is effectively horizontal rather than vertical, and horizontal permeability is typically about 1.5 times greater than vertical permeability in naturally undisturbed, horizontally layered soils.

The prefabricated drains are installed by specialized equipment called "stitchers," mounted on cranes or excavators, fitted with a mandrel to drive the drain from the surface to the desired depth, usually the bottom of the soft compressible strata (Figure 9.6). The installation is sometimes performed with the assistance of vibratory hammers (www.hbwickdrains.com), but most often is simply pushed into the ground hydraulically. The drains are usually laid out in a triangular or square pattern spaced at about 1–2 m (3–6 ft) apart. The designs are typically based on the most economical means of achieving a desired result projected from time–rate settlement curves and/or strength increases. These design curves are developed from radial consolidation theory (Barron, 1948; Hansbo, 1979), where time rate and estimated consolidation settlement are a function of horizontal coefficient of permeability (k_h),

Figure 9.6 Photo of PVD installation. *Courtesy of Hayward Baker/HB Wick Drains.*

drain spacing, and applied pressure (stress differential). Industry reports claim that typical applications of 8000 m per day can be accomplished with a single machine (www.cofra.co.uk) with installation rates of up to 1300 m (4000 l.f.) per hour (www.uswickdrain.com).

Once vertical drains have been installed at a site, a horizontal drainage layer must be provided to discharge the drained water. This is often done with a free-draining granular material, combined with horizontal strip drains placed to intercept the flow from the vertical drains (Figure 9.7).

Most commonly, vertical drains are used to strengthen subsurface soils prior to the application of loads such as highway embankments, bridge approaches, dams, buildings, and even airport runways. Vertical drains installed for the purpose of preconsolidation have even been used for underwater applications, port facilities, and near-shore marine construction (www.geotechnics.com; www.hbwickdrains.com) (Figure 9.8). One of the largest PVD projects to date was for the Virginia Port Authority, where over 4,000,000 m (12,700,000 l.f.) of drain was installed (www.uswickdrain.com). When considering projects such as these, each must be analyzed and designed individually, taking into account the many innovative ideas that have provided solutions for a wide variety of projects. For one case study, where an oil platform "island" was to be constructed over submerged soft clay river deposits in Alaska, drains were installed to depths of approximately 18 m (55 ft) through the compressible clay layers and into a permeable deposit of sand (www.hbwickdrains.com). The

Figure 9.7 Horizontal strip drain discharge system connected to PVDs. *Courtesy of Hayward Baker/HB Wick Drains.*

Figure 9.8 PVD install underwater (Port of Los Angeles). *Courtesy of Hayward Baker/HB Wick Drains.*

stitcher rig installed the drains from the frozen ice surface during the winter where the temperatures were near −40 °C (Figure 9.9). The soft sediments were drained to the permeable sand below and so were not impeded by ice capping at the surface. Using the ice as a working platform, construction of the island was completed before the spring thaw.

While PVDs have much less capacity than even a small-diameter sand drain, the low cost per length and ability to install to much greater depths often makes this a viable and economical choice. One reported difficulty with the use of PVDs is when the amount of consolidation settlement

Figure 9.9 PVD install through Arctic ice. *Courtesy of Hayward Baker/HB Wick Drains.*

exceeds about 5-10%. At that amount of deformation, the drains may "kink" and become ineffective (Holtz et al., 1988). Other problems that can produce inferior performance are *smear* of the drain fabric, which can occur during installation, and drain clogging resulting from ineffective filtering of fine-grained particles by the geotextile filter.

9.3 VACUUM-ASSISTED CONSOLIDATION

Vacuum-assisted consolidation (commonly referred to as simply "vacuum consolidation") is a method of preloading compressible fine-grained soils by applying vacuum pumps to the installed vertical and horizontal drainage system, either beneath an "airtight" membrane (Menard type; Figures 9.10 and 9.11) or through buried (embedded) horizontal pipes connected directly to the vertical drains (Cofra BeauDrain® type; www.cofra.co.uk). The Menard system requires significant care in creating an "airtight" seal, and the membrane often must be protected to ensure integrity of that seal.

This type of system effectively applies a differential near-atmospheric pressure throughout the full depth of the drainage system, resulting in an isotropic consolidation and faster drainage at greater depths. In fact, use of vacuum systems has been shown to speed up improvements, allowing, in some cases, additional loads to be applied within 2 weeks after starting the application (www.menardusa.com). Duration of completed applications is often within 4-6 months, which is significantly faster than traditional methods utilizing surcharge loads and drains alone.

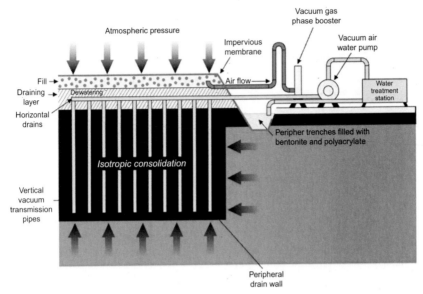

Figure 9.10 Vac. assisted consol schematic. *Courtesy of Menard USA.*

Figure 9.11 Field application of vac consol. *Courtesy of Menard USA.*

The idea of increasing the productivity of preconsolidation methods by use of vacuum pumps is not altogether new. In fact, several early attempts were made as early as the 1950s. The literature indicates that much was learned through both research efforts and attempted field applications over the past few decades to the point where it is now a predictable and reliable soil improvement method. Vacuum consolidation has been successfully applied for a variety of applications since the late 1980s, such as for power plants, sewage treatment plants, highway embankments, and airport runways (www.geopac.com). Most of the reported case studies have been outside of the United States, with some notable examples from France, Germany, the Netherlands, Australia, Japan, China, Thailand, Vietnam, and South Korea (www.menardusa.com; www.menardbachy.com.au; www.cofra.com; Bergado et al., 1998; Tang and Shang, 2000).

In conventional preloading schemes, an embankment fill is typically placed on the area to be treated. This adds stresses to induce consolidation settlements as well as increased shear stresses in the ground. This, in turn, could manifest in stress-induced outward lateral deformations of the foundation soil beneath the embankment or worse—bearing capacity failure. When vacuum consolidation is used in conjunction with vertical drains, the surcharge may sometimes be omitted (or limited) as the typical design pressure differential of 80 kPa induced by the vacuum system is equivalent to more than 4 m of a soil surcharge load (www.menardusa.com).

The added effective stress differential is also uniform (isotropic) throughout the depth of the drained material, as opposed to the added stress from surcharge loads, which will be reduced at depth as a function of stress distribution effects. In fact, there is greater stability added to the soil stratum being treated because the vacuum induces negative pore pressure (or increased effective stress), thereby allowing earlier application of increased surcharge or construction loads.

Cofra's BeauDrain® system (www.cofra.co.uk) uses a combination of vertical drains connected to horizontal vacuum pipe installed below the ground surface. This provides the same type of vacuum consolidation advantages as described above, but without the need for a sealed, airtight membrane. As a small compromise, these systems usually have a somewhat lower design pressure of −50 kPa. The drainage system is installed in one operation with specially designed equipment. This type of forced consolidation has been successfully applied in the Netherlands and Germany (www.cofra.co.uk).

9.4 INSTRUMENTATION AND PERFORMANCE MONITORING

In all cases of preconsolidation applications, it is critical to ensure that actual conditions be continuously monitored. As discussed earlier, consolidation and especially the time rate of consolidation are some of the most difficult geotechnical processes to accurately predict. Because of this, a combination of relatively simple methods may be used to monitor progress of the consolidation and compare to predicted results. Regular collection and analysis of data can be extremely important in cases where construction or soil improvement applications may need to be altered.

Instrumentation may include settlement markers either embedded at the original ground surface or at the top of the surcharge load (or constructed embankment), or both. It is important that settlement markers be measured from a static reference point not affected by the deformations occurring as a result of the forced consolidation.

Piezometers are also important instruments that can be used to indicate the progress of the consolidation process. Piezometers may be located at various depths beneath the load to obtain a profile of how pore water pressures are changing within the subsurface profile as a function of time. As a conventional load is applied, pore pressures will rise. As consolidation continues, the dissipation of excess pore pressures above initial or hydrostatic conditions can be monitored to evaluate the percent consolidation completed at various depth horizons.

Inclinometers may also be useful, as they can measure lateral deformations or displacements at depth, which in turn could indicate possible bearing or stability issues and/or account for some of the measured vertical settlement.

REFERENCES

Barron, R.A., 1948. Consolidation of fine-grained soils by drain wells. In: Transactions ASCE, vol. 113. pp. 718–724, Paper 2346.
Bergado, D.T., Chai, J.-C., Miura, N., Balasubramaniam, A.S., 1998. PVD improvement of soft Bangkok clay with combined vacuum and reduced sand embankment preloading. J. Geotech. Eng., Southeast Asian Geotechnical Society 29 (1), 95–121.
Hansbo, S., 1979. Consolidation of clay by band-shaped prefabricated drains. Ground Eng. 12 (5), 16–25.
Holtz, R.D., Jamiolkowski, M., Lancellotta, R., Pedroni, S., 1988. Behaviour of bent prefabricated vertical drains. In: Proceedings of the 12th ICSMFE, Rio De Janeiro, vol. 3, pp. 1657–1660.
Tang, M., Shang, J.Q., 2000. Vacuum preloading consolidation of Yaoqiang airport runway. Geotechnique 50 (6), 613–623.
http://www.cofra.co.uk (accessed 02.03.14.).
http://www.geotechnics.com (accessed 08.25.13.).
http://www.menardusa.com (accessed 08.30.13.).

CHAPTER 10

Electro-Osmosis (Electrokinetic Dewatering)

Applying an electrical current (electric potential) to a saturated soil will cause the water (and some positively charged adsorbed molecules) to flow toward the cathode, or negative terminal. If the water collected at the cathode is removed (usually by mechanical pumping), then the result is reduction of water content, which in turn results in consolidation of the soil mass, with corresponding strength gain and reduction is soil compressibility. This process is called *electro-osmosis*, or *electrokinetic dewatering*, which is the process of moving water and other positively charged ions through application of a direct electrical current. It has been shown that for certain low permeability fine-grained soils the application of an electrical gradient is more efficient in producing a water flow than a hydraulic gradient.

The principles of electro-osmosis were introduced as early as the 1930s by Casagrande (Hausmann, 1990) and have been utilized as a method of soil stabilization for a number of different types of projects and applications since that time. These projects include construction excavations, remediation for differential settlement, slope stabilization, stabilization of sediment deposits and mine tailings, and more. The concepts have also been used as an aid to pile driving and for remediation of contaminated ground. Holtz (1989) summarized a number of case histories where electro-osmosis was used to stabilize embankment foundations. The literature suggests that the most common classic application has been for slope stabilization, typically for emergency applications to active slides, either as a final fix or as an interim measure until a more permanent solution is implemented. The application of electro-osmosis has had mixed popularity since its inception due to costs and mixed results, but seems to be making a comeback as a more accepted alternative stabilization method, especially for cases where more conventional dewatering techniques have proven difficult and/or where speed of consolidation is critical.

10.1 PRINCIPLES OF ELECTRO-OSMOSIS

While Casagrande is often attributed with introducing the concept of soil improvement using electro-osmosis with his German patent in 1935, earlier

Soil Improvement and Ground Modification Methods

221

experiments conducted in Russia demonstrated the ability of applying direct current to cause flow of water through clayey soil as early as 1809 (Hausmann, 1990). Review of the literature suggests that the earliest field application may have been to stabilize the excavation of a long railroad cut in 1939 (Thomas and Lentz, 1990). Clay soil particles have a net negative charge due to their mineral composition and physical (geometric) make up. They tend to have very high specific surfaces, which essentially means large ratio of surface area with respect to volume. The end result is that clay particles have an electrical charge that naturally attracts positively charged molecules (cations) and dipolar water molecules (van der Waals interaction). A much more detailed discussion of clay mineralogy and the importance of the surface charge of clay particles is provided in Section 11.2.2.1. This is critical in understanding the physiochemical changes that result when chemical admixtures are combined with clayey soils, as will be described in Chapter 11. Therefore, in the field, saturated clayey soils will have a certain amount of water (known as the diffuse double layer) along with positively charged ions such as Na^+, K^+, or Ca^{2+}, bound to them by this inherent electrical charge. When a direct electrical current is applied to the ground, usually by insertion of two conductive metal electrode rods (one each connected to the positive and negative terminals of a power source), any free or loosely bounded water molecules or hydrated cations will be drawn by the current toward the negative pole (cathode). This process is depicted schematically in Figure 10.1. Understanding of this transport mechanism has also propagated attempts to introduce hydrated cations (chemical admixture) at the anode in order to alter soil properties.

The fundamentals of water flow through a porous media due to application of electrical current involves a number of variables. These include

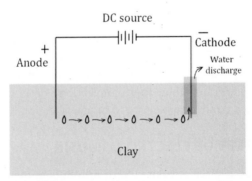

Figure 10.1 Schematic of electro-osmosis water transport and dewatering.

the (horizontal) coefficient of hydraulic permeability of the medium (soil), k_h, the electric potential applied, E (i.e., voltage), the coefficient of electro-osmotic permeability, k_e, in units of $(m/s)/(V/m) = m^2/sV$, and ionic content of the pore water, which affects the resistivity of the soil. Mitchell (1976) and others provided data showing that electro-osmotic permeability is relatively independent of soil type and constant with a typical average value of approximately

$$k_e = 5 \times 10^{-3} mm^2/sV$$

On the other hand, coefficient of hydraulic permeability varies widely depending on soil type and consistency (plasticity) and will significantly affect generation of pore pressures and time to develop electro-osmotic-induced pore pressures.

Prior to the 1970s, successful attempts at using electro-osmosis were often "hit or miss," as the details of the science behind the various factors and soil properties affecting the effectiveness of such applications had not been well developed. Since that time, much research and testing has been done to remedy the problem. Review of various case studies where electro-osmosis has been used or attempted has shown that a number of parameters can be employed to evaluate the most suitable conditions that would favor positive effects. In general, it has been found that for electro-osmosis to be effective, a soil should be above its plastic limit and saturated (Thomas and Lentz, 1990). Holtz et al. (2001) provided a summary of specific parameters that would be ideal for effective electro-osmosis applications. These are provided in Table 10.1. Shang and Mohamedelhassan (2001) developed a procedure for assessing the feasibility for using electro-osmosis as follows:

(1) Material characterization—measuring material properties related to the variability of electro-osmosis treatment.

Table 10.1 Soil Parameters with Favorable Effects for Electro-Osmosis

Soil Parameter	Unit	Range
k_h, (horiz) hydraulic conductivity	m/s	10^{-10}-10^{-8}
k_e, electro-osmotic permeability	m^2/sV	$\sim 10^{-9}$
K, electrical conductivity	S/m	0.01-0.5
E, electrical field intensity	V/m	20-100
c_v, coefficient of consolidation	m^2/s	0.01-1.0

after Holtz et al., 2001

(2) Measurement of electro-osmotic permeability—defines the flow rate in the material generated by electro-osmosis.

(3) Assessment of achievable parameters (e.g., the final solid content or shear strength), efficiency (treatment time), and power consumption.

Use of this procedure, along with assessment of the parameters provided in Table 10.1 and results of laboratory tests, has enabled a better evaluation of feasibility prior to attempted applications. Ultimately, however, the efficiency and economics of using electro-osmosis for any particular application will depend on the quantity of water that can be transported and needs to be moved (or removed) as a function of power consumed. Predictions of electro-osmosis efficiency can be made by a number of theoretical models as described by Mitchell (1976) and others.

10.2 APPLICATIONS/IMPROVEMENTS

Many successful applications of electro-osmosis have been reported over the years. The success will depend on the subsurface geology and soil properties as described in the previous section. Successful applications include emergency (and permanent) slope stabilization, dewatering, preconsolidation for settlement reduction, strength gain, reduction in shrinkage and swell, solidification of mine tailings and waste sludge, grouting assistance, and contaminant retrieval.

10.2.1 Dewatering

An obvious application of electro-osmosis is for dewatering and subsequent consolidation of the soil being treated. Dewatering consolidation as described here involves discharge of the water collected around the cathode. Most of the time, this is done by conventional pumping of a receiving well at the cathode to relieve any pore water pressure build up, while sealing or preventing the introduction of fluid at the anode. As water is withdrawn and consolidation occurs, the rate of flow decreases. Hausmann (1990) provides a detailed discussion of mathematical solutions for pore pressure distributions and percent consolidation, derived mostly from laboratory research. The strength increase resulting from dewatering and consolidation of clay soils has been successfully applied in a number of well-documented case histories. Two cases of successful field treatments involved improvement of sensitive clays in Norway (Hausmann, 1990; Mitchell, 1976). In one case, 2000 m^3 of soil was treated over a period of 120 days, resulting in a reduction in water content and increase in average shear strengths from about 10 to 60 kPa. In another case, the strength of a clay soil was more than doubled in about 90 days (Hausmann, 1990).

10.2.2 Electrostabilization/Electrohardening

While the most common successful applications of electro-osmosis have been seen as a dewatering technique for slope stabilization and consolidation, research testing has shown additional changes in important fundamental soil properties after treatment with electro-osmosis. These measured changes include strength gain, as would be expected after consolidation or dewatering, changes in liquid and plastic limits (Mitchell, 1976), and also reduction in swell potential (Thomas and Lentz, 1990). It should be noted that the greatest strength gains occur near the anode where the water is being drawn from, as long as no water is allowed to enter the system at that point. Little change in strength is observed at the cathode because there is minimal change in pore pressure and effective stress due to accumulation of the transported water at that point. Additional pumping at the cathode will achieve the same effects of dewatering as if there had been no application of electro-osmosis, but only after the additional water accumulated by the process has been removed. Some of the more notable successful applications of electro-osmosis reported in the literature involve stabilizing (consolidating and strengthening) soft and/ or sensitive clays prior to excavation as a result of strength gains. But, in fact, a number of researchers have also found that the strength increase achieved by electro-osmosis may be substantially higher than for a similar decrease in moisture achieved by normal consolidation (Hausmann, 1990). As reported by Mitchell in 1976, the measured strength gain in aNorwegian quick clay was on the order of 80% greater than could be attributed to the reduction in water content alone. And normal consolidation should have virtually no effect on Atterberg limit values. These differences have been attributed to additional physiochemical changes in the soil caused by the movement of ions, ion exchange, changes to the soil structure of the clay materials, and release of charged molecules from the anode, among other phenomenon. For example, the replacement of hydrogen or other ions on clay surfaces with aluminum tends to cause a reduction in the double water layer and create a more floc-culated structure. This added strength gain, sometimes referred to as *electrohardening*, can be enhanced by the choice of anode material and/or by the introduction of certain grout materials at the anode. This latter application falls under the guise of *electrogrouting* (also called electro-osmotic grouting or electrokinetic injection), described next.

10.2.3 Electrogrouting

When used to assist with grouting, electro-osmosis may function as a simple aid in transporting a material dissolved or suspended in a water solution by introducing the solution at the anode and drawing toward a strategically placed cathode, and/or it can also take advantage of the physioelectrical chemistry of the grout material and the soil into which it is being introduced. This has included the use of aluminum anodes designed to introduce additional aluminum ions into the soil or the introduction of such additives as sodium silicate, potassium chloride, calcium chloride, and aluminum acetate (Hausmann, 1990; Thomas and Lentz, 1990). Liu and Shang (2012) presented a study showing significant improvement of soft, very high water content, marine sediments when treated by electro-osmosis in combination with various chemical admixtures.

Studies also have demonstrated that electro-osmosis may be a good alternative to permeation grouting in silty soils where traditional hydraulic permeation grouting may not be cost-effective or practical. Thevanayagam and Jia (2003) describe a method to mitigate liquefaction potential in nonplastic, silty materials where other traditional mitigation methods, such as densification or drains, are either not practical (due to accessibility beneath existing structures or infrastructure), or ineffective (due to soil grain size and low permeability). Often, contractors will resort to grouting techniques when there is no vertical access to the subsurface. But for the case of silty soils, pumping capacity would be very high, making it cost-prohibitive. In Thevanayagam and Jia's study, the feasibility of injecting sodium silicate and colloidal silica grouts to strengthen the ground with the use of electro-osmosis was shown to be very promising. Other studies have reported using electro-osmosis-assisted, 2-shot injection grouting with sodium silicate as the primary agent and calcium chloride as a reagent.

10.2.4 Pile Driving and Capacity Enhancement

Another type of construction application using electro-osmosis is for assisting with pile driving operations. One case study reported in the literature describes the use of the pile as the cathode, resulting in a reduction in the required number of blows with the hammer by 33-42% (Nikolaev, 1962). Conversely, the generation of negative pore pressures at the anode has been shown to provide a temporary increase in capacity for friction piles. If the application is allowed to continue for an extended period, the treatment may result in a permanent increase of pile capacity of up to two times (Hausmann, 1990).

10.2.5 Contaminant Retrieval

In addition to merely drawing water molecules toward the cathode, individual cations are also drawn toward the cathode. This ion movement, sometimes called electromigration, can assist in treating/remediating soils with heavy metals, nitrates, sulfates, or other inorganic compounds (www. terrancorp.com), although specialized cathode materials may be required to make this effective.

A secondary benefit is that the applied electrical current results in heating of the soil. It is reported that temperatures can easily reach 80 °C. The result of heating can be useful in mobilizing volatile organics and may be useful for contaminated soil remediation. This effect will be discussed again in Chapter 13. The soil heating also increases electro-osmotic permeability by lowering the viscosity of the pore fluid in the ground (www. terrancorp.com). A number of studies have described effective removal of contaminants from saturated clays by electro-osmosis. Field demonstrations have shown that clayey soils heavily contaminated with chlorinated solvents, such as trichloroethylene (TCE or DNAPL) can be effectively treated by electro-osmosis. Other studies have described removal of phenol from contaminated clays in conjunction with chemical oxidation, where other extraction methods would be ineffective (Thepsithar and Roberts, 2006).

REFERENCES

Hausmann, M.R., 1990. Engineering Principles of Ground Modification. McGraw-Hill Inc, 632 pp.

Holtz, R.D., 1989. Treatment of Highway Foundations for Highway Embankments. NCHRP Synthesis of Highway Practice 147, TRB, National Research Council, Washington, DC, pp. 14–16.

Holtz, R.D., Shang, J.Q., Bergado, D., 2001. Soil improvement. In: Rowe, R.K. (Ed.), Geotechnical and Geoenvironmental Engineering Handbook. Kluwer Academic Publishers, pp. 429–462 (Chapter 15).

Liu, P., Shang, J.Q., 2012. Improvement of marine sediment by combined electrokinetic and chemical treatment. In: Proceedings of the 22nd International Offshore and Polar Engineering Conference, ISOPE, pp. 618–625.

Mitchell, J.K., 1976. Fundamentals of Soil Behavior. John Wiley & Sons, 422 pp.

Nikolaev, B.A., 1962. Pile Driving by Electro-osmosis. New York Consultants Bureau, 62 pp.

Shang, J.Q., 2011. Recent Development. Geotechnical Special Publication No. 217, ASCE, pp. 1–8.

Shang, J.Q., Mohamedelhassan, E., 2001. Electrokinetic Dewatering of Eneabba West Mine Tailings Dam. Geotechnical Special Publication No. 112: Soft Ground Technology, ASCE Press, Reston, VA, pp. 346–357.

Thepsithar, P., Roberts, E.P.L., 2006. Removal of phenol from contaminated kaolin using electrokinetically enhanced in situ chemical oxidation. Environ. Sci. Technol. 40 (19), 6098–6103.

Thevanayagam, S., Jia, W., 2003. Electro-osmotic grouting for liquefaction mitigation in silty soils. In: Proceedings of the Third International Conference: Grouting and Ground Treatment. ASCE Press, pp. 1507–1517.

Thomas, T.J., Lentz, R.W., 1990. Changes in Soil Plasticity and Swell Caused by Electro-Osmosis. ASTM Special Technical Publication 1095, American Society for Testing and Materials, pp. 108–117.

http://www.menardusa.com/MV-12%20years-en.pdf (accessed 08.27.13.).

http://www.terrancorp.com/content/case-electroosmosis-remediation (accessed 02.17.14.).

SECTION IV

Physical and Chemical Modification

CHAPTER 11

Admixture Soil Improvement

Admixture soil improvement refers to any improvement application where some material is added and mixed with existing or placed soil to enhance the engineering properties or engineering behavior of the soil. This chapter provides an overview of the improvement objectives, mixing methods, and some common applications for admixture treatments. There is also a discussion of the various materials used, including natural soils, chemical additives, and waste products, along with a discussion of applicability to different soil types. Included in the chapter are some case studies exemplifying some of the successful possibilities of utilizing admixture stabilization.

11.1 INTRODUCTION TO ADMIXTURE SOIL IMPROVEMENT

The engineering properties of soils can be dramatically enhanced through the addition (or subtraction) of materials to (or from) the soil. The mechanics of the improvements may be physical or chemical in nature. In most cases, changes in the soil properties are permanent. Improvement of soils through the use of admixtures (material added to the soil) is often called *soil stabilization*, as it may in many instances result in the soil being rendered more "stable" by being less susceptible to engineering property fluctuations (e.g., strength fluctuations, volume stability, moisture content change, etc.). Soil admixtures may include a wide array of materials such as natural soils, chemical reagents, binders, polymers, industrial by-products (waste or recycled materials such as fly ash, slag, shredded rubber, crushed glass, etc.), salts, poly-fibers, and bitumen/tar.

Soil stabilization with admixtures has been used for economical road building, conservation of materials, investment protection, and roadway upgrading. In many instances, soils that are unsatisfactory in their natural state can be made suitable for subsequent construction by treatment with admixtures, and/or by the addition of natural aggregate or other soil materials. Admixture improvement has also been used for repair of geotechnical failures by providing a rebuilt soil structure that is much stronger and more robust than the original construction. Admixture soil improvement is now routinely used for site and roadway rehabilitation as well as new construction projects.

Use of admixtures can improve engineered soil and in situ ground conditions so that significant cost savings may be possible. This can be achieved by requiring less costly foundation schemes, using a smaller volume of select fill material, utilizing lower-quality soils, and realizing economic savings over conventional excavation/replacement methodologies. This is especially important and useful for fine-grained soils, but also has numerous applications for coarser granular materials. Another driving force behind using admixture treatments is the shortage of available, conventional aggregates in many locales. Environmental concerns, regulations, and land use patterns have also severely impacted the availability of useable aggregate.

Physical improvements can be made by altering the soil gradation (or soil grain-size distribution) by adding or subtracting certain soil grain sizes, or by adding materials that physically "bind" soil particles together without causing any chemical reactions or changes to the mineralogical structure of the soil. Conversely, chemical improvements can be made by adding materials that intentionally cause reactions to occur, resulting in physio-chemical changes in the mineralogical structure of the soil. These changes can have pronounced improvements in the characteristics of the soil, even leading to a change in the fundamental classification of the soil.

11.1.1 Benefits of Admixture Soil Improvement

In so many parts of the world, poor soil conditions inhibit sound construction and development of quality infrastructure. For many people, transportation lifelines are severely impacted by poor subgrade soils and lack of quality fill or roadway materials. These conditions often exist in underdeveloped or developing areas where soil improvement engineering practice is sorely lacking or nonexistent. In many of these cases, relatively simple and inexpensive soil improvement techniques, using readily available admixture materials and equipment, can dramatically enhance conditions and reduce the degradation that otherwise would require continual repairs. Such improvements can lead to an improved quality of life and more efficient movement of needed supplies as well as the mobilization of emergency transportation.

Mixing admixtures into soil has been shown to be greatly beneficial in:
- Drying up wet soil
- Improving strength (including "solidification" of wastes for disposal assistance)
- Providing volume stability (reducing swell, controlling shrinkage)

- Reducing soil deformations (reduced compressibility/settlement concerns, minimizing differential settlement effects; increasing stiffness)
- Reducing erodibility (through increased surface strength and water repulsion)
- Improving durability to dynamic/repeated loads, including freeze-thaw (through increased intergranular strength and decreased degradation of aggregate)
- Permeability (or moisture) control (either reduced permeability for water conveyance or retention structures, moisture consistency, or water repulsion)
- Dust control

As a consequence, soil improvements have been responsible for:

- Improved working platforms and workability of soils
- Reduced thickness of roadway layers
- Slope stabilization
- Foundation/structural support
- Excavation support
- Liquefaction mitigation
- Reduced leakage/seepage from hydraulic retention/conveyance structures
- Stabilization of marine sediments
- Environmental (contamination) remediation

Increased soil strength allows steeper slopes to be constructed. Increased slope angles result in less volume of engineered fill required to attain a desired embankment height, less area (footprint) needed for the same embankment height requirements, economic savings from faster construction, and so on.

Stabilization projects are almost always site-specific, requiring the application of standard test methods, along with fundamental analysis and design procedures, to develop workable solutions. A number of standards for materials and testing related to soil stabilization with admixtures have been developed and are available from ASTM and others. A listing of some of the relevant ASTM test standards is provided at the end of this chapter.

11.2 ADMIXTURE MATERIALS

The materials that may be added to a soil for stabilization are wide-ranging and have a variety of properties, forms, and attributes. These generally range from naturally occurring soils (different in grain size distribution to those being treated) to chemical additives and even reused waste products. Class C fly ash, a by-product of coal combustion, has been widely used as a soil admixture either by itself or in addition to lime and/or cement. The type of admixture material to be used will depend on a number of variables including:

- Soil type to be treated
- Purpose of use
- Engineering properties desired
- Minimum requirement (or specification) of engineering properties
- Availability of materials
- Cost
- Environmental concerns

The selection of an appropriate additive may begin by following some general guidelines based on soil gradation and plasticity, which have been well-documented in the literature (e.g., Federal Highway Administration, 1992a,b; Joint Departments of the Army and Air Force, 1994). For example, while cement can be used with a variety of soil types, it is critical that it be thoroughly mixed with any fines fraction (grain sizes <0.074 mm). Therefore, in general, more plastic materials should be avoided as they would be difficult to mix with cement. Lime, on the other hand, will react with soils of medium to high plasticity, thereby reducing their plasticity and rendering them easier to mix, while also minimizing swell potential and increasing strength. Again, the choice or blend of admixture to use will depend on some or all of the variables listed above. A more detailed discussion of selecting an appropriate admixture(s) and the design of mixes is addressed later in Section 11.3.4.

In addition to conventional admixtures, there are many newer proprietary chemical additives that have been designed for the application and treatment of problematic and hard-to-mix soils with particular focus on environmental consciousness. Toxicity has been an issue for several chemical treatments, as will be discussed in greater detail in Chapter 12 on grouting.

11.2.1 Natural Soil Admixtures

Many of the engineering properties of the ground are a direct result of the distribution of soil grain sizes (grain size distribution) and the density of packing of the grains. As described in Chapter 5, most desirable engineering properties (particularly for granular soils) can be achieved with higher soil density. From the fundamental geometric perspective, a soil can achieve a higher density if the distribution of particle sizes is such that voids between successively smaller grains are filled with smaller and smaller grains. This leads to the notion that a well-graded soil (one that has a smooth distribution over a range of grain sizes) will be able to achieve the highest density. With increased density comes higher strength, higher stiffness, lower compressibility, lower permeability, increased durability, and so forth.

It has been suggested that an optimum particle size distribution (to achieve highest density) can be approximated by a simple expression for the percentage of each grain size based on the maximum particle size (Hausmann, 1990; NAASRA, 1986):

$$p = 100(d/D)^n \tag{11.1}$$

where p is the percent passing sieve with a nominal grain diameter, D is the maximum particle size, and n is the exponent (dependent on soil type; 0.45–0.50 typical for pavement layers).

Addition or removal of certain grain sizes and control of grain size distributions (where percentages of various grain sizes is controlled) can aid in achieving many desirable properties and can increase a soil's workability (ease to compact). This is often referred to as *gradation control*. In certain instances a more uniform gradation is desired, where there is a narrow range of grain sizes. This condition is often preferred for improved drainage (e.g., AASHTO, 2012). Uniformly graded soils can be found in nature (such as with beach sands and some alluvial and fluvial deposits), but often must be generated by screening or grading. Controlled soil/aggregate gradations are an important aspect of many different geotechnical applications depending on the intended use and desired properties. Some examples are for drainage, filtering, pavement layers, and for use with various admixtures.

Most roadway design guidelines as well as some engineered fills dictate specific gradations that may be nearly impossible to find in nature (e.g., AASHTO, 2012). In order to meet these specifications it is necessary to generate "select" materials by controlling the grain size distributions of the soils. This may be as simple as screening the material to limit maximum size and/ or limit the amount of fine-grained particles. In other cases, materials may have to be carefully graded and blended in order to achieve specified grain size distributions. There are many gradation specifications for different applications and for various admixture soil blends. These are readily available from FHWA, AASHTO, and others. An example of specified grading requirements, which may be achieved by grading the existing soil, is provided in Tables 11.1 and 11.2. This type of gradation control has been called *mechanical stabilization* by some (AASHTO, 2004), but in this text that term is used to describe physical manipulation of the soil such as compaction or other densification methods.

A "natural" soil additive that is commonly used either by itself or in conjunction with other admixtures is bentonite clay. This is very low

Table 11.1 Examples of Aggregate Grading Requirements

Ex.1—Aggregate Gradation Requirements for Open Graded Portland Cement Concrete Base (Adapted from National Lime Association)

Sieve Size	Percent Passing
1.5 in. (37.5 mm)	100
1.0 in. (25.0 mm)	95–100
½ in. (12.5 mm)	25–60
No. 4 (4.75 mm)	0–10
No. 8 (2.36 mm)	0–5
No. 200 (75 μm)	0–2

Ex.2—Untreated Base Grading Requirements (Adapted from HDOT (2005) (hidot.hawaii.gov/dot))

Sieve Size	Percent Passing by Weight		
	2½ in. Maximum Nominal	1½ in. Maximum Nominal	¾ in. Maximum Nominal
3 in. (75.0 mm)	100	—	—
2½ in. (62.5 mm)	90–100	—	—
2 in. (50.0 mm)	—	100	—
1½ in. (37.5 mm)	65–90	90–100	—
1 in. (25.0 mm)	—	—	100
¾ in.	45–70	50–90	90–100
No. 4 (4.75 mm)	25–45	25–50	35–55
No. 200 (75 mm)	3–9	3–9	3–9

Ex.3 (4.75 mm)—Fine Aggregate Grading Requirements (After HDOT (2005) (hidot.hawaii.gov/dot))

Sieve Sizes	Percent Passing by Weight
3/8 in. (9.53 mm)	100
No. 4 (4.75 mm)	95–100
No. 8 (2.36 mm)	80–100
No. 16 (1.18 mm)	50–85
No. 30 (600 μm)	25–60
No. 50 (300 μm)	10–30
No. 100 (150 μm)	2–12

FHWA Table 703-1. Gradation for Permeable Backfill

Sieve Sizes	Percent Passing by Weight
75 mm	100
19.0 mm	50–90
(4.75 mm)	20–50
No. 16 (1.18 mm)	0–2.0

Table 11.2 Examples of Grading Recommendations for Pavement Underlayers

Ex.1—Recommended Gradation for Bituminous-Stabilized Subgrades (Adapted from U.S. Army Field Manual 5-410)

Sieve Size	Percent Passing
3 in. (75 mm)	100
No. 4 (4.75 mm)	50-100
No. 30 (0.60 mm)	38-100
No. 200 (75 μm)	2-30

Ex.2—Recommended Gradation for Bituminous-Stabilized Base and Subbase (Adapted from U.S. Army Field Manual 5-410)

Sieve Size	Percent Passing for 1.5 in. max.
1.5 in. (37.5 mm)	100
1.0 in. (25.0 mm)	75-93
¾ in. (19.0 mm)	67-85
½ in. (12.5 mm)	57-75
3/8 in. (9.5 mm)	50-68
No. 4 (4.75 mm)	36-54
No. 8 (2.36 mm)	26-44
No. 16 (1.18 mm)	18-36
No. 30 (0.60 mm)	11-29
No. 50 (0.30 mm)	7-21
No. 100 (150 μm)	4-14
No. 200 (75 μm)	0-2

permeability clay that has been used as an admixture to lower permeability in naturally occurring soil (such as for landfill liners or covers and hydraulic conveyance/retention structure applications). Bentonite has also been applied in slurry form for hydraulic barriers (cutoffs) and as "driller's mud," although the latter is not really considered a soil admixture application.

11.2.2 Cement and Lime

The most widely used chemical stabilizing agents are cements (or modified cementitious chemicals), while lime is purported to be the oldest known stabilizing admixture, dating back thousands of years (i.e., Rome's Appian Way). Cement (ordinary Portland cement) and lime have several similarities for the purposes of admixture soil stabilization. This stems from the fact that they are both calcium-based chemical reactants. In fact, cement contains lime but also has its own source of additional reactants (pozzolans) whereas "pure" lime is limited in use to where other source(s) of reactant materials are present or added to the soil. In that light, the discussion of cement and lime stabilization will begin with the fundamentals of lime reactions and then be extrapolated and compared to the uses of cement as a soil stabilizer. As will be discussed, certain forms of fly ash may also contain a significant source of reactive calcium rendering it a useful cementing agent in its own right. Even nonreactive fly ash that is not self-cementing has proven to be a good pozzolan additive when blended with lime or Portland cement.

Cementitious stabilizers typically increase compressive strength, shear strength, tensile strength, and modulus of elasticity (soil stiffness), and those reactions can continue for months, continuing to improve those properties. Freeze-thaw and moisture resistance are also significantly enhanced by cementitious stabilization. Control of swell in potentially expansive soils is often a primary goal and objective of treatment with calcium-based admixtures. Cation exchange between monovalent cations—such as sodium and potassium commonly found in expansive clays, with higher valence calcium cations— can reduce the attraction of water molecules and therefore reduce swelling potential. Because of the importance of these reactions and their results, a brief discussion of clay chemistry and calcium reactions is provided.

11.2.2.1 Lime and Clay Mineralogy

Lime is one of the oldest soil stabilizing agents known. It is available in a number of different forms and therefore may be applied in a number of different manners. Lime has been used heavily over the past several decades for roadways, airfields, drainage canals, and foundation soils. While one of the most used admixtures for permanent, long-term stabilization (especially for poor-quality, fine-grained soils), lime has also been shown to be very effective in providing a rapid, short-term solution for enabling or expediting construction where wet soil conditions are present. In addition to drying wet soil, lime reduces plasticity and improves stability to provide a solid working platform for subsequent construction. Quality lime stabilization can be achieved with $\sim$2-8% lime. If more lime is needed to achieve the desired or required results, then another admixture type, such as cement, may be more economical.

Lime is generally a white to grey crystalline solid. Terminology for lime (as adopted from ASTM C51) depends on the amount of magnesium ($MgCO_3$) it contains. Three primary forms of lime are high calcium lime containing 0-5% $MgCO_3$ (magnesium carbonate), magnesian lime containing 5-35% $MgCO_3$, and dolomitic lime containing 35-46% $MgCO_3$.

Quicklime refers to "pure" calcitic calcium oxide (CaO) or *dolomitic quicklime* (CaO + MgO) where total oxides of CaO/MgO are equal to or >90%, meeting ASTM C977. Quicklime is a white, caustic, alkaline crystalline solid that is most commonly made by thermal decomposition of limestone or other calcium carbonate materials containing the mineral calcite ($CaCO_3$ or $MgCO_3$). Because of this, it is also sometimes called *burnt lime*. Quicklime is made by heating the source material to above 825 °C (1517 °F) in a process called *calcination*, which drives off CO_2 leaving calcium oxide:

$$CaCO_3 + MgCO_3 \rightarrow CaO + MgO + CO_2 \tag{11.2}$$

One of the advantages of using quicklime is the intense heat generated during hydration, which can reach temperatures above 150 °C. This basic hydration reaction is

$$CaO(s) + H_2O(l)Ca(OH)_2(aq) \tag{11.3}$$

$$(\Delta H_r = 63.7\,kJ/mol \text{ of } CaO;\ 490\,BTU/lb = 273\,cal/g) \tag{11.4}$$

The drying effect of adding quicklime to a wet soil can be attributed to two cumulative processes. First, the water content reduces due to hydration of the lime. Knowing that 1 kg CaO absorbs 0.32 kg of water through hydration, the decrease in moisture from hydration (Δw) is given by the following equation (Kitsugi and Azakami, 1982):

$$\Delta w = w_o - (w_o - 0.32a_s)/1 + 1.32a_s) \tag{11.5}$$

where w_o is the original soil water content and a_s is the mass ratio of lime to soil.

Second, moisture is also lost due to evaporation from the heat generated by hydration of quicklime. Hausmann (1990) showed that the additional moisture loss is equal to

$$\Delta w = 0.45a_s \tag{11.6}$$

Quicklime is volatile and must be kept sealed until use. It is perishable and must be "fresh" (typically <60 days) to be useful. Once exposed, quick-lime will spontaneously react with CO_2 in the air and ultimately revert back to a nonreactive form of $CaCO_3$. While very reactive and extremely useful for rapid stabilization, quicklime's caustic nature and special handling requirements often lead to the use of a more "user-friendly" *hydrated* or *slaked* lime. Hydrated lime is made by adding ~1% water to crushed granular quicklime. The material is still dry to the touch but with sufficient water to convert the oxides to hydroxides. Even though the product is already in the hydrated (aqueous) form of $Ca(OH)_2$, it is still available for further stabilizing reactions described below. Lime can also be applied as a slurry, which has the combined effect of enhancing distribution in mixing and adding water that is often necessary for proper compaction.

Hydrated lime is only found as a fine powder or as a slurry. Quicklime, however, is commercially available in a number of sizes (derived from ASTM Standard C51):

- Large lump lime—a maximum of 20 cm (8 in.) in diameter.
- Crushed or pebble lime—ranging from about 0.6 to 6.3 cm (¼ to 2½ in.).
- Ground lime—0.6 cm (¼ in.) and smaller.
- Pulverized lime—typical that most all passes a #20 sieve.
- Pelletized lime—2.5 cm (1 in.) sized pellets or briquettes, molded from fines.

An additional source of lime may be obtained as a by-product of various manufacturing processes, but it will be of lower and less-consistent quality, which may be effective for soil stabilization at a reduced cost. One such product is *lime kiln dust* (LKD) collected from the draft of the calcining process of lime production. LKD will typically contain only 18-30% total oxides with 7-15% alumina and silica oxides (www.dot.state.oh.us). LKD is best suited for the stabilization of silts and sands.

Whether quicklime or hydrated lime is used, numerous fundamental reactions can occur in reactive soils. *Reactive soils* are defined as soils that have a gain in unconfined compressive strength of at least 350 kPa (51 psi) (Thompson, 1970). The reactions between lime and soils include both *short-term reactions* and *long-term reactions*. It is necessary that sufficient lime be present to raise the pH in the soil pore water to enable these reactions to take place. For long-term cementation, adequate lime must be added to maintain available calcium to keep pozzolanic reactions going. This is typically evaluated by maintaining a high pH level of the pore water.

11.2.2.1.1 Clay Mineralogy

In order to understand and fully appreciate the stabilizing effects of lime treatment, it is important to understand the basics of clay mineralogy or, more specifically, clay chemistry and its interaction with water.

Clay minerals are made up of aluminum silicates together with water molecules and exchangeable cations (e.g., calcium, Ca^{++}; magnesium, Mg^{++}; sodium, Na^+; potassium, K^+). Most fundamental, elementary soil mechanics texts can provide a more detailed description of the molecular makeup of clay minerals, but there are a few notable details that must be understood as they pertain to soil stabilization reactions. First, clay particles are generally very thin compared to their lateral dimensions, which leads to very high specific surfaces (i.e., very large surface-to-volume ratios). Because of this, the chemistry of the clay particle surfaces is vitally important. Clay particles carry a net negative charge on their surfaces as a result of isomorphous substitution and a break in continuity of the mineral structure at the edges (Das, 2010). The magnitude of the charge is greater for those clays with higher specific surfaces. The net negative charge then attracts positively charged ions and the positively charged ends of dipolar molecules. Figure 11.1 shows an idealized schematic diagram of a clay particle.

Water molecules are dipolar, meaning that they have a preferential orientation with positive and negative charges at each end (Figure 11.2). In the presence of water, the net negative charge of a clay particle attracts the positive end of dipolar water molecules. As more water molecules attach themselves to a clay particle by attractive electrical charge, they form a layer surrounding the clay particle. When the positively charged ends of the water molecules are aligned with the negative charges of the clay surface, the outer edge of the water layer then provides a net negative charge that in turn attracts more positive charges, often in the form of another water layer. This configuration of water electrically attracted to the clay is termed the "double water layer," or "diffuse double layer" (Figure 11.3). This bounded water acts to effectively buffer between the clay particles, which affects several distinctive engineering properties of the material. The individual particles are kept from intimate edge-to-edge contact, known as a *dispersed* structure (Figure 11.4a). When the double water layer is significantly reduced, primarily due to a change in available net negative electric charge, the clay particles are allowed to have more contact with each other, forming a

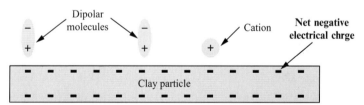

Figure 11.1 Clay particle schematic sowing attraction of positively charged ions and dipoles to negatively charged surface.

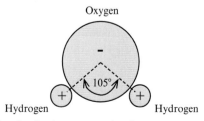

Figure 11.2 Schematic of a dipole water molecule.

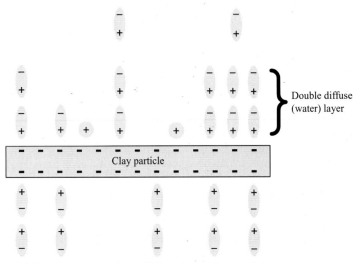

Figure 11.3 Representation of the diffuse double (water) layer.

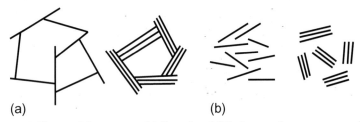

Figure 11.4 Clay particle structure (a) flocculated, (b) dispersed.

flocculated structure (Figure 11.4b). A dispersed structure (with much more attracted water molecules) tends to have lower peak strength, higher compressibility, be more ductile, have a higher swell potential, and have lower permeability. A clay soil with a flocculated structure will have higher peak strength, but will exhibit more brittle failure when ruptured and will have lower compressibility, lower swell potential, and higher permeability.

11.2.2.2 Soil-Lime Reactions
As soon as lime is introduced to a clay soil in the presence of water, reactions begin to occur. In the *short term* (occurring nearly immediately through a 24- to 48-h period) a number of physiochemical reactions may occur. If quicklime is used as the admixture, rapid hydration occurs. The hydration is an exothermic reaction, meaning that significant heat is generated (hence the caustic nature of quicklime to "burn" when it comes in contact with

moisture, as on one's hands). This hydration causes the lime to adsorb moisture, causing a tendency to dry the subject soil and consolidate the soil due to desiccation. Soil drying is also a simple consequence of adding a dry material into a moist soil if a dry lime is applied. Once mixed with the soil, the lime can then initiate *cation exchange*, replacing lower charge cations adsorbed on the surface of clay particles with high positive charge calcium cations (Ca^{++}). This will reduce the net negative charge, which in turn causes dissipation of the diffuse double layer and a lesser attraction of water.

As explained above, another result of the short-term reactions is *flocculation* of the clay particles. This is a change in the structural arrangement of the soil grains from a *dispersed* arrangement, where the bounded water (diffuse double water layer) effectively keeps the soil particles apart so that there is no edge-to-edge contact, to a *flocculated* structure, where the reduction in adsorbed water allows the particles to come in contact with each other forming *flocs*. Flocculation also leads to *agglomeration* of soil particles, which is the physical combining of particles to form what appear to be larger particles. This can sometimes result in providing the appearance and properties of lower fines content.

In short, these early reactions will dry wet soil, reduce plasticity, reduce attraction for water (which in turn reduces shrink and swell potential), improve compactability (workability), and provide a stable working platform through improved short-term strength. All of these may be possible even for very poor soil and site conditions.

Long-term reactions include *pozzolanic* reactions or cementation (Figures 11.5 and 11.6). This is where soils that contain a suitable amount of silica or alumina clay minerals (or added pozzolanic material such as fly ash), or the fine material already contained in Portland cement, react with the calcium and water to produce insoluble calcium silica hydrates, CSH (and/or calcium alumina hydrates, CAH, CASH). In addition, additional lime can react with moisture and carbon dioxide to form (or reform) calcium carbonate.

The resulting cementitious end products are permanent, formed from nonreversible reactions that may continue for days, months, and even years. The "ultimate" strength gain (typically measured after 28 days or longer periods of curing) and durability to resist repeated loading, freeze-thaw cycles, and prolonged soaking of the stabilized soil, have been shown to be affected by a number of variables. These include the soil temperature during the curing period, the uniformity of mixing, the delay time between mixing and compaction, the maintenance of proper compaction conditions (moisture and density), and the maintenance of adequate moisture during curing.

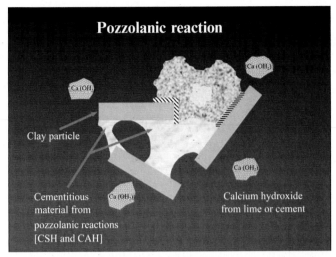

Figure 11.5 Schematic of complex soil-lime cementation resulting from pozzolanic reactions. *Courtesy of Carmeuse Natural Chemicals.*

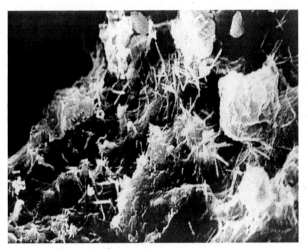

Figure 11.6 Micrograph of a reacted soil-lime mixture. *Courtesy of Carmeuse Natural Chemicals.*

11.2.2.2.1 Concerns of Using Lime and Cement Stabilization

A serious concern has been recognized for instances where a calcium-based stabilizer has been applied to a soil containing significant sulfates. While lime has been proven to minimize swell in many expansive clays, a number of notable cases have shown that long-term reactions with sulfate-rich soils can lead to the generation of secondary minerals such as ettringite and thaumasite. Crystallization of these minerals may cause expansion of the treated soils and damaging heave to overlying pavements over longer periods of time (e.g., several months or years). A number of studies have been conducted to identify the role of soil mineralogy and varying sulfate content on ettringite generation and have proposed a tolerable threshold level of sulfate, usually measured as "soluble sulfate" (Little et al., 2010). Sulfates are common in residual weathered soils, especially in semiarid and tropical environments.

Lime stabilization is also known to alter the compaction characteristics of a soil, resulting in an increase in the optimum moisture content and a reduction in the maximum density. This can be advantageous in cases where the natural soil has a relatively high natural moisture content, making it easier to achieve a particular compaction specification. This does indicate, however, that one must be dutiful in making sure that the soil being used for design and control testing is the actual, stabilized soil mix being used in the field. Where moisture content is critical to the desired engineering properties (as discussed in Chapter 5), care needs to be taken not to use too much lime, or the drying effect may cause the treated soil to dry to below a specified minimum level. As a general rule, one might expect the maximum density of a lime-stabilized soil to decrease on the order of ½-1% of the untreated dry maximum, per percent of lime added. Conversely, the optimum water content will usually increase 1-2% above the actual moisture content (or more!) with each percentage of lime added, depending on the type of lime used. An example of typical compaction curves for a soil with and without lime admixture is shown in Figure 11.7.

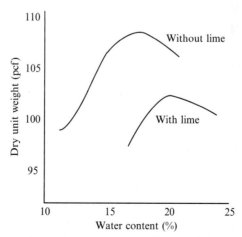

Figure 11.7 Typical moisture-density curve for soil and soil-lime mixtures.

11.2.2.3 Cement

For practical purposes, cement (typically ordinary Portland cement, meeting AASHTO M85 or ASTM C150) is most effective and economical for most granular soils. However, the use of cement may be relatively ineffective or economically inefficient for cohesive soils due to high dosage requirements and construction (mixing) difficulties, especially if the soil is wet and if excessive shrinkage properties are of concern. Ideal cement stabilization is most applicable to well-graded, granular soils, including gravelly soils and sands with only small amounts of silt or clay. Many countries have design guidelines that limit the range of applicable soils for cement stabilization by fundamental index and gradation properties. ASTM C150 specifications describe five types of Portland cement (Types I-V). The different types have controlled chemical components to aid in resisting sulfate attack, achieving high early strength, controlling hydration temperatures, and so on.

According to the FHWA (Federal Highway Administration, 1992a,b), the following definitions have been derived for cement stabilization:

Portland Cement: A hydraulic cement produced by pulverizing clinker consisting essentially of hydraulic calcium silicates, and usually containing one or more of the forms of calcium sulfate as an interground condition (ASTM C-1).

Cement-Stabilized Soil: A mixture of soil and measured amounts of Portland cement and water, which is thoroughly mixed, compacted to a high density, and protected against moisture loss during a specific curing period.

Soil-Cement: A hardened material formed by curing a mechanically compacted, intimate mixture of pulverized soil, Portland cement, and water. Soil-cement contains sufficient cement to pass specified durability tests.

Cement-Modified Soil: An unhardened (or semihardened) intimate mixture of pulverized soil, Portland cement, and water. Significantly smaller cement contents are used in cement-modified soil than in soil-cement.

Plastic Soil Cement: A hardened or semihardened intimate mixture of pulverized soil, Portland cement, and water, where the soil and cement is mixed at a high water content such that the material can be pumped. This mixture is often placed without compaction and is best suited for most soils except for clayey or organic soils.

The primary difference between categories of cement soil mixtures is the amount of cement added. These vary from only a few percent by weight for cement-modified soil, to as much as 6-10% (as much as 15% for clays) for soil-cement depending on the soil type. In general, the more fine-grained and higher plasticity soils will require more cement for stabilization. The quality and degree of stabilization will depend on a number of variables, including pulverization of the cement, pulverization of the soil, degree and quality of mixing, degree of compaction, adequate moisture, and proper curing.

11.2.2.3.1 Shrinkage

One of the major issues with cement soil stabilization is that soil–cement shrinks as a result of hydration and moisture loss, which can have deleterious effects where shrinkage cracks will reflect through overlying surfaces or where the loss of a continuous structural slab action or water tightness is desired. Preventative measures include limiting the plasticity (and thus the water affinity) of the soil, thorough pulverization of the soil, thorough mixing, and proper curing. For high cement contents, controlled or "guided" cracks can allow and design for planned shrinkage of the soil-cement, and help prevent undesirable random cracking. Cement stabilized soils can also exhibit changed compaction characteristics, but may not be as predictable as for lime treatment (www.wsdot.wa.gov).

Similar to the production of lime, a waste residue of the cement manufacturing process is cement kiln dust (CKD), which is a fine, powdery by-product of the production process. About two-thirds of the CKD generated is reused in the production of commercial cement. This material contains as much as 40% lime, which has led to its use as a soil stabilizer (www.wsdot.wa.gov). However, due to the lack of standards and variability of the material, it has not yet been fully accepted into the mainstream and/or accepted by many government agencies.

11.2.3 Fly Ash and Furnace Slag

Fly ash is used as a substitute or supplement for concrete, or simply where the natural soil lacks sufficient pozzolans. While it can be used by itself as a soil stabilizer or soil improvement admixture, fly ash is commonly used in conjunction with other admixtures, such as lime, cement, bitumen, and others, to enhance the improvement characteristics and/or economics of each. For example, fly ash used in addition to cement as an admixture will lower permeability, increase stiffness, and reduce shrink-swell tendencies. Additional discussion on use of admixture combinations is contained in Section 11.2.8.

Fly ash is typically a by-product of coal-fired electric generation facilities. While it is a readily available, inexpensive, and recycled material, only a relatively small amount of fly ash is utilized in the United States. In 2007, the United States produced ~131 million tons of coal combustion products. Of this amount only about 43% was used beneficially and nearly 75 million tons were disposed of. This is contrary to some European and Scandinavian countries, where nearly 100% of the fly ash generated is reclaimed and utilized. Incentives and mandates by state and federal highway agencies to use fly ash as a fill and soil stabilizer, and acceptance as a supplement/additive to cement, has helped increase the constructive use of fly ash in recent years.

Generally, there are two primary components of ash generated: top ash (fly ash) collected by cyclone or electrostatic precipitators, and bottom ash (or boiler/furnace slag), a granular aggregate collected by gravity. The principle components of fly ash are SiO_2, Al_2O_3, FeO_2, and CaO. Other materials may also be contained in smaller amounts.

Fly ash generally consists of silt and clay-sized, cohesionless particles with relatively low specific gravity (Gs), ranging from 2.2 to 2.7 (www.boralna.com). In contrast, the specific gravity of Portland cement is typically over 3.1. This attribute tends to allow the use of waste ash as a component of lightweight fill. Where the ash is rich in lime, the ash itself may be self-cementing. In these cases the cemented ash may be useful as a lightweight coarse aggregate.

The American Society of Testing and Materials (ASTM) defines two classes of fly ash, Class F and Class C fly ash, as determined by ASTM C618 (or AASHTO M295). The principle difference between the two is the difference in pozzolanic and calcium contents. *Class F* fly ash is produced from the combustion of bituminous anthracite and some lignite coals. It is pozzolanic but not self-cementing (onlinepubs.trb/org). Class F fly ash must have at least 70% pozzolanic materials (SiO_2, Al_2O_3, FeO_2), and provides a greater reduction in permeability of concrete. Class F fly ash also mitigates sulfate attack as well as corrosion of steel (reinforcement) and chemical attack. To produce cementitious end products, an active chemical additive such as lime or cement must be added. When non-self-cementing fly ash is used for stabilization, fly ash content will typically range from about 8% to 15% by weight. *Class C* fly ash will only have 50-70% pozzolanic material but will also have at least 20% CaO, allowing it to be self-cementing. Addition of Class C fly ash to cement will improve durability (www.boralna.com).

Since the passage of the Clean Air Act in the 1970s, electric utilities have adjusted operations to produce lower sulfur emissions. To this end, it has become common practice to burn low-sulfur subbituminous coals and combust with a fluidized bed of limestone. Both of these practices lead to the production of a self-cementing fly ash due to the presence of calcium oxide (CaO).

As mentioned previously, sulfates occurring naturally in soils can chemically combine with calcium hydroxide and hydrated calcium aluminum to form gypsum compounds and ettringite, which can result in subsequent expansion and heave. Therefore, use of Class F fly ash is often specified as a supplement to concrete, to provide sulfate resistance, while Class C fly ash will prove to be a better stabilizing agent for clay soils as long as sulfates are not present in significant amounts.

There are a number of major environmental benefits of using ash, including reusing a waste product, reducing the use of high CO_2 "footprint" cement, and minimizing the environmentally destructive need to quarry and/or transport expensive engineering fill materials. For every ton of fly ash used in place of Portland cement, about a ton of carbon dioxide is prevented from entering the Earth's atmosphere. Also, it takes the equivalent of 55 gallons of oil to produce a single ton of cement (www.coalashfacts.org).

Furnace slag (or ground granulated blast furnace slag, GGBFS) is the granular material formed during the processing of iron blast furnace slag generated from steel manufacturing. Sometimes referred to as *slag cement*, it is a cementitious material that can be substituted for equal parts of cement (fhwa.dot.gov). Specifications for use in concrete are provided by ASTM C989 and AASHTO, and when ground to cement fineness hydrates such as Portland cement. GGBFS has been used as a soil admixture for the treatment of clayey soils, sometimes in conjunction with (or as an alternative to) lime, for strength enhancement and mitigation of expansion (swell) for soils containing significant sulfates (www.industrialresourcescouncil.org). Nidzam and Kinuthia (2010) explained that blast furnace slag may be an ideal alternative to lime/cement admixtures in sulfate-bearing soils as it can provide the strength gain without the repercussions of generating secondary expansive materials described previously in Section 11.2.2.2.

11.2.4 Salts, Chlorides, and Silicates

A variety of salts have been successfully used as stabilizing agents for a number of different geotechnical applications. Calcium chloride ($CaCl_2$) and sodium chloride (NaCl, or common table salt) are two salts that are commonly used. Because most salts are soluble in water, they will be leached by rainwater if not protected, and therefore their effects are only temporary.

Calcium chloride (CaCl$_2$) is a by-product generated from the production of sodium carbonate chemical processing or it can be produced directly from limestone. This salt is an inorganic, hygroscopic (attracts water), calcium-based admixture. As moisture is "trapped," evaporation is reduced, making this admixture most commonly used as a dust palliative. Maintenance of consistent moisture content has also been shown to reduce the tendency for volume change. Calcium chloride has also been used as an aid to compaction, especially for gravels, which typically have difficulty retaining moisture. The calcium content may provide exchangeable Ca^{++} ions that can lower PI and improve strength in some soils. Sodium chloride has been similarly used to control moisture and improve compaction densities in sandy granular soils. The biggest drawbacks to using either of these salts (as well as a number of other salts) are concerns about promoting corrosion in metallic components of infrastructure, and environmental concerns with potable water and agriculture.

Potassium chloride (KCl) has also been used for modification of cohesive, expansive clay soils, with the objective of reducing swell potential. The objective is to attract and maintain water (moisture content) so that a soil is "preswelled" but will exhibit little additional volume change. ASTM D4546 is typically used to test for effectiveness of modification with potassium chloride, where a maximum of 1-2% swell is usually desired (www.haywardbaker.com).

Sodium silicates (e.g., Na$_2$SiO$_3$) have been used as dust palliatives, but also result in high soil pH, which may promote dissolution of silicates from soil particle surfaces making them available for cementation reactions. This may enhance stabilization with cementitious admixtures. Sodium silicates may therefore work best with silica sands and other lime-rich admixtures.

Investigations have been made for using phosphoric (polyphosphoric) acid (H$_3$PO$_4$) to modify asphalt binders to prevent rutting and brittle failures. The idea of using phosphoric acid as a stabilizer was first introduced in the mid 1900s as it was shown to increase strength and water resistance of soils. At present, there is no clear conclusion as to the benefits of using phosphoric acid as an additive to asphalt, as test results appear to be dependent on other factors.

11.2.5 Bituminous Admixtures (Asphalts, Bitumen, and Tar)

Bituminous admixtures include *asphalt cements/binders*, *cutback asphalt* (AASHTO M81 and AASHTO M82), and *emulsified asphalt*. As a group, these materials are often simply referred to as asphalts, although the terms "tar" and "bitumen" (oil tar) are often used interchangeably. These materials may be found naturally (as in tar sands), generated by the distillation of organic matter (e.g., wood, peat, crude oil processing), or as a by-product of coke production. Bituminous admixtures are most commonly used with the objectives of providing water repulsion and/or adding cohesive strength to soils. Tar has historically been used as a water repellant and wood preservative, while it is now used primarily in roadway applications. A significant difference between asphalts and the calcium-rich admixtures (cement, lime, Class C fly ash) is that no chemical reactions actually take place when asphalts are mixed with soil. The improvements to soils are basically a result of the physical binding of the asphalt-coated soil particles. Once cured, the result is a semihardened mixture of asphalt binder and soil particles providing added strength and stability to a soil mass, especially when well compacted. In addition to strength and stability improvements, asphalt stabilization provides a "waterproofing" action by coating the soil particles with a barrier that retards the absorption of moisture.

Asphalts are specified by viscosity grades (according to ASTM D3381) and performance grades based on "Superpave" specifications (www.fhwa. gov/pavements). Depending on the intended use and application, a variety of asphalt products may be used. Asphalt binders may be used as hot mix or cold mix applications.

Asphalt cements/binders (meeting ASTM D6373 or AASHTO M 20, M 226, M 320) are the norm for producing asphalt cements commonly used for roadway pavements. Asphalt cements are typically applied as a hot mix (typically at $>200\ °F = 182\ °C$), as the higher temperature aids in better mixing and improved workability. When hot, the asphalt mix (asphalt binder with controlled soil aggregate) is very malleable and easily compacted leaving a smooth wearing surface.

Cutback asphalt is a combination of asphalt cement and a petroleum solvent such as naphtha, kerosene, or heavy oil. The solvent reduces the viscosity of the asphalt at lower temperatures for use as tack/prime coats, fog seals, and slurry seals. After the cutback asphalt is applied, the petroleum solvent evaporates, leaving the asphalt cement residue distributed on the surface to which it was applied. Depending on the solvent used, the curing time (elapsed time for solvent to evaporate and deposit asphalt) can be controlled. AASHTO M 81 and M 82 provide specifications to control set times. The use of cutback asphalts has diminished due to environmental regulations and cost associated with the solvents. The use of cutback asphalts has generally been restricted to patching and repairs in colder weather (Roberts et al., 1996).

Emulsified asphalt is most commonly used (or associated) with cold mix asphalt concrete. It is produced by emulsifying asphalt in water, usually with a soap/detergent. The asphalt sets when the water evaporates. Asphalt emulsions have largely taken the place of cutback asphalts due to the environmental concerns with the cutback solvents. The electrical charge of the asphalt can be controlled (i.e., anionic emulsions or cationic emulsions) to improve adhesion properties, depending on the soil type being used. An important difference between cutback asphalt and asphalt emulsion is that cutback asphalt generally consists of about 85% asphalt cement and 15% cutter by weight, whereas asphalt emulsions consist of only about two-thirds asphalt cement. The result is that if one changes from using cutback asphalt to an asphalt emulsion, the residual asphalt content will be significantly less, which can lead to a less-than-expected volume of binder and underperformance of the asphalt concrete if not properly accounted for. It has been recommended that ~70% of aggregate particle height should be embedded in the residual asphalt for adequate performance (epg.modot.org).

Due to the viscosity of bituminous admixtures and the purely physical means by which they are combined with soil, the approach to mixing is very different than for chemical admixtures. Hot mix asphalt (HMA) is generally premixed in batches and applied in mixed form while the cutback and emulsions are most often applied to the soil in place.

11.2.5.1 Applications

Bituminous materials may be used to achieve a variety of soil stabilization objectives, although their use in pavements is by far their greatest use. In addition to conventional pavement layers, these materials are used for *tack coats* (essentially an adhesive between old and new asphalt concrete layers or for a new wearing surface), for waterproofing, as dust palliatives, for erosion control, for water conveyance structures (i.e., drainage ditches and culverts), and they may also be used to prevent loss of moisture by evaporation and hydration during curing of cement or lime-stabilized soils. Bituminous admixtures are also used as a stabilizer in underlying pavement layers from the subgrade to subbase and base layers. The type of bitumen to use depends on the type of soil to be stabilized, the method of construction, and so forth. As soil gradation will obviously affect the engineering properties and performance of pavement layers, different gradations are recommended for each layer. For example, Table 11.2 provides recommended gradations for road-way pavement under layers.

The term *soil-bitumen* (or soil-asphalt) has been used to refer to a water-proofed cohesive soil, typically employing the use of ~4-7% bitumen. Asphalt concrete usually refers to stabilized granular soil, where sand or well-graded aggregate is mixed with between 3% and 10% bitumen to provide a durable water- and abrasion-resistant structural layer. When a *wearing layer* is to be applied to a new or existing pavement, a uniformly graded coarse sand or fine gravel aggregate may be placed over a bituminous tack coat.

Asphalt *seal coats* are a thin layer of asphalt material applied to pavement surfaces for added wear protection or waterproofing. These sometimes include the use of modifiers or fillers (e.g., sand, aggregate, latex, polymers, etc.). Seal coats are often used as a periodic application to "renew" and protect the pavement surface to extend the wear and life of a pavement.

Strength is typically included in specifications of the engineering properties for asphalt cements. This is most commonly evaluated by unconfined compression tests, although bearing ratio tests have also been used. Strength may also be evaluated after a period of soaking. Soaked strength is usually considerably less than that of as-compacted material. While adding bitumen to soil will provide added cohesive strength up to a point, too much of this admixture has been shown to actually decrease unconfined compressive strength depending on aggregate type (pedago.cegepoutaouais.qc.ca; Hausmann, 1990; www.virginiadot.org). Ductility (flexibility), durability, impermeability ("waterproofing"), and fatigue resistance are also important properties evaluated and sometimes specified for asphalt-soil combinations.

Water resistance is most often evaluated by the degree of water absorption for an asphalt-stabilized soil. Liquid asphalt mixed at 5-6% into various soil types has been shown to have absorption ratios generally <2% (Fang, 1991). In general, soil absorption of <1-2% is considered "waterproofed."

Fatigue testing has shown that compacted asphalt concrete will fail at significantly lower loads after repeated loading cycles (fatigue strength). ASTM has developed a standard test procedure (D7460) to evaluate this phenomenon. Testing has shown that fatigue strengths may be only 55-60% of the peak strengths reported. These reduced strength values must be taken into consideration for actual pavement designs. Superpave specifications were generated as a result of the Strategic Highway Research Program (SHRP) to address rutting, fatigue, and thermal distress of asphalt concrete pavements (ftp.dot.state.tx.us).

11.2.5.2 Typical Problems with Asphalt Stabilized Soils

One of the greatest concerns with asphalt-stabilized mixtures is a process referred to as *stripping*, where the asphalt material separates from the soil particles. This is particularly prevalent with aggregates. In many instances, hydrated lime or other liquid antistripping additives can aid in minimizing this phenomenon. In addition to stripping, asphalt is known to deteriorate with age, drying, and repeated loading (fatigue). Typical deterioration of asphalt pavements may be expressed as crocodile cracking, potholes, heaving, raveling, and rutting. While all asphalt cements will deteriorate over time, a number of factors are known to expedite deterioration. including construction quality, temperature extremes, frost heaving, and the presence of water in underlying soil layers. At high temperatures asphalt binders can soften, leading to rutting of pavements under heavy loads. At the same time, high heat and strong sunlight can cause asphalts to oxidize, which results in pavements becoming drier, stiffer, and more brittle. This, in turn, leads to cracking. In regions susceptible to freezing, spring thaws can create a situation where water can be trapped between the pavement above and the frozen ground below. The saturated soil creates a weak zone, which may not provide enough support for the pavement under significant loads, thereby leading to the creation of potholes. This situation has led some jurisdictions to enact "frost laws," which restrict loads and/or speed limits during spring months (en.wikipedia.org).

11.2.6 Polymers and Resins

Resin-based admixtures have been around for many years. Resins are typically available as liquids, foams, or gels. Many ionic, chemically derived resins, such as acrylics, acetates, lignosulfonates, and epoxies, were found to have significant environmental impacts due to toxicity, and have been taken off the market. In their place, several new proprietary additives have come on the market over the past two decades; these promote naturally derived components that are environmentally considerate and can result in a stabilized soil with natural soil appearance, ideal for shallow unpaved surfaces for visually "sensitive" applications such as parks and natural recreation areas. The largest use of polymers and resins are for grouting applications. The topic of grouting and discussion of materials used will be covered in some detail in Chapter 12. Here only shallow mixes or applications will be covered.

Most of the available surface polymer and resin products were derived for applications of dust control and solidifying and/or strengthening near-surface soils. The resins typically provide a physical bonding of soil grains. These types of products have been used extensively for rapid improvement of unpaved rural roadways so critical to infrastructure and quality of life for developing regions, as well as an attractive option for military maneuvers (roads, airfields, helipads, etc.). Resins have also been widely used to provide erosion resistance for soil structures and other unpaved surfaces subjected to high erosional forces. The ease of application is often an advantage to using these materials. These types of admixtures are available as premixed solutions (requiring no water) or as dry powder, which may be blended directly with the soil or be prepared at the site in portable batch plants.

Widely used polymer additives include products from Chemilink, Soilworks, Midwest (Soil-Sement, Road Oyl, Roadbond), DirtGlue Enterprises, and many others. These materials are primarily polymeric and sometimes organic "biomaterials"-based. They may stabilize through an integration of chemical, biochemical, and physical reactions.

11.2.6.1 Ecoalternatives

Many of the new additives now on the market are touting "ecofriendly" or "green" engineering with synthesized organic and biodegradable materials (e.g., Soilworks's Soiltac). These additives have a number of advantages over more traditional materials and have been developed to work with most soil types. While providing similar improvements such as strength gain and reduced swell obtainable from lime application, there are advantages over lime treatment for clay soils, including reduced water usage, no need for remixing, reduced energy consumption, reduced carbon footprint, much lower permeability, and no adverse reaction to sulfates (www. roadbondsoil.com).

Some polymeric admixtures have been designed for improvement of particular soil types and conditions, such as SS-100 from Chemilink. This is a polymer-modified cementitious chemical binder designed specifically for improvement of soils in tropical areas (e.g., Southeast Asia). These regions often have problematic conditions such as high and/or frequent rainfall, high water table, lack of good and/or economical construction materials, and unsuitable, weak, peaty, or swampy soils. More conventional admixtures such as lime, cement, and fly ash may have limited use in tropical regions due to surface cracking for shallow and/or surface applications, and ineffective reactions with organic soils. Polymer-modified materials, such as those available from Chemilink and others, have been used widely for large, high-profile projects, including reclaimed land, throughout Southeast Asia for airfields (Singapore, Malasia), seaports (Indonesia, Malaysia), as well as for roadways and building foundations (www.chemilink.com).

11.2.7 Fibers

The use of natural fibers for soil stabilization dates back thousands of years to when straw was mixed with clay. Several attempts at using fibers to stabilize soils have been made in recent years. Both synthetic (typically polyester, polyethylene, polypropylene, or fiberglass) and natural fibers have been used for this purpose. There are advantages to using natural fibers (such as coir or papyrus), as they tend to be low-cost, locally available, and made of biodegradable "eco-friendly" materials (Sivakumar-Babu and Vasudevan, 2008; Adili et al., 2012).

Fiber reinforcement has been utilized for both sands and fine-grained soils as well as some asphalt applications. The fibers provide tensile strength to soils, which in turn adds shear strength to soil masses. Typical improvements reported in the literature include an increase in peak shear strength ($\sim$20-50%), increased stiffness, limited reduction in postpeak shear resistance, resistance to desiccation cracking in clay soils (up to 80% reduction), increased durability (up to 33% increase in fatigue strength), and increased liquefaction resistance. Generally, these improvements are primarily due to the added tensile resistance of fibers mixed with natural soils. Synthetic fibers are typically between 0.6 and 5 cm (¼-2 in.) long fibrillated or tape strands that can be used in conjunction with other admixtures where additional improvement is needed. Fibers are blended with existing soils with rotary mixers similar to the type used for shallow mixing of lime, cement, or fly ash, then compacted in lifts with conventional compaction equipment.

Fiber reinforcement was chosen by the Louisiana Department of Transportation and Development (LADOTD) as an effective and cost-efficient solution to recurring sloughing and slope failures in highway embankments of very weak soils also subject to desiccation cracking (www.landfilldesign.com). The locally available borrow soils used for many embankments have very low long-term strengths where there is little to no overburden pressure. Desiccation and weathering reduce the cohesive strength of these soils so that rainfall tends to trigger slope failures. The fiber reinforcing adds enough tensile strength and shear resistance to prevent future failures. Some additional work has been done on adding fibers along with other soil admixtures such as cement, fly ash, or other chemical stabilizers (Collins, 2011).

Work at the Alaska University Transportation Center (AUTC) described strength gains of more than two to three times using fibers alone in sands (SP, SM) and uniform silts. This may be very useful for subgrade applications, but low abrasion resistance deemed it unacceptable for surface layers (Collins, 2011; Hazirbaba et al., 2007).

11.2.8 Combined Materials

In many cases, a combination of stabilizing materials may be used. One material may provide effective treatment of a particular attribute of the soil, which another admixture may not. An admixture may also provide a pretreatment of the soil, enabling more effective treatment by additional admixture materials. This is often seen where lime is first used with clay soils to make them more friable and less plastic, therefore making the soil easier to mix with cement or asphalt. While using multiple admixtures in combination may at first appear to be less economical, the end result may achieve the desired level of engineering properties that is not possible with the application of a single admixture material.

Lime/fly ash and lime/cement/fly ash are common combinations used. Where soils may be wet and/or plastic clays are encountered, but the soil is insufficiently reactive, the lime will reduce moisture and plasticity making the soil more workable, while fly ash will provide a source of pozzolanic material (silica and alumina) to enhance cementitious reactions. If a higher strength and/or stiffness are desired, cement may also be added. When using combinations of admixture types, quality of mixing becomes even more critical. Therefore, whenever possible, central plant mixing is usually recommended.

Combining polymers and bentonite with soil has proved to be an effective hydraulic barrier with many advantages, including self-sealing capabilities, ease of installation, and minimal degradation (Liao, 1989). Where bituminous admixtures are used to stabilize soil to increase strength, small additional amounts of cement may be blended in to attain required strength specifications.

11.2.9 Other Recycled Materials

Other waste materials have been utilized for soil improvement. Of particular note is recycled rubber from waste tires. Recent experimental test results (Moghaddas-Tafreshi and Norouzi, 2012) showed an increase in bearing capacity of more than 2.6 times with an optimum 5% rubber content. Other studies have examined recycled plastic waste from water bottles mixed with the soil to improve strength and reduce compressibility (Sivakumar-Babu and Kumar-Chouksey, 2011). Unconfined compressive strengths in the laboratory were nearly doubled with a 1% plastic mixture. *Recycled (crushed) glass* has also been given consideration as a soil admixture for engineered fill and as a drainage material. Wasted *crushed concrete* is now commonly used as a recycled material for new roadway construction either as an aggregate for new concrete mixes or as a substitute gravel in subbase and base layers. In order to be used, the recycled concrete must be crushed and graded and must be free of contaminants, including trash, wood, paper, and other such materials. This technique has been termed "rubblization" by some (en.wikipedia. org), referring to making "rubble" from waste concrete. Larger pieces of crushed concrete have also been used for erosion control, riprap, fill for gabions, and landscaping stone. Foundry sands have also been used to improve workability and drainage by mixing them with fine-grained soils to effectively change their gradation.

Another industrial by-product that has emerged for use in recent years is the result of combusting municipal solid waste (MSW) (mostly residential and commercial trash). This creates a waste stream of both top and bottom ash, collectively called *municipal solid waste ash* (MSWA). This process has the benefits of reducing the volume of the waste stream transported to and dumped into landfills, as well as supplying a supplemental source of power for municipalities. The composition and quality of the ash generated is dependent on the waste stream delivered to the combustion facilities. Depending on the region, this can be very irregular, making the applicable use somewhat limited. In some cases the residual ash may be considered toxic, containing heavy metals and other pollutants, which renders its use unacceptable. In other areas, where the waste stream is more uniform and toxicity levels are within acceptable limits, the waste ash (MSWA) may be a viable soil admixture that may enhance soil properties such as drainage and/or permeability. Some studies have shown promising results of utilizing MSW ash, both as a soil admixture (or supplement) and as a component of CMU blocks for use in developing regions. MSWA has been used for landfill cover material in some municipalities (Lee and Nicholson, 1997). Other studies have shown that some MSWA may be capable of an environmental application by binding up hydrocarbon contaminants, thereby improving the quality of leachates (Nicholson and Tsugawa, 1996).

Recently, research and field tests have shown that *steel slag fines* blended with dredged coarse media can be used not only as an extremely competent and cost-effective engineered fill, but can successfully immobilize a wide range of heavy metals (Ruiz et al., 2013). Steel slag fines are the by-product of commercial scale crushing and screening operations passing a 9.5 mm sieve at steel mills that convert bulk slag into construction aggregates. This results in a double bonus of recycling waste materials while providing an important environmental benefit.

While each of these waste materials may not be biodegradable, their use could be considered an environmental solution for recycling the massive amounts of these types of wastes.

11.3 APPLICATION METHODS AND MIXING

For the most part, the degree or quality mixing of the admixture into the soil will have a direct correlation to the level of improvement attained. To a certain degree, better or more intimate mixing can be achieved with increased cost of mixing method, although there are certainly other considerations, including project size, scope, areal extent (e.g., a small or concentrated structure footprint vs. a long stretch of highway), and so on.

There are a number of methods for adding and/or mixing materials into the ground to achieve the benefits available from soil stabilization with admixtures. These can be divided into four general categories primarily based on the type of mixing equipment used, and does not necessarily depend on what type of materials are being added. The four categories are:
- Surface mixing
- Layering (or surface placement) and quicklime piles
- In situ mixing (in situ soil mixing, ISS; shallow soil mixing, SSM; and deep soil mixing, DSM)
- Grouting (primarily jet grouting for admixture applications)

11.3.1 Surface Mixing

Surface mixing essentially covers all types of mixing where admixture materials are applied at the ground surface or in layers (lifts) of placed engineered fills. This type of application serves well for projects where applied surface loads are relatively small or where moderate improvement of the surface soils is adequate to provide the needed benefit in supporting (or resisting) the applied load. Scenarios typical of this type of loading conditions are roadways, hydraulic structures, beneath slab construction, (light) shallow foundations, paved parking areas, athletic fields, and so forth. Shallow surface mixing has also been widely used for treatment and upgrading of roadway/pavement layers for major highways and freeways as well as airport runways. One application for these cases has been to treat expansive, high-plasticity clays. A few very notable and historical application cases are the Dallas-Fort Worth and Denver International airports. These projects utilized extensive soil improvement with lime for stabilization of existing clay subgrade soils. At the time of each of these projects, they were considered the world's largest lime stabilization projects. A lime-treated subbase and base treated with a combination of lime, cement, and fly ash were used for the Houston Intercontinental airport.

The equipment needed for surface mixing applications ranges from very simple, to those that provide more elaborate specialized processes. This depends on type of application, desired mix quality, and cost. For low-use secondary and unpaved roads, traditional disk harrows have been used for mixing and blending of the admixture materials, followed by grading and compaction with conventional roadwork equipment. For more intimate blending, *traveling mixers* (such as asphalt recyclers) can be utilized. These types of recyclers (shown in Figure 11.8) provide a good quality mix that is often ideal for roadway stabilization or rehabilitation for small to moderate-sized projects. Traveling mixer equipment typically incorporates an internal pugmill for mixing material that has been drawn up into the machines. Traveling mixers can process in excess of 300 cubic meters of soil per hour. This makes them ideal for larger projects such as state and federal highways that may extend for significant distances.

For large-footprint projects that do not extend for great distances (such as concentrated highway projects or large commercial facilities) where high-quality mixing and a greater depth of improved soil is required, a central mixing plant is sometimes used. In this type of application the soil may be excavated and transported to a mixing plant where the admixture material(s) are intimately blended with the soil and then transported back to the site for placement and compaction. This is obviously the most costly type of surface admixture stabilization application, but will provide a very controlled and high-quality product.

11.3.1.1 Admixtures in Roadway (Pavement) Designs

Admixtures have played an important role and have been used extensively in roadway applications for several different pavement "layers." There are many benefits to using admixtures in subsurface pavement layers, including cost savings in materials through reduction in required layer thickness, reduction in deterioration rate of pavement layers, reduced maintenance, and improved drainage characteristics, to name just a few.

Figure 11.8 Typical surface mixing equipment. (a) Disk harrow (Courtesy of EPA). (b) Recycler/traveling mixer (Courtesy of Bomag).

Some states have adopted abbreviated procedures for recommending mix designs based primarily on soil classification alone (e.g., Oklahoma DOT, Alaska DOT, Ohio DOT). But many state as well as federal agencies (i.e., state DOTs, FHWA, AASHTO) have accepted detailed practices for mix designs and utilizing admixtures for roadway applications. Some industry groups (e.g., National Lime Association, Portland Cement Association, American Coal Ash Association) have also contributed to design guidelines. These include design procedures from FHWA, AASHTO, and the National Lime Association, as well as several individual state DOT guidelines (Federal Highway Administration, 1992b). For example, the Mechanistic-Empirical Pavement Design (MEPD) procedure was developed by the National

Cooperative Highway Research Program (NCHRP) as a design and analysis tool to provide a more realistic characterization of in-service pavements. It can enable engineers to create more reliable pavement designs for new and rehabilitated pavements (www.fhwa.dot.gov).

One of the more common applications of using (typically lime and/or cement) admixtures in roadways is in treatment of subgrade soils. As the quality and competency of the subgrade is improved, the overall quality and capacity of the finished roadway is improved. The increased strength, stiffness, and durability achievable with common admixtures allows for reduced thickness requirements for subsequent base layers. The design charts and guidelines used to determine required minimum pavement thicknesses are usually based on measured strength parameters of the stabilized layer(s). This provides for obvious savings in materials, construction time, and ultimately cost for initial construction as well as reduced future maintenance.

In addition, lime has been used to improve clay-contaminated base aggregate and calcareous base aggregates. Benefits of lime stabilization for calcareous bases have been demonstrated in the United States, South Africa, and France (onlinepubs.trb.org). It seems that the use of lime has the potential to improve many regions worldwide where calcareous aggregates are much more prevalent than higher-quality, igneous rock aggregates. Hydrated lime is also sometimes used in asphalt blends as it can significantly improve cohesion when it reacts with contained fines. This is related to mitigation of striping as described earlier in Section 11.2.5.

11.3.1.1.1 Recycled Roadway Materials

It is generally acceptable to use up to 25% reclaimed asphalt concrete pavement in "new" pavement systems provided that it meets ASTM or AASHTO specifications. Full-Depth Reclamation (FDR) is described by the Portland Cement Association (PCA) and others as a process of recycling the entire thickness of existing roadway layers. The old asphalt and base materials are pulverized, mixed with cement and/or lime and water, and compacted to produce a strong, durable base that may be finished with an asphalt or concrete surface. Full-depth recycling may also incorporate some subgrade soils into the mix to be stabilized. This may affect the choice of stabilizing agent to be used. Recycling saves money and natural resources by recycling the old asphalt and base materials. There is no need to haul and dispose of the old pavement materials. When utilizing reclaimed asphalt concrete pavements there is little or no waste, and no need to haul in new aggregate (which can be expensive due to availability/quality), although some new aggregate may be added to upgrade the mix to meet requirements if necessary. Cost of rehabilitating a recycled base is estimated to be 25–50% less than removal, disposal, and replacement with new pavement materials (www.cement.org). The recycled base is generally stronger, more uniform, and more moisture resistant than the original base. Ideally, this results in a longer lasting, low-maintenance life for the new pavement structure. Recycled base materials can also be treated with other chemical stabilizers with similar improved results. Again, the choice of stabilizer to use will be dependent on the existing roadway materials involved and the desired (required) engineering properties.

The design and construction of recycled bases is relatively simple. The design of stabilized base thickness can be determined with the assistance of available guides such as the *AASHTO Guide for Design of Pavement Structures* (AASHTO, 1998) and AASHTO HM-32-M (AASHTO, 2012). Field cores or test pits should be made to determine the existing thickness of pavement layers and for the collection of material samples for evaluation of properties in laboratory tests.

Other construction wastes, such as waste concrete, may also be suitable for use as base and subbase course materials for both construction of new pavement and repairing of existing pavement after simple mechanical treatment.

11.3.2 Layering (Surface Placement) and Quicklime Piles

One method of lime treatment that does not require mixing with the soil is placing and compacting quicklime in predrilled holes. This type of application has been used successfully to stabilize soft clay, primarily in Asia (India, China, Japan, Taiwan, Singapore, and Malaysia) (Moseley, 1993). The basic premise is that strength is gained by absorbing pore water from the surrounding soil while expanding the volume of the lime "pile" on the order of 85%. The result is to consolidate the surrounding soil, increasing its strength and adding lateral pressures while the quicklime pile itself also hardens. Another application of lime treatment without mixing involves the application of relatively thin layers of quicklime between placed clay layers, sometimes referred to as the "lime sandwich method." In some cases, a filter fabric is also used at the soil-lime interface.

11.3.3 In Situ Mixing

In the context of this text, *in situ* mixing covers a variety of application methods that call for the admixture material to be blended with the soil "in place." This includes *shallow soil mixing* (SSM) and *deep soil mixing* (DSM) up to ~100 ft (or more) depths without excavation. Some have generically called these processes *in situ soil mixing* (ISS). There is some discrepancy (or intraindustry debate) as to what defines shallow versus deep soil mixing, as the process is essentially the same. SSM has been suggested as being limited to 10–15 m (~30–50 ft) depths and may create larger diameter individual columns than those typical for deep soil mixing applications.

Some newer in situ mixing techniques for "wall" construction with specialized equipment are also introduced here. Grouting (particularly jet grouting) can also blend admixture materials with the ground in situ, but these applications as previously mentioned, will be addressed in more detail in a separate chapter. Major advantages of in situ mixing are that poor soils can be utilized without excavation or disposal, low volume of spoils is typical (dependent on soil type), no dewatering is required, and there is generally reduced noise and vibrations during construction.

The admixture materials used for in situ mixing may vary depending on the variables discussed previously, and may include (but are not limited to) lime, cement, fly ash, furnace slag, and so forth. A mixing tool capable of injecting an admixture at or near the auger tip (either under air pressure or as a fluid/slurry; see Figure 11.9), is rotated into the ground while simultaneously mixing the admixture material into the soil, thus blending the materials together. Figure 11.10 shows a schematic diagram of the in situ mixing process. Deep soil mixing was first developed and used extensively in Japan and Europe in the 1960s, 1970s, and 1980s before really taking hold as a worldwide soil mixing technology. Early applications were primarily for construction of foundation support and earth retention structures (Andromalos et al., 2012). Applications have now expanded to include other construction components, waste soil (dredging materials and sludge) disposal, and environmental remediation. Deep soil-mixed elements have replaced many conventional techniques such as sheet piles, slurry walls, or concrete cutoff walls.

Dry soil mixing (applying a dry admixture) is best suited for soils with moisture content >60% and near the liquid limit, and is most applicable to clays, high organic soils, peats, and other weak soils (www. haywardbaker.com). This method is limited to about 60 foot depths, and has even been effectively used for underwater applications. Wet soil mixing is better suited for drier soils, where the admixture materials would need additional moisture for cementation reactions and can reach depths of up to 100 ft.

Soil mixing equipment now available is quite varied and can be highly specialized. This may range from a single mixing paddle used to create a

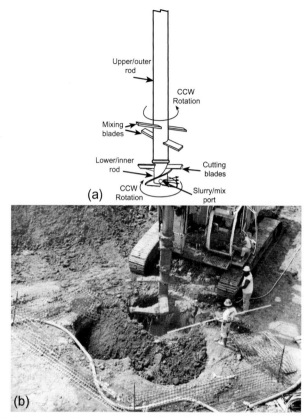

Figure 11.9 Examples of in-situ soil mixing equipment. (a) Schematic of a typical bidirectional mixing tool. (b) Photograph of a single-axis mixing tool (Courtesy of Hayward Baker).

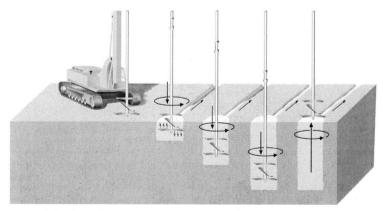

Figure 11.10 In-situ soil mixing process. *Courtesy of Hayward-Baker.*

soil-mixed column, to a system of several overlapping auger-type (multiaxis) mixers (Figure 11.11). These machines are used to construct soil-mixed "panels" for earth retaining wall support and hydraulic cutoffs, as well as other applications. Most methods described in the literature involve first rotating the mixing tool to the maximum design depth to loosen the soil. The tool is then slowly withdrawn as the admixture is injected and blended with the soil to the diameter of the mixing tool. Mixing tool diameters can vary from ∼0.3 to 3.5 m (1-12 ft). Some contractors have incorporated water jets to help cut and blend the native soil with the admixture. Another type of equipment uses a rotating mixer mounted on the end of a mechanical

Figure 11.11 Multi-axis deep soil mixing equipment. *Courtesy of Malcolm Drilling.*

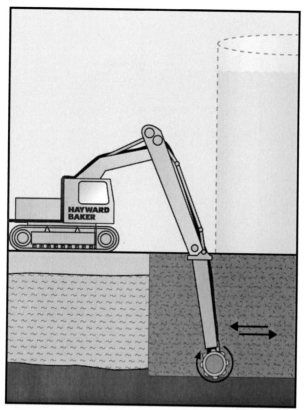

Figure 11.12 Mass soil mixing equipment drawing. *Courtesy of Hayward-Baker.*

arm; in this way, an entire mass of soil may be mixed rather than just in vertical columns. A schematic of *mass soil mixing* is provided in Figure 11.12.

Quality control and quality assurance are maintained through real-time monitoring during the construction process, coring and lab testing of soil-mixed column samples, load testing, and even excavation of test columns for visual inspection. Detailed visual inspection is also possible by lowering a camera down a corehole.

The outcome of most in situ soil mixing operations is ultimately the generation of single or overlapped "piles" of up to 46 m (150 ft) in length (www.geocon). These type of soil-mixed piles have been successfully used to improve bearing capacity, reduce settlements, and mitigate liquefaction deformations, as well as provide "drainage" and aid in consolidation of soft, wet clays. If a continuous line of overlapped piles is constructed, the resulting wall (Figure 11.13) may be utilized for excavation support, wall bearing, and

Figure 11.13 Excavated soil-lime column wall. *Courtesy of Hayward-Baker.*

as a hydraulic barrier (cutoff walls, discussed next). Soil–mixed piles have also been used to improve stability of slopes and embankments (including levees).

11.3.3.1 Cutoffs and Slurry Walls

Slurry and soil–mixed cutoff walls were described in Chapter 7 as a means of seepage control and positive hydraulic barrier. In fact, while deep mixing has been and continues to be popular for this type of application, primarily with cement and bentonite admixtures, jet grouting has taken the place of deep mixing for many of these applications.

An in situ soil mixing method that is becoming popular for creating soil-mixed continuous cutoff or retaining wall structures is termed *cutter soil mixing*. New equipment has been devised for this specific purpose. Several specialty geotechnical contractors have now developed such equipment (e.g., Malcolm Drilling Company, Nicholson Construction, Bauer Foundations, etc.). This method uses a tool with two sets of counterrotating, vertically mounted, cutter wheels that can cut through difficult soils, including stiff clays, gravels, and cobbles (Figure 11.14). A slurry of admixture is injected between the cutting/mixing wheels. This type of equipment uses directional cutting/mixing wheels that can create rectangular "panels." When overlapped, the panels form continuous walls. An advantage of this type of equipment is that the housing of the hydraulic motor is located within the cutting heads, allowing direct application of energy rather than

Figure 11.14 Cutter soil mixing apparatus. *Courtesy of Malcolm Drilling Company, Inc.*

at the top of a Kelly bar, as is common with more conventional deep mixing equipment. Sensitive location instrumentation located within the cutter head can be computer-controlled in real time to assure alignment and complete overlap between installation points of soil-mixed panels (www.malcolm. com). This equipment is reportedly capable of creating soil-mixed walls of 40 and up to 60 m (~195 ft) deep (pure.ltu.se; www.bauerfoundations. com). The powerful cutting/mixing heads of this type of equipment allows for effective mixing in a wider range of soil conditions than is typically feasible with most other deep mixing methods.

11.3.3.2 Deep Mixing Case Study—New Orleans Levee LPV111

The largest deep mixing operation ever undertaken to date was done as part of the rehabilitation and upgrading of the New Orleans levee system to withstand the 100-year hurricane storm surge. In this case, it had been shown that weak and compressible foundation soils would be incapable of either handling the added loads of a raised and enlarged levee or maintaining an acceptable level of settlement. A greater lateral support was needed to resist storm surges. Therefore deep mixing was used to treat subsurface soils to strengthen the ground and reduce compressibility. Approximately 1.4 million m^3 (49.4 million ft^3) of the ground was treated in this manner for this project.

For the most part, dual-axis, deep mixing rigs were utilized for the project to mix soil elements to a maximum depth of 20.5 m (67 ft). 1.6 m diameter (5.25 ft), deep mixed elements were arranged in buttresses (or panels) perpendicular to the levee alignment at a maximum spacing of 4.7 m (15.5 ft), with an additional element placed in between buttresses to aid in resisting differential settlement (Schmutzler et al., 2012). Extensive quality assurance and quality control were implemented, including monitoring of grout mixes and independent testing to confirm quality control test results.

11.3.4 Mix Design Procedures

In order to prepare a proper mix design, a series of steps should be followed. These steps begin with identifying the properties of the soil to be improved. There are some soils that are known to be problematic when calcium-based stabilizers are being considered. It is important that the amount of soluble sulfates in the soil be predetermined (such as with ASTM C1580). Stabilization with calcium-based additives may not be suitable for high-sulfate soils as expansive secondary minerals may result. Unfortunately, this important lesson was learned the hard way, as a number of disastrous case studies demonstrate. This was discussed briefly in Section 11.2.2.1, Lime and Clay Mineralogy. Appreciable organic content, usually taken as >1% by weight, has also shown to be difficult to adequately stabilize, or otherwise may require excessive lime or cement.

In choosing a suitable admixture for soil stabilization, two soil characteristics must first be considered: grain size distribution and soil plasticity. Grain size distribution is typically determined by sieve analyses (e.g., ASTM D6913; ASTM C136), while plasticity can be determined from the results of Atterberg limits tests (ASTM D4318). The National Lime Association spearheaded an effort to develop a definitive lime stabilization mixture design and testing procedure (MDTP) for use by design engineers, specifying agencies and laboratory personnel (www.lime.org). If there is >25% passing the #200 sieve (fines), and if soil plasticity, reported as plasticity index (PI), is >10, then lime treatment should be considered first. For coarser-grained and low-plasticity soils, other (or additional) stabilizers will likely be more effective and efficient. Often, lime has been used with the addition of fly ash for roadway projects. Studies in Mississippi reported that pavement structures have been successfully treated using HMA directly over lime-treated subgrades and over lime-fly ash stabilized bases (Little and Yusuf, 2001).

Design charts employing simplified procedures have been commonly used for preliminary choice of appropriate admixtures for roadway and other designs. These charts have been based on common performance test values such as CBR, M_r, LRFD, and others (Federal Highway Administration, 1992b).

11.3.4.1 Laboratory Testing

In most cases, once preliminary selections have been made for the choice of admixture(s) to be utilized, more detailed designs are made by testing for desired (or required) engineering properties of controlled laboratory mixes. These lab mix designs allow for refinement on amounts of admixture(s) to be used in the field. The lab tests that are typically performed include:

- Soil–lime pH determination, to identify the approximate optimal percentage of lime (minimal amount) required for stabilization. This is a procedure known as the *Eades-Grim* test (ASTM D6276), and defines the amount of lime needed to raise the pH of the soil pore water to about $pH = 12.4$. This is also the amount of lime needed to maintain the elevated pH necessary for sustaining the reactions required for both short-term and long-term (pozzolanic) reactions. This test is typically performed on soil–lime mixtures at various percentages of lime (by dry weight) from between 0% to 10%. While this test will indicate the proportion of lime needed to maintain elevated pH levels to sustain stabilizing reactions, it does not provide any reliable information pertaining to the reactivity of a particular soil or magnitude of strength gains. Caution should be used in assuring accurate measurements, as pH meters have variable resolution and may also need to be periodically calibrated.

- Optimum moisture content (determined by ASTM D698, Standard Proctor Test (AASHTO T-99); ASTM D1557 (AASHTO T-180)), to determine the moisture content required to achieve the maximum dry unit weight, and moisture-density relationship for the modified soil at various admixture percentages. The optimum moisture content of lime-treated soils generally increases with increasing lime content. The dry unit weight (density), coincidentally, decreases with increasing lime content. These changes must be considered when specifying compaction levels for field applications.

- Unconfined compressive strength (UCS, ASTM D5102), where a minimum of 700 kPa (100 psi) is commonly required along with freeze-thaw and wet-dry durability. The unconfined compressive strength test is also used to evaluate the reactivity of a soil to admixture stabilization. It has been suggested that a soil may be deemed *reactive* if there is a strength gain of at least 350 kPa (50 psi).

- Expansion (swell) test, to determine the swell potential of a compacted specimen. This test may be useful in indicating the amount of expected swell, or the change in swell potential, for varying amounts of admixture. This test can be run as a separate test (as in ASTM D4546) or in conjunction with other tests from specimens prepared similarly to the compaction test or UCS test.

For LKD, the requirements should be modified to reflect differences in the stabilization expectations from lime.

A mellowing period of 24 h is recommended to allow hydration and early reactions to occur prior to compaction. After compaction, the treated soil should be "cured" for 7-, 14-, or 28-day tests. In the field, an initial mix may be made with only part of the total amount of lime to be added. This *pretreatment* of the soil provides for improved mixing and better ultimate results. The remainder of the admixture (if any) is then mixed in and the soil compacted.

Quality assurance and quality control (QA/QC) are critical components of design and application with soil admixtures. A good design will incorporate not only careful preparation and testing of mix designs in the laboratory, but also requires strict attention to monitoring all aspects of the construction process followed by field control inspection through observation of processes and testing of the final product. This last step may involve shallow and/or deep field tests, as well as additional testing of specimens taken from the field.

Guidelines are available for design and mixing specifications that typically include test standards and detailed requirements for contractor qualifications, quality control/quality assurance, and so forth. An example specification guide for soil mixing is included as an appendix at the end of this chapter (courtesy of Geo-Solutions).

11.3.5 Grouting

First, it should be made clear that grout is not a specific admixture material. Grouting is a method of adding material to the ground, usually by injecting a fluidized material under pressure. In the simplest form, grouting has been used to infill void space in an open-graded or coarse soil (or fractures in rock) to reduce seepage. Pumping a slurry of fine-grained soils into gravelly soil and rockfill dams to improve their capacity to retain water has been utilized for many years. For the purposes of discussion in this chapter on admixture stabilization, grouting generally refers to *jet grouting*, where admixture materials are mixed with the surrounding soil in a manner similar to deep soil mixing. The differences are that jet grout mixing involves introduction of the admixture by high-pressure jets capable of cutting through the existing ground and creating an intimate blend of the admixture with the surrounding ground. This often results is significantly larger-diameter treated columns. The effectiveness of jet grouting also depends on the soil type it is being applied to, as well as the desired outcome of the jet grouted mass. A distinct advantage of jet grouting over many of the mixing methods introduced earlier in this chapter is the ability to mix with a wider range of soil types and conditions. This is primarily due to the ability of the high-pressure jets to cut through stiff, dense, or coarse soils.

11.4 STABILIZATION OF WASTES AND CONTAMINATED SOILS

Soil mixing for environmental remediation was first introduced in the 1980s and 1990s. It is now increasingly used in the United States and elsewhere as a technically sound and cost-effective treatment for a variety of waste and contaminant applications (Andromalos et al., 2012). For environmental remediation applications, soil mixing is performed with large-diameter augers (typically 1-3.5 m = 3-12 ft diameters) fitted with mixing paddles and grout ports to bind contaminants in the soil or contain the contaminants from migrating (www.geo-solutions.com). The primary goals of soil mixing for environmental applications are to solidify contaminated soils by increasing strength and reducing hydraulic conductivity, and to allow removal and/or prevent further spread of contaminants. As more contaminated sites are "rehabilitated" for productive use with more environmental forethought and oversight, and as cleanup of other contaminated sites is addressed to contain contaminants, admixtures have become a key tool in remedial work for these situations. For contaminated site remediation, the type and amount of admixture to be used will depend on testing of soil samples for pollutants and soil properties. If solidification is the desired result, then common admixtures such as lime, cement, fly ash, and kiln dust can be batched and mixed with the soil. If fixation is the goal, there are a number of commercially available proprietary materials that will react with a variety of pollutants producing varying degrees of fixation.

RELEVANT ASTM STANDARDS

C51—11 Standard Terminology Relating to Lime and Limestone (as used by the Industry), V4.01

C136—06 Standard Test Method for Sieve Analysis of Fine and Coarse Aggregates

ASTM C150/C150M-12 Standard Specification for Portland Cement, V4.01

C400—98(2006) Quicklime and Hydrated Lime for Neutralization of Waste Acid, V4.01

C593—06(2011) Specification for Fly Ash & Other Pozzolans for use with Lime for Soil Stabilization, V4.01

C595/C595M—13 Standard Specification for Blended Hydraulic Cements, V4.01

C618—12a Standard Specification for Coal Fly Ash and Raw or Calcined Natural Pozzolan for Use in Concrete, V4.02

C977—10 Specifications for Quicklime and Hydrated Lime for Soil Stabilization, V4.01

C989/C989M—13 Standard Specification for Slag Cement for Use in Concrete and Mortars, V4.02

C1097—07 Specification for Hydrated Lime for Use in Asphalt, V4.01

C1529—06(2011) Specification for Quicklime and Hydrated Lime for Environmental Uses, V4.01

C1580—09 Standard Test Method for Water-Soluble Sulfate in Soil, V4.02

D558—11 Standard Test Methods for Moisture-Density (Unit Weight) Relations of Soil-Cement Mixtures, V4.08

D559—03 Standard Test Methods for Wetting and Drying Compacted Soil-Cement Mixtures (Withdrawn)

D560—03 Standard Test Methods for Freezing and Thawing Compacted Soil-Cement Mixtures (Withdrawn)

D698—12 Standard Test Methods for Laboratory Compaction Characteristics of Soil Using Standard Effort (12,400 ft-lbf/ft^3 (600 kN-m/m^3)), V 4.08

D1557—12 Standard Test Methods for Laboratory Compaction Characteristics of Soil Using Modified Effort (56,000 ft-lbf/ft^3 (2700 kN-m/m^3)), V 4.08

D1632—07 Standard Practice for Making and Curing Soil Cement Compression and Flexure Test Specimens in the Laboratory, V4.08

D1633—00(2007) Standard Test Methods for Compressive Strength of Molded Soil Cement Cylinders, V4.08

D2487—11 Standard Practice for Classification of Soils for Engineering Purposes (Unified Soil Classification System), V 4.08

ASTM D3381—09 Standard Specification for Viscosity-Graded Asphalt Cement for Use in Pavement Construction, V4.03

D4318—10 Standard Test Methods for Liquid Limit, Plastic Limit, and Plasticity Index of Soils, V 4.08

D4546—08 Standard Test Methods for One-Dimensional Swell or Collapse of Cohesive Soils, V 4.08

D4972—01(2007) Standard Test Method for pH of Soils, V4.08

D5102—09 Unconfined Compressive Strength of Compacted Soil-Lime Mixtures, V4.08

D6236—11 Standard Guide for Coring and Logging Cement or Lime Stabilized Soil, V4.09

ASTM D6249—06(2011) Standard Guide for Alkaline Stabilization of Wastewater Treatment Plant Residuals, V4.01

D6276—99(2006) Using pH to Estimate the Lime Requirement for Soil Stabilization, V4.09

D6373—07e1 Standard Specification for Performance Graded Asphalt Binder, V4.03

D6913—04(2009) Standard Test Methods for Particle-Size Distribution (Gradation) of Soils Using Sieve Analysis, V 4.09

D7460—10 Standard Test Method for Determining Fatigue Failure of Compacted Asphalt Concrete Subjected to Repeated Flexural Bending, V4.03

Reference: ASTM Book of Standards, ASTM International, West Conshohocken, PA, www.astm.org.

APPENDIX EXAMPLE SPECIFICATION GUIDE FOR SOIL MIXING WITH ADMIXTURES

GUIDE
TECHNICAL SPECIFICATIONS
SOIL MIXING

{This technical specification is to be used to guide the writer in the contract requirements for Soil Mixing (SM) construction for a specific site. Included are _____ to be filled in with project specific data. Also included are [ ] which denote options to be considered for specific design requirements.

Parenthetic remarks { } are included when appropriate to provide the writer with additional, non-essential information.}

{NOTE: Soil Mixing is used for a variety applications and known by a number of different names: This specification is geared to soil mixing that uses a single large diameter mixing tool to treat soils up to 100 ft deep without excavation. Typical applications include: soil or sludge stabilization, waste treatment, retaining walls, soil improvement, dewatering, foundation improvement, etc. Other names for soil mixing include: auger mixing, deep/shallow soil mixing (DSM/SSM), in situ soil mixing, in situ soil stabilization (ISS), deep mixing method, soil cement columns/piles, cement soil mixing, rotary mixing, etc.}

A.1 Scope of Work

This section of the specifications includes requirements for Soil Mixing and related work as indicated on the drawings and as hereinafter specified. The work consists of furnishing all plant, labor, equipment, and materials and of performing all operations as required to construct the [treated area or retaining wall or sludge stabilization or, foundation improvement, etc.] using the soil mixing method.

A.1.1 Reference Standards

Following is a list of standards, which will be referenced in this specification. Such referenced standards shall be considered part of these specifications as if fully repeated herein.

Reference	Title or Description
API Spec 13A	API Specification for Oil-Well Drilling-Fluid Materials
API RP 13B-1	API Recommended Practice Standard Procedure for Field Testing Water-Based Drilling Fluids
ASTM C 150	Specification for Portland Cement
ASTM D 422	Particle-Size Analysis of Soils
ASTM C 989	Ground Granulated Blast-Furnace Slag for Use in Concrete and Mortars
ASTM D 1633	Unconfined Compressive Strength of Soil-Cement
ASTM D 4380	Density of Bentonite Slurries
ASTM D 4832	Preparation and Testing of Soil-Cement Slurry Test Cylinders
ASTM D4972	Standard Test Method for pH of Soils
ASTM D 5084	Hydraulic Conductivity Using a Flexible Wall Permeameter

A.1.2 Abbreviations and Definitions

A. API—American Petroleum Institute

B. ASTM—American Society for Testing and Materials

C. Owner—The Owner as referred to herein is _____

D. Owner's Representative—The Owner's Representative or the Engineer is _____ (or individuals) designated by the Owner to act on its behalf in the execution of these specifications

E. Soil Mixing Specialist—An individual who has had proven and successful experience in soil mixing construction

F. Grout—A stable colloidal suspension of powdered cement, bentonite, additives and/or other similar materials in water. The terms "grout" and "slurry" is used interchangeably in these specifications

G. Injection Ratio—A volumetric ratio of grout to soils (e.g., 100 gal/cubic yard) to be mixed in a SM column. The grout injection ratio is determined for each column based on the column dimensions, soil density, pattern of treatment, and desired application rate

H. SM Column—One completed insertion, injection and mixing of the soil with the mixing auger to the design depth. This creates a column of treated soil the same diameter as the mixing auger. The column may be primary (through virgin soils) or secondary (connecting primary columns) or tertiary, etc.

I. Working Platform—The working platform is the leveled, surface of stable soils from which the SM equipment operates

J. Overlap Ratio—The ratio between the overlap distance (measured along the column diameter) and the diameter of the column. For example; a pattern of columns with a 15% overlap ratio has an overlap of 1.2 ft between two 8 ft. diameter columns

K. Mixing Pass—Operation of the mixing auger/tool from the top of the column to the bottom. Generally, a number of passes are required to completely mix a SM column

L. Mixing Auger/Tool—The special tool that attaches to the Kelly bar and is inserted into the ground to mix the soils. The tool may be fitted with ports for injecting grout, mixing paddles, auger blades, etc.

M. Soil Mixing (SM)—A soil improvement technique used to construct in situ soil structures or treat soils in place, without excavation or dewatering. Soil mixing uses a large drill used to insert a larger diameter {typically 4–8 ft diameter} tool into the ground while injecting and mixing a grout with the soil. Stabilized soil columns are created that may be joined together to by overlapping to form retaining walls, foundation elements or to treat a large block of soil or sludge

N. Swell—The excess material resulting from adding grout to the in situ soils. The swell is typically a mixture of soil and grout similar to the materials in the SM column

A.2 Submittals

The following information shall be submitted at least [4 weeks] prior to construction.

A.2.1 Qualifications of Contractor

The Contractor or his subcontractor or consulting advisor shall submit evidence that he is experienced and competent to construct the project using the Soil Mixing method. The evidence shall include references from at least five similar and successful projects constructed over the last 10 years. Project descriptions shall include at a minimum the dimensions of work, type of grout, and equipment description. This evidence will insure that the Contractor will have sufficient competent experienced personnel and proven methods and equipment to carry out the operations specified.

In particular, a SM specialist shall supervise the construction, grout preparation, column mixing, and quality control. [This individual shall be an experienced engineer with at least 5 years of experience in SM construction on at least five similar projects shall plan and direct the methods of the work.] This individual shall be knowledgeable of: (1) the proper mixing methods employed to mix, control and test grout, (2) SM construction equipment and tools, (3) in situ mixing injection ratios, overlaps and overlap ratios, and (4) testing for SM quality control. The SM specialist shall have at least [5 years] of experience and [five] projects in the successful construction of Soil Mixing.

[The company name, key contact, and qualifications of the Contractor's off-site laboratory shall be submitted. The laboratory will have previous experience with soil mixed materials, experienced laboratory technicians, and modern laboratory testing equipment.]

A.2.2 Work Plan

The Contractor shall submit a detailed operating plan describing his proposed construction equipment, procedures, and schedules. This shall include, but not be limited to, the Contractor's plan for:

A. Listing of supervisory personnel: Name and experience of the various persons, their role and primary responsibilities

B. Equipment set-up and site use layout: including storage areas, grout mixing plant location, haul roads, and work platform

C. Soil Mixing equipment specifications, including maximum depth capability of SM machine, dimensions of the mixing auger/tool and capacity of grout mixing plant

D. Construction means and methods: Listing of equipment and capabilities, construction steps, handling of excess grout and swell, layout, overlap control, control of drainage, spills, wastes, etc.

E. A layout drawing showing the geometry, overlaps, and dimensions of the column overlapping pattern

F. A quality control plan describing all testing, sampling, reporting forms, methods, responsible persons, non-conformance procedures, and all other means to ensure the quality of the work

G. Schedule: A bar chart schedule showing all major activities and durations

A.2.3 Design Mix

A pre-construction, laboratory design mix program is required to determine appropriate materials and material proportions for the required SM performance. {In the case of contaminated sites, a contaminate stabilization (e.g., SPLP, TCLP, etc) may also be required.}

In addition to the Work Plan, the following specific information shall be submitted prior to the start of SM construction:

A. [Sampling Plan. A description of the methods and locations of all samples used in the design mix testing. {Generally, test borings and/or test pits are used to obtain soils samples.} Mixing water, [groundwater or leachate,] cement, bentonite clay, native soils, etc. should also be obtained and tested.]

B. [Contaminate Stabilization testing report, including the results of SPLP, TCLP, ANS 16.1, or other leachate generation test. {Note: this could require about 2 months to complete}]

C. Laboratory SM grout mix and SM soil mixtures, including grout ingredients, soils mixed, and injection ratios. The SM mixture submittal shall report moisture content and unconfined compressive strength [and hydraulic conductivity] on at least [4] samples of each proposed design mix. {Note: laboratory testing may required 45-90 days to complete.}

D. Source of all imported material, including, mix water, cement, bentonite and any additives. Shipment of materials to the site shall be accompanied by the vendor's written certification of the quality or specification of the material and MSDS

{Note: Due to the time and uncertainty in laboratory testing, it is often advised that the owner or Engineer subcontract this work to an experienced soil mixing technical advisor early in the project schedule to avoid delays and uncertainty in construction.}

A.3 Materials

A.3.1 Grout

Grout shall consist of a stable colloidal suspension of cement, bentonite or other additives in water. The purpose of the grout is to assist in loosening the soils for mixing and to modify the soils for improved strength [impermeability, reduced leachate generation, etc]. The grout shall be pumpable and workable with the SM injection equipment.

A.3.2 Cement

Cement used in preparing grout shall conform to ASTM C 150 and/or C 989. The cement shall be adequately protected from moisture and contamination in storage on the jobsite. Reclaimed cement or cement containing lumps or deleterious matter shall not be used.

A.3.3 Water

Fresh water, free of excessive amounts of deleterious substances that adversely affect the properties of the grout shall be used to manufacturer grout. It is the responsibility of the Contractor that the grout resulting from the water shall always meet the standards of this specification.

A.3.4 Additives

Ad mixtures may be used to enhance the workability or final properties of the treated soil. {Common additives include bentonite, fly ash, lime, set retarder, etc.}. Additives may be added to the water or the grout. Propriety chemicals may be approved based on the results of pre-construction tests. The owner's representative shall approve all additives used.

A.3.5 SM Material

The material formed by mixing the grout with the native soils shall have an unconfined compressive strength of [typically 25-150 psi minimum] at 28 days [and a permeability of 1×10^{-6} cm/s {and SPLP, TCLP or other required properties}].

A.4 SM Construction

The [treated area] shall be constructed using SM to the lines, grades, and cross sections indicated on the drawings. The SM structure shall be essentially vertical with a pattern of overlapping columns and shall extend through the overburden [and key [1] ft into the designated layer or] [to refusal] [or a minimum depth of _____.] A generalized description of the soil profile through which the SM is to be constructed is provided on the boring logs attached to this specification.

A.4.1 Tolerances

The following tolerances shall apply to the SM dimensions and construction.

A. The SM columns shall be essentially vertical. The working platform and/or crane shall be leveled to be plumb within [3%] of vertical and/or the Kelly bar shall be measured to be within [3%] of vertical
B. The depth of the SM columns shall be measured or surveyed to within [6 in.] of the desired elevation. The depth shall be measured from the surface to the bottom of the mixing auger/tool
C. The SM wall shall follow the designed alignment within [1] ft of the centerline. The SM wall may vary from the designed alignment to accommodate equipment limitations if approved by the Engineer
D. The SM pattern of overlaps shall be surveyed and staked to ensure that the overlap ratio is constructed as designed. The center of each SM column shall be constructed within [6] inches of the designated location
E. Construction will not be permitted when the air temperature is below [20 °F] or when severe weather conditions may compromise the quality of the work
F. The injection ratio shall be calculated and checked for each SM column. The injection ratio may be corrected for previous overlaps in the same column. In all cases, the minimum injection ratio shall be observed. There shall be no maximum injection ratio

A.5 Equipment

A.5.1 SM Machine

The SM machine consists of a drill that turns a hollow Kelly bar. The top of the Kelly bar is attached to a grout swivel to permit the injection of grout through the bar, and the bottom of the Kelly bar is attached to the mixing auger/tool. The mixing tool has a series of holes or jets to permit the discharge of grout into the soil. The mixing tool may be configured with mixing paddles or special teeth to be capable of blending the soil and grout into a homogeneous mixture. The power source for the drill shall be sufficient to maintain the required penetration rate and mixing speed from a stopped position at the depths specified.

A.5.2 Grout Mixing Plant

The grout mixing plant shall include the necessary equipment including a high shear mixer capable of producing a colloidal suspension of cement and additives in water and pumps, valves, hoes, supply lines, and all other equipment as required to adequately supply grout to the mixing tool. Positive displacement grout pumps shall be used to transfer the grout to the mixing/auger. The grout pump shall be capable of pumping the required distance and elevations to provide an adequate supply of grout to the mixing tool. The plant shall be equipped to accept dry or liquid additives in measured amounts. Storage tanks [or ponds] shall be provided (as needed) to store to allow for an adequate supply of batches or continuously mixed grout to the SM machine. Grout shall be agitated until fully mixed and recirculated in the storage tanks to maintain a homogeneous mix and prevent flash set. Grout meters or calibrated tanks shall be provided to measure injection volumes.

A.5.3 In situ Sampling Tool

A special sampling tool will be provided by the contractor for obtaining samples of the wet, mixed soil, at depth in the SM column. The sampler shall consist of a weighted chamber, which can be opened and closed from the surface to obtain mixed soil and grout. The sampler may be attached to the SM machine or supported by a second machine.

A.6 Execution of Work

A.6.1 Alignment

A. The [contractor or engineer] shall accurately stake the alignment of the proposed soil mix construction, as shown on the drawings

B. Two sets of control lines (e.g., east-west and north-south) shall be established by survey outside the limits of the work. The center of each SM column shall be measured {or established by string lines between the control lines} from these control lines based on a drawing of the overlap pattern. [Alternately, a multiple laser grid system or other remote survey device may be used]. The SM work shall advance stepwise, using primary, secondary, etc. column and overlapping portions of previously completed columns to ensure a proper overlap and continuity

A.6.2 Column Depth

A. The depth of the SM columns wall shall be determined by the lines and grades shown on the Drawings and based on pre-construction soil borings. The Engineer may observe the power usage of the SM machine as an aid in verifying the proper depth

B. The total depth of penetration shall be measured and recorded on each column. The depth may be observed by pre-measured marks on the Kelly bar or survey of a fixed point on the Kelly bar. The depth of each column shall be measured from the bottom of the auger/mixing tool

A.6.3 Obstructions

If obstructions including boulders, bedrock or other potentially damaging materials are encountered the SM operator shall stop drilling until the nature of the obstruction is known. Obstructions, which cannot be penetrated, may be remediated by removal, grouting or other acceptable means. Obstructions, which reduce the drilling rate to less than one foot of penetration per minute for at least 5 min, may be acceptable as refusal upon approval of the owner's representative.

A.6.4 Grout Plant

The grout plant shall consist of a slurry mixer, transfer pumps, storage tanks, metering, proportioning (or weighing) equipment and other equipment, as needed. The proportioning equipment may use meters, weights or weight-volumes to ensure proper proportions. The density [and viscosity] of the grout shall be monitored and recorded, as per the quality control plan to verify grout proportion. Weighing equipment shall be calibrated to with 2% at the beginning of the project and verified monthly thereafter.

A.6.5 Soil Mixing and Penetration

Each soil column shall be penetrated while simultaneously injecting grout and then mixed by repeated passes of the mixing auger. The number of mixing passes shall be monitored and optimized [and recorded] for each column to ensure adequate mixing. The mixing rotation speed shall be adjusted to accommodate drilling conditions based on the degree of drilling difficulty. Additional mixing or passes may be required to evenly distribute the grout throughout the column. After the initial penetration, the rotation speed shall be increased to maximize mixing.

A.6.6 Injection Rate

The grout injection rate shall be monitored and recorded for each column and adjusted as necessary for minimum drilling resistance and to accommodate the design mix. The minimum injection rate shall be calculated for each stroke based on the volume of unmixed soil in the column {whether primary or secondary, etc.}, the density of the soil, and the volume of grout required to achieve the design mix proportions. The flow of grout through Kelly bar shall be verified periodically by observing the flow out of the mixing auger as it is suspended in the air above column. Any blockage in the tool or Kelly bar shall be cleared prior to injection and mixing.

A.6.7 Swell Management

The contractor shall place, regrade, and otherwise manage excess materials resulting from the SM treatment. These materials shall be placed over the SM materials (or other designated area) in a manner so that the final SM material properties are [acceptable to the owner {e.g., stable, able to drain, etc.}. The owner shall not be responsible for retreatment of swell that is improperly managed.

A.7 Clean-up and Treatment for Top SM Construction

Upon completion of the SM construction, the surface shall be covered in accordance with the capping details shown on the Drawings. {Note: typical cap for a soil-cement SM wall is a layer of soil, but if the strength is adequate, the SM may be left uncovered.} After completion of the capping, all remaining grout and SM swell shall be leveled and the surface shall be cleaned as directed by the owner's representative.

A.8 Quality Control

The Contractor shall maintain his own quality control for the SM wall construction under the direction of the SM Specialist. Testing requirements are summarized specified herein.

A.8.1 SM Area Continuity and Depth

The Contractor shall be responsible for demonstrating to the satisfaction of the Engineer that the work is continuous and the minimum specified depth. The owner's representative will be available onsite to verify these measurements. SM continuity shall be assured by an overlapping pattern of the SM columns constructed in accordance with these specifications.

A.8.2 Materials

All permanent materials shall be certified by the manufacturer to comply with the specified standard. Certificates of Compliance with the specification shall accompany each truckload of materials received on site.

A.8.3 Grout

A series of tests shall be conducted at the mixer or holding tank containing fresh grout ready for injection into the soil. Tests shall include density [viscosity, pH, etc] [or weight of each material] to ensure that the specified design mix is properly prepared for injection into the soils.

A.8.4 Soil Mixed Materials

Samples of the soil mixed materials shall be obtained with the in situ sampler, formed into test cylinders, cured and tested. A series of test cylinders shall be made once every [day or 500 cubic yards]. The soil mixed material shall be tested at 7 and 28 days after sampling and exhibit an unconfined compressive strength of [25-150] psi at 28 days [and a permeability of 1×10^{-6} cm/s or other required properties].

A.8.5 Documentation

All quality control records, test, and inspections shall be documented by the contractor and available for review by the owner's representative. The contractor shall record all measurements and test results for submittal to the owner's representative each day.

A.9 Measurement and Payment

The treated area (structure, retaining wall, improved soil, stabilized soil, etc.) shall be measured based on the actual volume of soil mixing constructed. Measurements shall be made of the area and depth of the constructed works from the working platform as approved by the owner's representative during construction. Separate payment will be made for the remediation of obstructions removed during the soil mixing construction. Payment for the Soil Mixing shall be made at the contract unit price ($/cy) for the treated area (structure, retaining wall, improved soil, stabilized soil, etc.). Payment for obstructions removed shall be made at the contract hourly rate ($/h).

REFERENCES

AASHTO, 1998. AASHTO Guide for Design of Pavement Structures, 4th ed., with 1998 Supplement. American Association of State and Highway Transportation Officials, Washington, DC, 700 pp.

AASHTO, 2004. Mechanistic–Empirical Pavement Design Guide, Interim Edition: A Manual of Practice Guide for Mechanistic–Empirical Design of New and Rehabilitated Pavement Structures. Publication MEPDG-1, American Association of State and Highway Transportation Officials, Washington, DC.

AASHTO, 2012. AASHTO HM-32-M, Standard Specifications for Transportation Materials and Methods of Sampling and Testing, 32nd ed. American Association of State and Highway Transportation Officials, Washington, DC.

Adili, A.A., Azzam, R., Spagnoli, G., Schrader, J., 2012. Strength of soil reinforced with fiber materials (papyrus). In: Soil Mechanics and Foundation Engineering, vol. 28, No. 6. Springer, Heidelberg, pp. 241–247.

Andromalos, K.B., Ruffing, D.G., Peter, I.F., 2012. In situ remediation and stabilization of contaminated soils and groundwater using soil mixing techniques with various reagents.

In: SEFE7: 7th Seminar on Special Foundations Engineering and Geotechnics, Sao Paulo, Brazil.

Collins, R.W., 2011. Stabilization of Marginal Soils Using Geofibers and Nontraditional Additives (M.S. Thesis). Fairbanks: University of Alaska.

Das, B.M., 2010. Principles of Geotechnical Engineering, seventh ed. Cengage Learning, Stamford, CT, 666 pp.

Fang, H.-Y., 1991. Foundation Engineering Handbook. Kluwer Academic Publishers, Boston, MA, 923 pp.

Federal Highway Administration, 1992a. Soil and Base Stabilization and Associate Drainage Considerations: Volume I, Pavement Design and Construction Considerations. Publication Number FHWA-SA-93-004. Washington, DC, 160 pp.

Federal Highway Administration, 1992b. Soil and Base Stabilization and Associate Drainage Considerations: Volume II, Mixture Design Considerations. Publication Number FHWA-SA-93-005. Washington, DC, 208 pp.

Gray, D.H., Ohashi, H., 1983. Mechanics of fiber reinforcement in sand. J. Geotech. Eng.-ASCE 109 (3), 335–353.

Hausmann, M.R., 1990. Engineering Principles of Ground Modification. McGraw-Hill Inc., New York, 632 pp.

Hazirbaba, K., Connor, B., Davis, D., Zhang, Y., 2007. The Use of Geofiber and Synthetic Fluid for Stabilizing Marginal Soils. Final Report, INE Project No. RR.07.03. University of Alaska, Fairbanks, Alaska University Transportation Center.

Joint Departments of the Army and Air Force, 1994. Soil Stabilization for Pavements. Report TM 5-822-14/AFMAN 32-8010.

Kitsugi, K., Azakami, H., 1982. Lime-column techniques for the improvement of clay ground. In: Proceedings of the Symposium on Recent Developments in Ground Improvement Techniques, Bangkok, pp. 105–115.

Lee, M.T., Nicholson, P.G., 1997. An engineering test program of MSW ash mixed with quarry tailings for use as a landfill construction material. In: Wasemiller, M.A., Hoddinott, K.B. (Eds.), ASTM Special Technical Publication 1275. American Society for Testing and Materials, Philadelphia, PA.

Liao, W.A., 1989. Polymer/bentonite/soil admixtures as hydraulic barriers. SPE Drilling Engineering, Society of Petroleum Engineers, Richardson, TX 4 (2), 153–161.

Little, D., Nair, S., Herbert, B., 2010. Addressing sulfate-induced heave in lime treated soils. J. Geotech. Geoenviron.-ASCE 136 (1), 110–118.

Little, D.N., Males, E.H., Prusinski, J.R., Stewart, B. Cementitious Stabilization. State-of-the-Art Report of TRB Committee A2J01: Committee on Cementitious Stabilization. http://onlinepubs.trb.org/onlinepubs/millennium/00016.pdf (accessed 24/11/12).

Moghaddas-Tafreshi, S.N., Norouzi, A.H., 2012. Bearing Capacity of a Square Model Footing on Sand Reinforced with Shredded Tire—An Experimantal Investigation. Construct. Build Mater. 35, 547–556, Elsevier.

Moseley, M.P., 1993. Ground Improvement. Blackie Academic & Professional, London, 218 pp.

NAASRA, 1986. Guide to Stabilisation in Roadworks. National Association of Australian State Road Authorities, Sydney, Australia.

Nicholson, P.G., Tsugawa, P.R., 1996. Stabilization of diesel contaminated soil with lime and fly ash admixtures. In: Fang, H.-Y., Inyang, H. (Eds.), Proceedings of 3rd International Symposium on Environmental Geotechnology. Technomic Publishing Company, Lancaster, PA.

Nidzam, R.M., Kinuthia, J.M., 2010. Sustainable soil stabilisation with blastfurnace slag. Proc. ICE 163, Thomas Telford, London, UK (3), 157–165.

Roberts, F.L., Kandhal, P.S., Brown, E.R., Lee, D.Y., Kennedy, T.W., 1996. Hot Mix Asphalt Materials, Mixture Design, and Construction. National Asphalt Pavement Association Education Foundation, Lanham, MD.

Ruiz, C.E., Grubb, D.G., Acevedo-Acevedo, D., 2013. Recycling on the waterfront II. Geo-Strata, ASCE. July/August.

Schmutzler, W., Leoni, F., Bertero, A., Leoni, F., Nicholson, P., Druss, D., Beckerle, J., 2012. Construction operations and quality control of deep mixing at levee LPV111 New Orleans. In: Proceedings of the 4th International Conference on Grouting and Deep Mixing, ASCE, pp. 682–693.

Sivakumar-Babu, G., Kumar-Chouksey, S., 2011. Stress-strain response of plastic waste mixed soil. Waste Manag. 31 (3), 481–488, Elsevier.

Sivakumar-Babu, G., Vasudevan, A., 2008. Strength and stiffness response of coir fiber-reinforced tropical soil. J. Mater. Civil. Eng. 20 (9), 571–577.

Thompson, M.R., 1970. Suggested Method of Mixture Design Procedures for Lime-Treated Soils. ASTM Special Technical Publication 479, pp. 430–440.

http://www.acaa-usa.org/associations/8003/files/rptRegs-Stds.pdf (accessed 20/12/12).

http://armypubs.army.mil/eng/DR_pubs/DR_a/pdf/tm5_822_14.pdf (accessed 20/12/12).

http://www.asphaltinstitute.org/public/engineering/design/index.dot (accessed 20/12/12).

http://www.astmnewsroom.org/default.aspx?pageid=998 (accessed 14/01/13).

http://www.bauerfoundations.com/en/products/soil_mixing.html (accessed 07/11/12).

http://www.boralna.com/flyash (accessed 28/02/14).

http://www.cement.org/pavements/Portland Cement Association (accessed 12/04/12).

http://www.chemilink.com/files/pdf/C-P34/C-P34-FullPaper.pdf (accessed 28/02/14).

http://www.coalashfacts.org (accessed 20/12/12).

http://www.dirtglue.com (accessed 29/10/12).

http://www.dot.state.oh.us/Divisions/Engineering/Geotechnical/Geotechnical_Documents/2009%20Workshop%20Mix%20Design%20Procedures%20for%20Lime%20Stabilized%20Soil.pdf (accessed 18/12/12).

http://ftp.dot.state.tx.us/pub/txdot-info/cst/superpavebindspec.pdf (accessed 14/01/13).

https://engineering.purdue.edu/NCSC/PPA%20Workshop/2009/PPA_pdf/Vanfrank%20-%20Utah%20Trials.pdf (accessed 12/01/13).

http://pure.ltu.se/portal/files/42050076 (accessed 03/03/14).

http://epg.modot.org/files/3/3c/409_Minnesota_Seal_Coat_Design.pdf (accessed 12/01/13).

http://www.fhwa.dot.gov/infrastructure/materialsgrp/ggbfs.htm (accessed 17/01/13).

http://www.fhwa.dot.gov/pavement/materials/pubs/hif11031/hif11031.pdf   (accessed 28/02/14).

https://www.fhwa.dot.gov/resourcecenter/teams/pavement/pave_3PDG.pdf   (accessed 01/03/14).

http://www.fhwa.dot.gov/publications/research/infrastructure/structures/97148/bfs1.cfm (accessed 06/03/14).

http://flh.fhwa.dot.gov/resources/pse/specs/fp-03/fp-03met-div700.doc (accessed 17/01/13).

http://www.fibersoils.com/GroundedSuccess.html (accessed 17/12/12).

http://www.geo-solutions.com/equipment/soil-mixing.php, (accessed 17/12/12).

http://www.geosyntheticsolutions.com/documents/fibers.pdf (accessed 17/12/12).

http://www.graymont.com/pdf/Soil_Stabilization_Brochure_En_05_06.pdf   (accessed 10/11/12).

http://hawaii.gov/dot/highways/specifications2005/specifications/specspdf/specspdf-700-715/703A__Aggregates__Print.pdf (accessed 20/12/12).

http://www.haywardbaker.com/WhatWeDo/Techniques/GroundImprovement (accessed 11/01/13).

http://www.industrialresourcescouncil.org/Applications/SoilStabilization/tabid/443/Default.aspx (accessed 17/01/13).

http://www.itc.nl/~rossiter/Docs/FM5-410/FM5-410_Ch9.pdf (accessed 13/01/13).

http://www.landfilldesign.com/materialselect/prodhighlights/geofibers/AGS/ProjHighlight/I-10%20slope%20failure.pdf (accessed 10/01/13).

http://www.lime.org/documents/publications/free_downloads/tech-brief-2006.pdf (accessed 08/11/12).

http://www.lime.org/uses_of_lime/construction/soil.asp (accessed 04/11/12).

ftp://ftp.mdt.mt.gov/research/LIBRARY/FHWA-SC-06-07.PDF (accessed 11/01/13).

http://www.midwestind.com (accessed 29/10/12).

http://www.okladot.state.ok.us/materials/pdfs (accessed 20/12/12).

http://onlinepubs.trb.org/onlinepubs/millennium/00016.pdf (accessed 10/01/13).

http://pedago.cegepoutaouais.qc.ca/media/0260309/0378334/SCGC-BON/ Documents/TR149-Abd-El-Naby.pdf (accessed 13/01/13).

http://www.roadbondsoil.com (accessed 29/10/12).

http://www.virginiadot.org/business/resources/Materials/MCS_Study_Guides/bu-mat-Chapt2AP.pdf (accessed 13/01/13).

http://en.wikipedia.org/wiki/Asphalt_concrete (accessed 12/01/13).

http://www.wsdot.wa.gov/research/reports/fullreports/715.1.pdf (accessed 11/11/12).

CHAPTER 12

Ground Modification by Grouting

Grouting is a method often applied as a soil and ground improvement method whereby a flowable (pumpable) material is injected into the ground under pressure to alter the characteristics and/or behavior of the ground. This chapter provides an overview of soil and ground improvement technologies by various methods of grouting. While used for many decades in various forms, grouting technology has evolved to the point where, generally, it is now applied only by specialty contractors. Except for a few historically common applications, such as prior to construction of dam foundations and abutments, grouting has been used most often as an expensive remedial measure after project problems have occurred. As stated by the Federal Highway Administration, "Grouting as a means of stabilizing soils has more often been used in the U.S. in shaft sinking and to repair collapses than as a routine method because it is an expensive and time-consuming process that is not perfectly reliable even when very great care is exercised" (www.fhwa. dot.gov). Today, grouting techniques have become more common in project designs, as they seem to be effective methods for preventing or mitigating potential future problems, or for serving as a primary component of construction. When used in this manner, the applications may be more cost-effective than other solutions. In many cases, grouting methods may be one of the only feasible solutions, especially when working in and around the constructed environment and existing infrastructure.

12.1 FUNDAMENTAL CONCEPTS, OBJECTIVES, AND HISTORY

Grouting can be defined as the injection of flowable materials into the ground (usually) under pressure to alter and/or improve the engineering characteristics and/or behavior of the ground. Modification of the ground by filling voids and cracks dates back more than two centuries (ASCE, 2010; Karol, 2003; Weaver and Bruce, 2007). Technically, this would include reported sluicing of permeable rockfills and gravels well before the 1800s. A detailed history of injection grouting, starting as early as 1802, is documented by Weaver and Bruce (2007) and Karol (2003), as well as other references.

12.1.1 Improvement Objectives

The general objectives of grouting are to improve strength and stability, and to control and/or reduce permeability (seepage). While historically most often used as a remedial measure, grouting is now included in more new design work for a wide range of applications. The types of improvements attainable by grouting include increasing bearing capacity and stiffness, reducing permeability and/or groundwater flow, excavation support, underpinning, stabilization for tunneling, and even densification for liquefaction mitigation. A number of different methodologies or "types" of grouting are available, depending on the site-specific variables and requirements, including soil type, soil groutability, and porosity. These different grouting methods will need to be closely coordinated with the wide variety of grout materials available for use. The different types of materials most commonly utilized are covered in Section 12.2. An overview of commonly applied grouting methods is described in Section 12.3.

12.2 GROUT MATERIALS AND PROPERTIES

12.2.1 General Description and Properties

Grout is any material used to fill the cracks, fissures, or voids in natural (or man-made) materials. It does *not* refer to any particular type of material. Grout materials span a wide range of properties, from very low viscosity "fluids" to thick mixtures of solids and water (Karol, 2003). The type of grout material used for a project will depend on a number of variables, including specific project requirements, soil type, material travel expectations, required set times, and so forth. In general, grout material types can be separated into three general categories: (1) particulate (cement) grouts, where solid particles are suspended in a fluid, (2) chemical grouts, where materials are fully dissolved in a fluid, and (3) compaction grout, which is typically a thick, low-slump concrete mix, and so may technically be classified as a particulate grout, although not in a "fluid" form. A major difference between the first two categories is that penetrability of a particulate grout is a function of particle size and void opening size, while the penetrability of chemical grout is primarily a function of the solution's viscosity. Other materials have been used that do not seem to fall into either of these broad categories. These might include materials that are neither cementitious, nor chemical in nature. Examples of these types of materials are hot bitumen (sometimes used to plug high-volume seepage through rock formations) or organic matter used as filler.

It is helpful and instructive to define some terminology that describes properties of grout materials affecting their function and applicability for various uses:

Rheology is the science of flow of materials (www.en.wikepedia.org). It is characterized by fundamental material properties, including viscosity, cohesion, and internal friction (Weaver and Bruce, 2007). The ability of the grout material to flow into and through the groundmass to be treated is fundamental to the process and integral to design.

Grout stability refers to the ability of a grout to remain in a uniform mixture or solution without separation. This includes the *mixture's* ability to not separate or "bleed." Bleed refers to the settlement of particles from the suspension fluid after the material is injected. The grain size, shape, and specific gravity (Gs) of suspended particulate grout particles will be directly related to the amount of bleed. The settlement rate is directly proportional to the difference between the Gs of the particles and the suspension fluid. An unstable grout often leads to incomplete sealing of voids or fractures.

Viscosity is a measure of the ability of a fluid to flow or deform, and corresponds to the notion of "thickness" of a fluid (www.en.wikepedia.org). Obviously, viscosity will have a profound effect on the ability of a grout to penetrate or permeate through the ground or soil mass. This ability of a low viscosity grout may tempt a contractor to use a higher water to cement ratio (*w:c*) to allow (ensure) that the materials migrate to at least their design location, but may result in poor overall results. It has been suggested that a *w:c* ratio of greater than 3:1 should not be used (Weaver and Bruce, 2007). The use of additives such as *superplacticizers* (described in the next section) may enable the use of stable grouts by reducing their viscosity. *Cohesion* of a grout material will also impede its ability to flow freely.

Grouts designed with very low viscosity (and slow set times) may travel to greater distances within the ground and more widely disperse the grout material into smaller voids and cracks. These materials are called *high mobility* grouts. These types of grouts are used most often for remedial seepage control and grout curtains. Grouts that are intended to remain close to their point of application may also be designed by using lower water to cement ratios and, in some cases, by using "quick set" reagents that restrict their ability to flow beyond a certain distance from point of injection. This may be useful for conditions where there is running groundwater, or where there is a tendency for the grout materials to dissipate into surrounding voids. These materials are referred to as *low-mobility* grouts. For certain applications, very low-mobility grout with low slump is used to fill large voids, displace and/or densify loose soil, and remediate settlement distress.

Grout *particle grain size* will obviously affect the size of voids into which a grout can penetrate. As a general rule, if D_{85} of the grout particles is $>1/3$ of the average void or fracture size of the material being treated, then the openings may become blocked (a process known as "blinding") and intrusion of the grout will be incomplete. Mitchell (1981) proposed groutability ratios for the soil grain size and grain size for the particulate constituents of a cement type grout:

$$N = D_{15s}/D_{85g} \qquad (12.1)$$

$$Nc = D_{10s}/D_{95g} \qquad (12.2)$$

where N and Nc are the groutability ratios for the soil to be grouted, D_{15s} is the grain size relating to 15% finer for the soil, D_{85g} is the grain size relating to 85% finer for the grout particles, D_{10s} is the grain size relating to 10% finer for the soil, and D_{95g} is the grain size relating to 95% finer for the grout particles

Weaver and Bruce (2007) suggest that good results could be obtained for $N > 24$ or $Nc > 11$. Similarly, a groutability ratio (GR) for fissured rock was presented as:

$$GR = \text{width of fissure}/D_{95g} \qquad (12.3)$$

A GR >5 is considered a good indicator of fissured rock groutability.

Pressure filtration is a term used to describe the effect of separation (water loss) that occurs when a grout is forced into the soil through small soil voids, much like pressing the grout against a geotextile filter. This can lead to a buildup of a cementitious "cake" around the perimeter of a grout hole, prohibiting any additional grout take. To enhance penetrability of a grout, a low-pressure filtration coefficient is desirable. The values of pressure filtration coefficients are primarily a function of the type and stability of mixes, and secondarily of the water to cement ratios. Details of pressure filtration coefficients and different grout mixes can be found in grouting references such as Weaver and Bruce (2007).

12.2.2 Cement Grouts

Generally, grouts that consist of a flowable mixture of solids and water are termed *suspended solids grouts*. The most common suspended grout is Portland cement, often with a variety of additives. Portland cement is manufactured from a combination of lime, silica, alumna, and iron, which, when prepared as a chemically reactive agent, will by itself, or in combination with a soil mixture, provide a strong, permanent, water resistant, structure.

Cement grouts are commonly used with water to cement ratios of about 0.5-4. At lower $w{:}c$ ratios, the grout will tend to be more uniform, but also more difficult to inject due to high viscosity. Balanced stable cement grouts (commonly used in dam foundation grouting) may include a number of additives to generate a homogeneous balanced blend of water, cement, and additives to produce a product with zero (or near zero) bleed, low cohesion, and good resistance to pressure filtration (www.laynegeo.com). Typical types of additives may include:

(1) Superplasticizers, to reduce grout viscosity and inhibit particle agglomeration. This reduces the need to use higher water to cement ratios.

(2) Hydrated bentonite (or sodium montmorillonite), used at ~1-4% by weight of water, to stabilize the grout, increase resistance against pressure filtration, and reduce its viscosity.

(3) Type F fly ash or silica fume, used at up to 20% by dry weight of cement as a pozzolanic filler, to improve the particle size distribution, and to increase durability of the cured grout by making it more chemically resistant.

(4) Welan gum, used at about 0.1% by dry weight of cement, a high molecular-weight biopolymer used as a thixotropic agent to enhance resistance to pressure filtration and increase cohesion (www.layne.com).

Microfine cements are cement materials that have been pulverized to attain finer grain sizes, thereby enabling greater penetration into smaller fractures and pore spaces. This also keeps solid particles in suspension much longer and can result in improved seepage control. These improved qualities come at a significantly higher cost, up to eight times as much as Portland cement (Karol, 2003). Grain size distributions of microfine cements are typically about an order of magnitude smaller than common Portland cements. Microfine cements typically contain up to 25% blast furnace slag crushed or milled to a very fine particle size. This material is also known as ground blast furnace slag, or GBFS. Other microfines may contain up to 100% slag fines. These materials have played an important role in enabling the use of particulate cement grouts to treat medium- to fine-grained sands, which otherwise would have required more costly (and often environmentally sensitive) chemical grouts. A number of definitions exist pertaining to the grain size of a microfine cement, from $d_{\mathrm{max}} < 15\ \mu\mathrm{m}$, $d_{95} < 30\ \mu\mathrm{m}$, to ultrafine cements with $d_{\mathrm{max}} < 6\ \mu\mathrm{m}$. Some issues with microfine cements arise from agglomeration of grains, which may form large lumps or create flash setting (Weaver and Bruce, 2007). This problem can be alleviated by carefully controlled mixing, wet grinding, or the use of additives to enhance penetrability, as described above.

12.2.3 Chemical Grouts

Grout materials that are in full solution are generally termed *chemical grouts*. These include variations of sodium silicates, chrome-lignins, acrylamides, acrylates, and a variety of polymers and resins. Resins are true solutions of organics in water or solvent without suspended particles, and tend to be the most expensive. They are used where situations require very low viscosity, rapid gain in high strength, and high chemical resistance. "Relative costs" for common categories of chemical grouts were proposed by Koerner (2005):

Silicates	0.2-1.2
Acrylamides and lignosulfates	1-8
Resins	10-80

Chemical grouts often contain reagents that chemically react with the soil, causing the mixtures to solidify and harden with time. Others are mixed in place where they undergo *polymerization* with a second catalyzing agent (and can be applied as a two-shot injection). The types of components and reagents can be proportioned and mixed to control viscosity, strength, and durability. One distinct advantage of chemical grouts is the ability to very precisely control set times to within a few seconds. These set or "gel" times may be designed from seconds to hours, depending on the application and desired control. Adjustments can be made to set times by careful control of mixture proportions. Some additives, including water and calcium chloride (even including suspended solids, i.e., cement and bentonite) may be blended with these grouts to modify certain properties, such as dilution, freeze resistance, strength, and better set time control.

One serious issue with some of the chemical grouts is the concern about toxicity. Probably the most notable example is the use of acrylamides, first developed in the early 1950s. Some of the main advantages of acrylamides is the very low viscosity and corresponding ability to penetrate finer-grained soils, ability to accurately control set time (at which point the material would very rapidly change from liquid to solid), good strength, excellent waterproofing capabilities, and chemical resistance. Acrylamide was banned in Japan in 1974 after some cases of water poisoning, and was recommended for a ban after a U.S. government memorandum reported 56 cases of poisoning (Karol, 2003). It was voluntarily withdrawn from the market in 1978 by its U.S. manufacturer, but never banned. As a result, the use of imported acrylamide products has continued.

Acrylate grouts first came on the market in the early 1980s in response for a need to replace the toxic acrylamides (Karol, 2003). While not providing quite as much desirable strength, viscosity, and set time control as the acrylamides, acrylates are "relatively" nontoxic.

Polyurethane (and urethane) grouts have become popular, as they can be manufactured to quickly react with water, making them suitable for applications with flowing water conditions. These types of materials form an expanding foam and are often used in structural defects (i.e., cracks, joints) in structural floors or walls, or used to fill voids.

Some other chemical grouts include lignosulphates, formaldehydes, phenoplasts, and aminoplasts. While no longer widely used in the United States due to toxicity concerns, these types of grouts are still used regularly in Europe.

12.3 TECHNIQUES, TECHNOLOGY, AND CONTROL

Techniques or methods of grouting can generally be divided into category types based on the way in which the grout material is transmitted into the ground. Figure 12.1 depicts five typical grouting category types. These will each be described in Section 12.3.1.

Technology of grouting has evolved along with practice, experience and the development of more advanced equipment over the years. The technology of actually getting the materials placed in the ground to the desired locations is described in Section 12.3.2. This will include

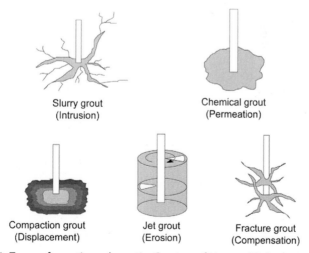

Slurry grout
(Intrusion)

Chemical grout
(Permeation)

Compaction grout
(Displacement)

Jet grout
(Erosion)

Fracture grout
(Compensation)

Figure 12.1 Types of grouting schematic. *Courtesy of Hayward Baker.*

methodology, equipment, point(s) of application, pressures used, and control of where the grout materials end up.

12.3.1 Types/Methods of Grouting

This section provides an overview of the most common categories of grouting application methods used. While there may be some amount of overlap, or in some cases use of multiple methods for a particular project, the distinction between grouting application methods is a function of how the grout material interacts or is placed in the ground. Different grouting methods are also applicable to different ranges of soil grain sizes, as depicted in Figure 12.2.

Slurry Grouting (Intrusion) involves injecting a material so that it intrudes into existing soil formations by following preferred paths of voids or fractures without necessarily disrupting the preexisting formations. The amount of penetration available will be a function of the grout mobility, particulate grain sizes, and sizes of the voids in the ground to be treated. It is generally applicable to coarser soils, such as gravels and coarse sands, as well as fractured rock, but with specialized microfine materials and low viscosity, slurry grouts can be applicable to somewhat finer-grained, sandy soils.

Chemical Grouting (Permeation) generally refers to the use of commercially available agents that will permeate through existing pores and voids of a soil mass. As a general rule, chemical grouts are complete solutions, in that

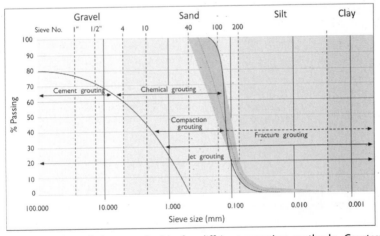

Figure 12.2 Soil gradations applicable for different grouting methods. *Courtesy of Hayward Baker.*

there are no particulate solids in suspension. As such, chemical grouts may be able to permeate into finer soil gradations (medium to fine sands and silty sands) and may contain dissolved materials that react directly with the soils being treated. As an example, certain chemical additives may stabilize expansive soils. Chemical grouting is commonly applied through sleeve ports of a grout pipe placed in a predrilled hole. Sleeve pipe injection will be discussed later in Section 12.3.3.

Compaction Grouting (Displacement) is a technique used mainly for treating granular material (loose sands), where a soil mass is displaced and densified by a low-slump mortar (usually a blend of water, sand, and cement) injected to form continuous "grout bulbs." Compaction grout will typically have no more than 2.5-5 cm (1-2 in.) slump, as measured by a standard concrete slump cone (ASTM C143). A relatively newer grouting technology only developed in the 1950s, compaction grouting is the only major grouting technology developed in the United States (ASCE, 2010). It is also the only grouting method designed specifically to *not* penetrate soil voids or blend with the native soil. It is a good option for improving granular foundation materials beneath existing structures, as it is possible to inject from the sides or at inclined angles to reach beneath them. The grout can also be applied by drilling directly through existing floor slabs. Compaction grouting improves density, strength, and stiffness of the ground through slow, controlled injections of low-mobility grout that compacts the soil as the grout mass expands. Compaction grouting is commonly used to increase bearing capacity beneath new or existing foundations, reduce or control settlement for soft ground tunneling, pretreat or remediate sinkholes and abandoned mines, and to mitigate liquefaction potential (Ivanetich et al., 2000). Compaction grouting can be applied to improve soils equally well above or below the water table. The technology can be applied to a wide range of soils; in most cases, it is used to improve the engineering properties of loose fills and native soils that are coarser than sandy silts (ASCE, 2010). When applied in stages from deeper to shallower, columns of overlapping grout bulbs can be formed, providing increased bearing capacity and reduced settlements (Figure 12.3). One caution that must be exercised when applying compaction grouting is to ensure that there is adequate confinement pressure to prevent disruption of overlying features. As a result, monitoring of surface displacements is often a critical component for compaction grouting. For some shallow applications, the soil may be grouted from the top down to provide confinement and prevent surface heave from the grout pressures applied below.

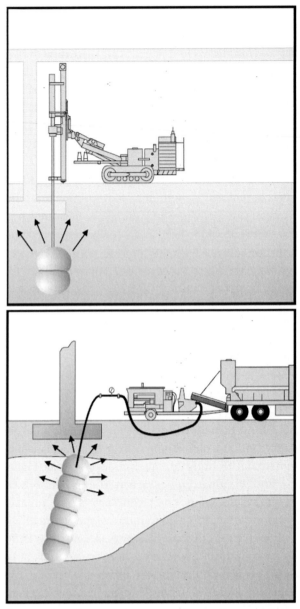

Figure 12.3 Construction of compaction grout columns. *Courtesy of Hayward Baker.*

A version of compaction grouting commonly used to remediate settlement problems beneath foundations and/or slabs is a method sometimes referred to as *"mud jacking"* or *"slab jacking."* In these instances, low-mobility grout is used to slowly lift whole structures or components (such as distressed floor and/ or basement slabs) while carefully monitoring pressures and displacements.

The use of injected *expanding polyurethane* has some similarities to using low-mobility compaction grouts in that it is often used for filling of voids and releveling of distressed slabs. But grouting with expanding polyurethane also has a number of advantages, including its light weight, accurate control of set times, variable expansion characteristics, flexibility, and very good water shutoff capabilities. As mentioned previously, expanding polyurethane has been used to remediate small local deficiencies such as voids behind retaining structures or beneath slabs. These applications are often not accessible for larger grouting equipment.

Jet Grouting (Erosion) is a method that involves injecting the grout material under very high pressures (300-600 bars) through high-velocity jets (600-1000 ft/s) (Figure 12.4) so that they hydraulically cut, erode, replace, and mix with the existing soil to form very uniform, high-strength, soil-cement columns (Figure 12.5). As such, jet grouting could be considered a form of deep mixing with the advantages of generally higher compressive strengths and more uniform soil treatment. Originally developed in Japan in the early 1970s, jet grouting soon spread to Europe and then to the United States in the 1980s, where it has now become very popular for a wide range of applications (Figure 12.6). Typical applications involve drilling to the maximum design depth, followed by injection of grout (and other fluids) while the drill stem/grout pipe is rotated between 10 and 20 rpm (Karol, 2003), and then slowly raised to form a relatively uniform column of soil-cement. There are generally three types of jet grout systems in common use: the single jet or monofluid system, the two-fluid system, and the three-fluid system (Figure 12.7). In single-fluid systems, grout alone is injected from nozzles located above the drill bit and can create a grouted mass to a radial distance of around 40-50 cm (15-20 in.) in cohesive soil and 50-75 cm (20-30 in.) in some granular deposits. The radial distance will be a function of the volume of grout placed as well as pressure and soil type. The two-fluid system combines air jetting with the grout mixture, which can assist in increasing the radius of influence by several inches. The three-fluid system adds a water jet in addition to the grout and air, which helps to cut and erode the existing soil, enabling an even larger radius column, but this also generates a larger volume of wet spoil that must be

Figure 12.4 Jet grout pipe application. *Courtesy of Yogi Kwong Engineers.*

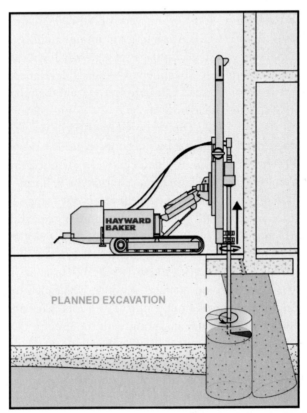

Figure 12.5 Schematic of jet grouting application. *Courtesy of Hayward Baker.*

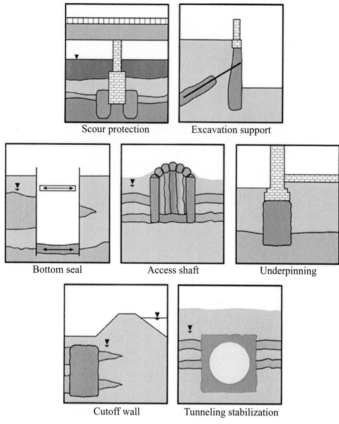

Figure 12.6 Typical jet grouting applications. *Courtesy of Hayward Baker.*

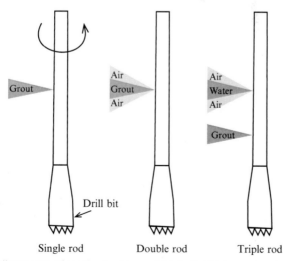

Figure 12.7 Illustration of single, double, and triple fluid jet grout systems.

Figure 12.8 Overlapped jet grout columns. *Courtesy of Yogi Kwong Engineers.*

collected at the surface. Note that, for all system types, the drill bit is larger in diameter than the stem rod to allow an annulus for return of spoils.

Constructed in various configurations, overlapped columns can create seepage barriers, cutoff walls, excavation support, and stabilization of large "blocky" masses of soil (Figure 12.8 and 12.9). The grouted columns can be installed at considerable angles, enabling application beneath existing structures where vertical drilling is not feasible. More recently, jet grouting has been used to stabilize very loose and difficult ground conditions prior to tunneling and microtunneling. Jet grouting has even been used to encapsulate radioactive waste in situ with a special hot wax (www.layne.com). Jet grouting can be performed above and below the water table and can be applied to a wide range of soil types, from cohesionless to plastic clays, as depicted in Figure 12.2. Available equipment now includes multiaxis rigs with up to 30-m (100-ft) drill lengths for higher efficiency production (Figure 12.10).

Fracture Grouting (Displacement) (also called *claquage*) involves utilizing high-pressure systems that intentionally disrupt the preexisting ground formations by a method often referred to as *hydrofracture*. Grout is typically injected with sleeve pipes (Section 12.3.3). Here the high-pressure injection actually creates interconnected fractures in the ground filled with grout to provide reinforcement as well as some densification (consolidation). This process is typically performed in repeated stages of injection to ensure the interconnection of multiple fractures. When used in conjunction with construction in soft soils, fracture grouting may be used to provide intentional

Figure 12.9 Continuous jet grout wall underpinning an existing building. *Courtesy of Hayward Baker.*

controlled heave to compensate for settlement. When used in this manner, the process is referred to as *compensation grouting.*

12.3.2 Grouting Technology and Control

Improvements in grouting technology, materials, and equipment have had a dramatic impact on the increasing use of grouting applications and improving efficiency (reducing costs). A number of variables must be carefully monitored and controlled to ensure that applications are successful. A number of these control variables are described here.

12.3.2.1 Injection Pressure

There are some general rules of thumb pertaining to appropriate grout-injection pressures to be used. The widely used rule in the United States

Figure 12.10 Triple axis jet grouting for rehabilitation of 17th Street levee in New Orleans, LA. *Courtesy of Layne Christensen.*

is that the injection pressure should be "1 lb/ft^2 per foot of depth," at least for the top several feet of ground. In Europe the "rule" is 1 kg/cm^2 per meter of depth. These limits are effectively limiting injection pressures to overburden stresses. There is some controversy as to the rationale and adequacy of adhering to these limits and whether these limits may have been responsible for the poor performance of many grouting projects (Weaver and Bruce, 2007). Certainly, for higher-strength ground and fractured rock, the strength of the material can support much greater pressures than would be provided by the overburden pressures. A number of grouting practitioners (primarily Europeans) have advocated using higher pressures so that existing fissures will open to accept the grout, and smaller voids in finer-grained soils will be penetrated. In fact, the pressure(s) used will depend greatly on the type of grout material being injected and the method of grout-ing being performed. For example, lower pressures may be appropriate for intrusion or permeation of materials, where it is undesirable to disturb the preexisting ground structure, while up to 20,000 kPa (3000 psi) may be warranted for intended fracturing (hydrofrac) or water sealing deep in a rock foundation.

It should also be understood that, when a groundmass is subjected to higher hydrostatic pressures, as when a reservoir behind a dam is filled, exist-ing fractures and/or voids will expand as a result. It is only prudent that pres-sures used to grout these groundmasses should be higher than the expected hydrostatic pressures, or seepage will be inevitable. In fact, Lombardi (2003) recommended that injection pressures on the order of two to three times the anticipated hydraulic head be applied.

12.3.2.2 Set Times

Control of where the grout material finally ends up may be adjusted by adding dispersants, retarders, or accelerators to the mix. Fast (quick) set times may be desired to limit the radius of injected materials, particularly in strat-ified soils with more permeable lenses or in gravelly soils and fractured rock with wide fissures. Fast set times may also be necessary if being applied where there is moving groundwater that would otherwise tend to transport the grout away from the area intended for treatment. Set times for various grout mixtures may be evaluated by ASTM C191 or C953.

12.3.2.3 New Technology

Grout injection control systems continue to evolve, with several "smart" systems now routinely used for many field applications. Most of these new systems involve continuous monitoring and data acquisition of variables such as precise injection location, grout flow, volumes, grout mix, and pres-sure. These are often aided by automated, computer-controlled interfaces and/or graphic displays, which can greatly improve the efficiency and qual-ity of grouting applications.

12.3.3 Grouting Equipment

There are several types of equipment required for introducing grout material into the ground. Much of this depends on the grouting method applied (Section 12.3.1) and the desired results for the particular application. In addition to drilling equipment, some of which is integrated with the grout injection pipes, there are a number of critical components that must be carefully designed to meet the requirements of each application.

12.3.3.1 Batch and Pumping Systems

Virtually all grouting applications rely on pumps to place the grout and provide the required pressures for various grouting methods. As described in Section 12.3.2, these pressures may vary widely from a few thousand to tens of thousands of kPa. For cement grouts, the mixture of cement, water, and any other additives must be blended, continuously agitated, and pumped into the ground before the material sets. In these cases, the water is the catalyst and fluidizer, and must be part of the batch. Ideally, the pump system should have a volume capacity to batch all of the grout needed for a single injection process.

The advantage of two-part chemical grouts is that the two portions may be pumped or added separately, allowing the use of shorter and more controlled set times. These pump systems often have accurate (and sometimes adjustable, computer-automated) metering of the component volumes for control of catalyst concentrations and set times. Therefore, the critical criteria for a pumping system are adequate volume, pressure capacity, and control of mix proportions (if not prepared in a single batch). A large variety of commercial pumping configurations are readily available.

12.3.3.2 Packers

In order to maintain grouting pressures and control where the grout is injected into the ground, tight "seals" must be utilized. These seals may be mechanically tightened where the grout hole meets the insertion pipe, or against the pipe wall or hole at a desired depth (downhole packers). Balloon packers are generally hydraulically or pneumatically inflated membranes, which provide a seal above and/or below a grout injection point to control the injection location within a grout hole or grout pipe location. Use of multiple packers may be desirable to isolate the injection point to specific subsurface horizon(s).

12.3.3.3 Pipes

There are a variety of grout pipe configurations available, depending on the type of grouting application. Single point, "push-in" or *lance*-type driven pipes may be used for certain applications in a wide range of soil conditions. Single point pipes are also commonly inserted in drilled (or jetted) holes, especially for significant depths and hard or difficult-to-penetrate soils and rock. Many single point applications use readily available standard commercial pipe, hollow drill rods, or drill casing.

For more control over the precise depth at which the grout enters the ground, *sleeved pipes* may be used. Sleeved pipes (also known as *tubes-á-manchette*) were first introduced in the 1930s in France (Weaver and Bruce, 2007). The use of sleeved pipes requires a predrilled hole into which the pipe is inserted, and the annulus between the pipe and hole is filled with a weak grout slurry. The sleeved pipe typically consists of a PVC pipe with perforated holes at regular intervals. The holes are covered on the outside of the pipe with a rubber sleeve (Figure 12.11). During application, a desired depth interval is isolated by a double packer system and the grout pressure between

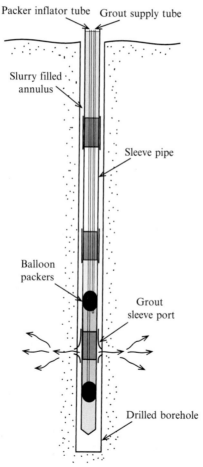

Figure 12.11 Schematic of a grout sleeve pipe (tube-á-manchette).

packers forces the grout past the rubber sleeve, through the weak grout, and into the surrounding ground. The use of sleeved pipes has an additional advantage in that specific horizons may be regrouted by repositioning the injection point.

12.3.3.4 Monitoring

As mentioned earlier in Section 12.3.2, real-time computer monitoring of pressures, volumes, and injection locations is now commonplace and has greatly improved efficiency and quality, as well as provided a good record for later review. In addition, control of mixes is critical, and periodic manual tests often are still performed to evaluate apparent viscosity (Marsh funnel test or ASTM D4016), specific gravity (Baroid mud balance), bleed (ASTM C940), cohesion, and other parameters important to quality assurance and quality control. Some non-ASTM test methods are provided by API Recommended Practice 13B-1 (1990).

12.4 APPLICATIONS OF GROUTING

Grouting can be used for a wide range of applications as mentioned throughout Section 12.3.1. But, as stated at the beginning of this chapter, the general objectives of grouting are to improve strength and stability, and to control and/or reduce seepage. This section will describe some typical applications that are used to achieve these goals, as well as a few case studies exemplifying the versatility of grouting.

12.4.1 Water Cutoff/Seepage Control

As described in Chapter 7, slurry walls are likely the most common type of cutoff wall used, particularly when a "positive" cutoff is required (such as for geoenvironmental applications). But grouting is also a commonly used (and generally less expensive) method for seepage remediation and preventative seepage. Grouting applications in the United States include dam foundations as early as the 1890s to the 1930s (Weaver and Bruce, 2007). Many of these early applications incurred problems or provided inadequate results, requiring additional remedial grouting. Weaver and Bruce (2007) reported that the first construction of a *grout curtain* in the United States was for the Estacada Dam in Oregon in 1912. Between the 1930s and 1980s, many seepage cutoffs and grout curtains were installed with varying degrees of success. Several other notable cases provided insight into the effectiveness of grout curtains. From these early experiences, which were often well documented, much was learned and implemented. Over years of practice, improvements in technology and a better understanding of the design parameters have improved, so that many positive success stories have now been reported.

Grouting for water cutoff may utilize a number of different grout methods, usually depending on project requirements, the subsurface materials, and geologic/hydrologic conditions. This may include intrusion, permeation, jet grouting, or fracture grouting. The applicability of these methods was outlined earlier. When cement grouts are used for water cutoff applications, they are often blended with bentonite or other clay material to aid in reducing permeability of the grouted mass. Sodium silicates and acrylate gels are some of the most utilized chemical grouts materials for hydraulic barriers, and provide "modest performance at modest cost" (Mitchell and Rumer, 1997).

For many years, intrusion, permeation, and fracture grouting have been used for preparing dam sites by tightening up fractured or permeable abutment materials and bedrock. Jet grouting, albeit somewhat more expensive, tends to provide a more uniform and more effective barrier, usually recommended with two to three overlapping rows. Remedial foundation grouting for seepage control of dams and levees has been a major use of grouting.

12.4.1.1 Case Studies

In Dearborn, MI, chemical grouting was used to preclude artesian inflow (including hydrocarbons, methane, and hydrogen gasses within the groundwater) into two 37 m (120 ft) diameter by 46 m (150 ft) deep sewer overflow shafts. Acrylamide permeation grouting was used in the contact soils, while a combination of acrylamide and traditional cement grout was used in the underlying bedrock. This was one of the largest acrylamide grouting projects ever undertaken in North America.

An example of high-profile, remedial dam foundation grouting is the Dworshak Dam, located east of Lewiston, Idaho. This is the third highest dam in the United States, where increased seepage flows exceeded 19,000 l/min (5000 gpm). Material also was being washed out, suggesting some erosional degradation. The solution was to reestablish (reconstruct) the grout curtain in the underlying weathered/fractured rock. In another case, grouting was employed to construct a remedial cutoff through 18 m (~60 ft) of embankment material plus an additional 18 m into underlying, highly fractured bedrock. For this application, 106,000 l (28,000 gal) of balanced-stable grout was injected into 409 grout holes (www.layne. com). For a detailed guide to design and other considerations for dam foundation grouting, refer to in-depth texts on the subject, such as Weaver and Bruce (2007).

12.4.1.2 Horizontal Seepage Barriers

Installation of interconnected, short jet grout columns at depth can provide a suitable hydraulic barrier and excavation base support in the form of a horizontal panel (Figure 12.12). When deep excavations or shafts are constructed well below the water table, a large hydrostatic pressure is exerted on the base as well as the sides of these openings. Compressive forces on the sidewalls of deep shafts may be easily handled by the arched shape of the shafts. Deep rectangular excavations may require additional reinforcement (i.e., tiebacks described in Chapter 15). The high fluid pressure on the bases of these excavations promotes seepage as well as stress. A notable case of a large, deep-shaft water cutoff by jet grouting was a 42 m (137 ft) diameter, 50 m (163 ft) deep excavated shaft for a sewer pump station in Portland, OR, where a jet grouted cutoff plug was installed to a 100 m (335 ft) depth (www.layne.com; Figure 12.13).

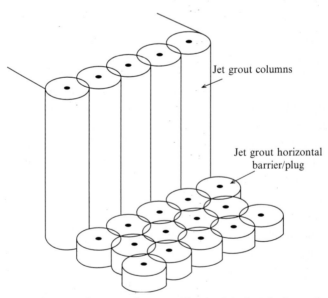

Jet grout columns

Jet grout horizontal barrier/plug

Figure 12.12 Illustration of jet grout horizontal barrier (plug) at the base of a jet grout supported excavation.

Figure 12.13 Deep shaft with jet grouted cut-off plug for sewer pump station in Portland, OR. *Courtesy Layne Christensen.*

12.4.2 Ground Support

Jet grouting has been used for a range of ground support applications, including earth retention, excavation base support (as described above), shallow foundation support, underpinning, scour protection (and remediation) around bridge piers, and stabilization for tunneling. Stabilization in this context refers to improvement with the general objective to keep soil in place. This may include applications for erosion resistance, and retaining caving or running sand during tunneling or excavating. For some tunneling cases, horizontal jet grouted elements have been used to form a strong supporting arch of treated soil to support tunneling beneath. Single point slurry or permeation grouting prior to excavation has also been used to support (and prevent seepage from) tunnel roofs.

12.4.2.1 Case Studies

Layne (www.layne.com) reported successful remedial slope stabilization of a poorly compacted highway fill along Rt. 243 East of Manassas, VA, by densification and shear resistance gained from grout columns, using 5-7 cm (2-3 in.) slump compaction grout.

Jet grouting performed with multiaxis machines was utilized to stabilize and support an 800 m (2600 ft) long excavation of a cut and cover for a Bay Area Rapid Transit (BART) subway station in Fremont, CA. Over 8000 jet grouted 2 m (7 ft) diameter columns were installed to depths of 20 m (65 ft), treating over (150,000 yd^3) of jet grouted soil for excavation support and base seal (www.layne.com).

As part of the $14.3 billion project to rebuild and strengthen the greater New Orleans levee system after devastating failures caused by Hurricane Katrina in 2005, jet-grouted columns were installed to strengthen the levees behind the floodwalls along the 17th Street Canal. This project involved installation of 76 cm (30 in.) thick, 6×12 m (20×40 ft) deep shear panels spaced at 3 m (10 ft) centers along the levee alignment with average 3500 kPa (500 psi) strengths, to provide stability against 100-year flood levels. Figure 12.10 shows the triple-axis, multidirectional jet equipment used to efficiently create the shear panels involving more than 77,000 m^3 (100,000 yd^3) of jet grouting.

12.4.3 Ground Strengthening, Displacement, and Void Filling

Chemical (permeation) grouting has long been known to add strength to granular soils by means of bonding grains together. The strength gain can be represented as an apparent cohesion. While the strength gain from chemical grouting may not be very large, at shallow depths or where confining stress is low, the increase in strength may be significant enough to prevent caving, sloughing, and/or raveling of loose granular materials.

Compaction grouting has become more common as a means of strengthening soft/loose ground by displacement densification and creation of relatively strong, cemented inclusions. Previously described in Section 6.1.5, compaction grouting has been used for increasing bearing capacity, reducing settlements, releveling floor slabs, and mitigating liquefaction potential.

A somewhat newer approach has been to increase capacity of deep foundations by compaction grouting. Installation of compaction grout columns adjacent to deep foundations exerts an increased lateral stress, which in turn provides significant enhancement of side resistance (Figure 12.14).

12.4.3.1 Case Studies

As described earlier, compaction grouting has been used for "leveling" or "jacking" of distressed slab construction or settled foundations. Figure 12.15 depicts a large-scale project where compaction grouting was used to remediate settlement of a 20,000 m^2 (215,000 ft^2) continuous 2.1 m (7 ft) thick floor slab of a dry dock at the Puget Sound Naval Shipyard. Carefully controlled compaction grouting raised the floor slab back to a level position where up to 13 cm (5 in.) of settlement had occurred.

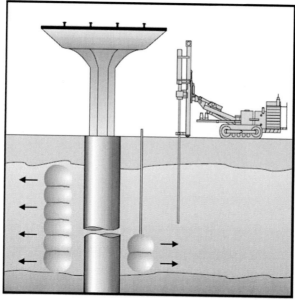

Figure 12.14 Compaction grout to improve deep foundation capacity. *Courtesy of Hayward Baker.*

Figure 12.15 Re-leveling of the distressed floor slab of Puget Sound Naval Shipyard dry dock. *Courtesy of Layne Christensen.*

12.4.3.2 Sinkhole Remediation

Compaction grouting has also become a solution for sinkhole remediation and prevention, as well as filling of abandoned mine shafts and other subsurface voids. Low-mobility grouts have been used to stabilize karstic materials prior to construction and to fill active sinkholes of all sizes (Figure 12.16). Figure 12.17 shows an application of low-mobility grout to seal the throat of a sinkhole measuring ~90 m (300 ft) in diameter.

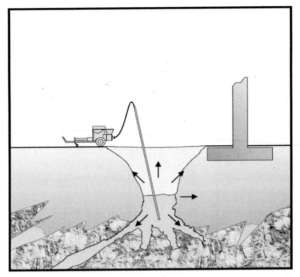

Figure 12.16 Schematic of compaction grouting to remediate sinkholes. *Courtesy of Hayward Baker.*

Figure 12.17 Compaction grouting to remediate large sinkholes. *Courtesy of Moretrench.*

12.4.4 Other Grouting Applications

12.4.4.1 Grouted Anchors, Nails, and Micropiles

As will be covered in Chapter 15, which outlines in situ reinforcement, conventional soil nails, ground anchors, and mini/micropiles are usually set with grout. Grout is also commonly used for sealing piezometers in boreholes to isolate them from infiltration, sealing sheetpile interlocks, and rehabilitation of sewer lines.

12.4.4.2 Pile Installation Assistance

As part of the fortification for the New Orleans Hurricane and Storm Damage Risk Reduction System (HSDRRS), a $1.5 billion, 2400 m (7800 ft) long storm surge barrier was constructed between the Inter Coastal Waterway (ICWW) and the Mississippi River Gulf Outlet (MRGO) to block wind- and storm-generated flooding such as occurred during Hurricane Katrina. The barrier was designed to consist of relatively large 1.7 m (5.5 ft diameter) cylindrical piles with 46 cm (18 in.) square precast concrete piles in between. The difficulty of placing the smaller square piles between and adjacent to the large-diameter piles was solved by inserting the square precast concrete piles into fluidized jet-grouted columns (Figure 12.18). This solution provided for gap closure and water seal between the large cylindrical piles. More than 30,000 m (100,000 ft) of jet-grout assisted piles were installed with more than 14,000 m^3 (18,300 yd^3) of grout to depths of 30 m (100 ft) (www.layne.com).

12.4.4.3 Pressure Grouted Piles

A twist on drilled shafts has been introduced by some contractors by using pressure grout to fill augered holes to as deep as 40 m (130 ft). These are known as *auger pressure grouted piles*. High-strength grout is pumped under pressure through the hollow shaft of continuous flight augers, producing concrete shafts with design capacities of over 180 metric tons (200 tons). Some of these deep foundation alternatives have been load tested to over 900 metric tons (1000 tons) (www.berkelandcompany.com). Over 3000 auger pressure grouted piles with up to 22 m (72 ft) lengths were used to construct the new San Francisco 49ers stadium in Santa Clara, CA.

RELEVANT ASTM STANDARDS

C143/C143M—12 Standard Test Method for Slump of Hydraulic-Cement Concrete, V4.02

C150/C150M—12 Standard Specification for Portland Cement, V4.01

C191—13 Standard Test Methods for Time of Setting of Hydraulic Cement by Vicat Needle, V4.01

C940—10a Standard Test Method for Expansion and Bleeding of Freshly Mixed Grouts for Preplaced-Aggregate Concrete in the Laboratory, V4.02

Figure 12.18 Jet grouted assisted pile installation and closure IHNC MRGO. *Courtesy Layne Christensen.*

C953—10 Standard Test Method for Time of Setting of Grouts for Preplaced-Aggregate Concrete in the Laboratory, V4.02

D4016—08 Standard Test Method for Viscosity of Chemical Grouts by Brookfield Viscometer (Laboratory Method), V4.08

Reference: ASTM Book of Standards, ASTM International, West Conshohocken, PA, www.astm.org.

REFERENCES

American Petroleum Institute (API), 1990. Recommended Practice Standard Procedure for Field Testing Water Based Drilling Fluids. Recommended Practice 13B-1, API, Washington, DC.

ASCE/G-I 53-10, 2010. Compaction Grouting Consensus Guide: ASCE/G-I 53-10. ASCE Publications, New York, 79 pp.

Bruce, D.A., Dreese, T.L., Heenan, D.M., 2008. Concrete walls and grout curtains in the twenty-first century: the concept of composite cut-offs for seepage control. In: USSD 2008 Conference, Portland, OR, 35 pp.

Burke, G.K., 2007. Vertical and horizontal groundwater barriers using jet grout panels and columns. In: Grouting for Ground Improvement. ASCE, New York, pp. 1–10, Geotechnical Special Publication 168.

DePaoli, B., Tornaghi, R., Bruce, D.A., 1989. Jet grout stabilization of peaty soils under a railway embankment in Italy. In: Foundation Engineering: Current Principles and Practice. ASCE, New York, pp. 272–290, Geotechnical Special Publication No. 22.

Hausmann, M.R., 1990. Engineering Principles of Ground Modification. McGraw-Hill Inc., New York, 632 pp.

Ivanetich, K., Gularte, F., Dees, B., 2000. Compaction grout: a case history of seismic retrofit. In: Advances in Grouting and Ground Modification. ASCE, Reston, VA, pp. 83–93, Geotechnical Special Publication 104.

Karol, R.H., 2003. Chemical Grouting and Soil Stabilization. Marcel Dekker, Inc., New York, 558 pp.

Koerner, R.M., 2005. Designing with Geosynthetics, fifth ed. Pearson Education Inc., Upper Saddle River, NJ, 796 pp.

Lombardi, G., 2003. Grouting of Rock Masses. ASCE, Reston, VA, Geotechnical Special Publication 120, pp. 164–197.

Mitchell, J.K., 1981. Soil Improvement State-of-the Art. In: Proceedings of the 10th International Conference on Soil Mechanics and Foundation Engineering, ICSMFE, Vol. 4, pp. 509–565.

Mitchell, J.K., Rumer, R.R., 1997. Waste Containment Barriers: Evaluation of the Technology. In: In Situ Remediation of the Geoenvironment. ASCE, Reston, VA, pp. 1–25, Geotechnical Special Publication 71.

Weaver, K.D., Bruce, D.A., 2007. Dam Foundation Grouting. ASCE Publications, Reston, VA, 473 pp.

http://www.berkelandcompany.com (accessed 01/03/14).

http://www.fhwa.dot.gov/bridge/tunnel/qa.cfm (accessed 30/11/13).

http://www.haywardbaker.com (accessed 18/01/14).

http://www.layne.com (accessed 15/11/13).

http://www.layne.com/en/projects/ (accessed 04/1213).

http://www.moretrench.com (accessed 18/01/14).

http://www.nicholsonconstruction.com (accessed 13/12/13).

http://www.soilfreeze.com (accessed 09/12/13).

http://en.wikipedia.org/wiki/Rheology (accessed 04/12/13).

CHAPTER 13

Thermal Treatments

This chapter provides an overview of soil stabilization methods with thermal treatments. Thermal treatment refers to the modification and/or stabilization of soils by application of (1) heat (typically by way of combustion of fossil fuels) for improving properties of clayey soils and (2) artificial ground freezing for the temporary treatment and stabilization of soils and fractured rock. These two approaches are obviously very different in many respects and will, therefore, be addressed separately herein.

13.1 TYPES OF THERMAL TREATMENTS

Heat treatment has been utilized for soil stabilization for many years. It includes burning petroleum products directly in soil borings and surface heating from the close proximity burners of traveling heaters. In general, heating is an effective method of soil treatment for fine-grained (clayey) soils only. The high temperatures cause permanent physical reactions in the clay minerals, as well as a drying effect by evaporation of water. The increased costs and environmental concerns of using petroleum products have rendered many of these types of processes extinct, although, recently, heating has made a comeback for limited applications in the remediation of contaminated soils. Heating the soil at a moderate temperature assists the vapor extraction of volatile organic compounds. Soil vapor extraction performance can be enhanced or improved by injecting heated air or steam into the contaminated soil through the injection wells. Heating the soil to extremely high temperature is the in situ *vitrification* by which electrical current is used to heat and melt the soil in place (Terashi and Juran, 2000). The technique is effective for soils contaminated with organic, inorganic, and radioactive compounds. Heating, or more properly "firing" of clays to make bricks could also be considered a soil heat treatment.

Ground freezing is a technique that provides a temporary increase of strength and shut off of water seepage. It is used around open cuts and excavations, small and large diameter shaft excavations, underpinning of existing structures, and tunneling. Often, it may be the best choice for sinking deep excavations and shafts that extend below the water table. It has also found a

319

significant environmental purpose in handling and/or containing contaminated ground or wastes, arresting landslides, and stabilizing underground collapses in emergency situations.

Another method, called *active freezing*, is used in northern latitudes to maintain frozen ground in permafrost zones beneath heated structures where passive systems using insulation alone are not considered adequate to maintain a frozen state. Thawing of permafrost beneath heated buildings may result in unwanted settlements and/or loss in bearing strength. Mageau and Nixon (2004) describe this type of system, which utilizes natural and forced ventilation through air ducts and ventilated granular pads to remove heat from beneath structural foundations. Active freezing has also been used to aid in maintaining frozen conditions for artificially frozen ground for stabilizing soil or for water cutoff.

13.2 HEAT CAPACITY OF SOILS

In order to better understand the mechanics of thermal treatments, one needs to understand the basics of heat energy. For example, ground can be artificially frozen when heat energy is removed. Here we should also review and/or define some terminology and units.

Heat energy is transferred as a result of a temperature difference, where energy is transmitted from a body with higher temperature to one with lower temperature. Heat energy is commonly defined in units of joules (J) or calories (cal). A calorie is defined as the amount of heat required to change the temperature of one gram of liquid by 1 °C. The Joule is the official SI unit of heat energy:

$$1\,cal = 4.184\,J$$

The transfer of energy due to temperature difference alone is called *heat flow*. The SI unit of power, the watt, is used for reference to heat flow. A watt is defined as 1 J/s.

In reality, the total heat energy of a fluid is dependent on both temperature and pressure. This total energy is called *specific enthalpy*, and refers to the total energy of a unit mass. The unit of measure most commonly used is kJ/kg.

Specific heat is the amount of heat required to change the temperature of 1 kg of a substance by 1°. The units of measure would then be kJ/kg K (K = kelvins). *Heat capacity* can then be defined as the heat required to change the temperature of a whole system by 1°.

The heat capacity characteristics of soils, water, ice, and other earth materials must be carefully evaluated for each specific application in order to properly design and monitor successful applications.

13.3 HEAT TREATMENT OF SOILS

Heat treatment has been utilized as a method of ground modification by improving engineering properties of fine-grained soils. Heat can affect clay chemistry and has the ability to alter clay mineralogy through *diagenesis*, allowing for improved engineering properties of these materials. Granular soils are generally unaffected by the application of heat at temperatures less than 1000 °C, with the exception of drying, which has little effect on engineering properties of these soil types.

Heat treatment of clayey soils results in permanent, irreversible changes as a consequence of both the drying effect and changes in the actual mineral structure of these soils. A number of significant improvements can be made by utilizing heat treatments, although examples described in the literature indicate that the energy and associated fuel consumption is relatively high. Some efforts were made to apply heat treatment to stabilize clay slopes in the former Soviet Union (Turner and Schuster, 1996). Because of the cost of and conscious awareness toward reducing consumption of nonrenewable energy sources and concerns of pollutants, the current and future use of heat treatment for soil modification is likely to be restricted to (biological) control and treatment of contaminated soils. One area in which heat treatment may still be viable is in the production of *Ferroclay* building blocks. These may range from earth/mud bricks, still utilized in third-world construction, to fully fired bricks (Hausmann, 1990).

13.3.1 Improvements and Applications of Ground Heating

Generally, improvement of engineering properties of clayey soils occurs with an application of at least 400 °C. Improvements, including decreased compressibility, reduced plasticity, reduced swelling potential, lower optimum moisture content, and increased strength, have been detailed in the literature (Abu-Zreig et al., 2001). Case studies have shown strength increases of up to 10-20 times. Heat treatment has been applied to soil through a variety of techniques, including combustion of fuel in boreholes, surface treatments by traveling "burners" in close proximity to the ground surface, and through "baking or firing" of clay blocks (forming a range of construction elements from crude mud blocks to conventional bricks, as described above).

In situ improvement at depth has been successful only where there is a source of relatively low-cost fuels. As a result, this approach has all but disappeared, given the rise in fuel costs and other environmental considerations. Surface treatment by means of traveling heaters can successfully treat to a limited depth of existing, in situ surface soils or layers of engineered fill. One note of caution is to beware of possible ground movement resulting from expansion of water followed by consolidation upon cooling.

13.4 GROUND FREEZING

The principle of ground freezing is that when the moisture (pore water) in the soil freezes, the soil particles are bound together, creating a rigid structure with considerable strength and stiffness. Ground is artificially frozen when heat energy is removed from it. This is accomplished by introducing a lower temperature medium that causes a flow of heat energy from higher to lower temperature, thereby reducing the heat (cooling the soil). Understanding the relatively simple mechanics involved points to the fact that, unlike heat treatments, artificial freezing may be applicable to a wide range of soil types, grain sizes, and ground conditions. Fundamentally, the only requirement is that the ground has sufficient soil moisture (pore water). Ground freezing and associated improvements and/or stabilization is possible only if continuous artificial cooling is maintained. It is, therefore, of critical importance to understand that ground freezing is always *only* a temporary stabilization technique. As a result, consideration should be given to back-up systems as a part of initial planning and design. However, once ground is frozen, some time will be needed for it to thaw, so relatively short power interruptions are not necessarily critical.

The first reported use of ground freezing was in South Wales in 1862 in conjunction with a mine shaft excavation (Schaefer et al., 1997). The strength of frozen soil may be on the order of 1–10 MPa, although it depends on a variety of factors, such as soil type, water content, rate of freezing, and maintained temperature of the frozen soil. An important attribute is that frozen soil becomes a nearly impermeable material. The technique is currently used for the temporal increase of strength and temporal shut off of water seepage around open cuts, shaft excavations, and tunneling. There have been a number of specialty symposia on ground freezing that provide an overview of applications, including the International Symposium on

Ground Freezing that has been held periodically since 1978. In addition, the increasing number of specialty contractors providing ground freezing services has provided even more available literature and case studies.

13.4.1 Improvements and Applications of Ground Freezing

The fundamentals of ground freezing have been known and used since the 1880s for the mining industry. The principle improvements of freezing the ground are typically either strengthening or stabilizing the ground, controlling seepage, or a combination of both. Frozen ground can have increased shear strengths of up to 20 times that of unfrozen soil (or nearly twice that of concrete) by combining the inherent soil shear strength with that of ice. Seepage is controlled by the formation of a frozen barrier of the pore water acting as an effective cutoff if sufficient pore water is available. One caution and/or concern is the disruption of soil structure and associated volume change due to expansion of the pore fluid upon freezing. Another issue has been with deformations and loss of soil strength upon thawing of the frozen soil mass.

Because successful ground freezing fundamentally relies only on there being enough moisture in the ground, it is applicable to virtually all earth materials, making this method more versatile for temporary water cutoff than many others. Figure 13.1 demonstrates the range of applicability of freezing compared to other common cutoff methods.

Ground freezing has been successfully used for temporary construction elements (e.g., excavations (see Figure 13.2), cofferdams, underpinning of existing structures, stabilization for tunneling, etc.), incipient or active slope failure stabilization, containment (or exclusion) of contaminated groundwater, hazardous wastes and toxic "spills," undisturbed sampling of cohesionless soils, and so forth. At the same time, frozen ground provides a hydraulic barrier for temporary seepage control of construction dewatering applications. As such, freezing eliminates the need for costly construction of both structural shoring systems and dewatering (hydraulic barrier) systems. In addition, freezing can provide a hard, durable working surface even in soft and/or wet soils. Figure 13.3 shows a freezing project for excavation of a deep shaft.

Where accessibility, space limitation, and "sensitive" infrastructure exist, ground freezing has been demonstrated as a workable solution. Examples of this are excavations adjacent or in close proximity to historic structures.

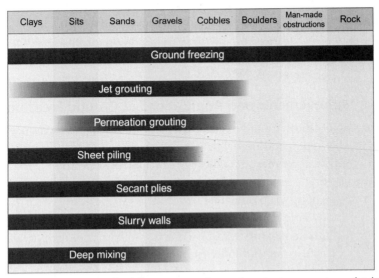

Clays	Sits	Sands	Gravels	Cobbles	Boulders	Man-made obstructions	Rock

Ground freezing

Jet grouting

Permeation grouting

Sheet piling

Secant plies

Slurry walls

Deep mixing

Figure 13.1 Freezing applicability compared to other improvement methods for ground support. *Courtesy of Moretrench.*

13.4.2 Ground Freezing Techniques

Freezing is typically induced by insertion of equally spaced pipes circulating supercooled brine (often < -25 to $-30\,°C$) or, more expensive but much quicker, by injection of liquid nitrogen (LN_2), which boils at $-196\,°C$. In the case of using brine, the solution is circulated down a central tube and back up through the annulus to extract heat from the surrounding soil (Figure 13.4). A strong saline (usually calcium chloride) solution has a much lower freezing point that that of typical pore water and will therefore remain fluid even at temperatures as low as $-35\,°C$. The pipes are usually placed in a row or "line" to provide a continuous wall or temporary "structural" element to support higher loads and/or provide a hydraulic barrier for groundwater cutoff (Figure 13.5). *Freezewall* is a term sometimes used to describe a continuous wall of frozen soil columns. As previously indicated, this type of "frozen wall" construction can be utilized to provide both wall support and a hydraulic barrier. In some cases, freezing of a larger mass of soil may be desired to temporarily stabilize unsafe or unstable conditions. This turns the ground into the consistency of soft rock, which can then be excavated, drilled, or tunneled through by conventional or more modern techniques.

Using liquid nitrogen for ground freezing is more costly due to the expense of the nitrogen (which is expended and, therefore, must be regularly replenished to maintain freezing), but due to the extremely low

Figure 13.2 Frozen ground for excavation shoring. *Courtesy of SoilFreeze.*

Figure 13.3 Freezing around the periphery of a deep-shaft construction. *Courtesy of Moretrench.*

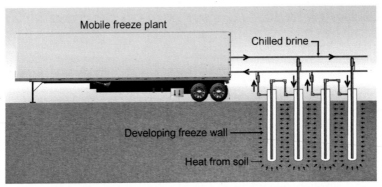

Figure 13.4 Schematic example of freezing by circulating super cooled brine. *Courtesy of Moretrench.*

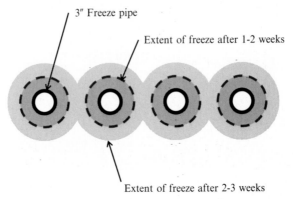

Figure 13.5 Example of how the frozen zone surrounding freeze pipes eventually joins to form a continuous strong, impermeable "wall." *Courtesy of SoilFreeze.*

temperatures generated ($-196\,°C$ or $-320\,°F$), freezing will be very rapid. In addition, the necessary cooling equipment is substantially less involved and, therefore, less costly than a brine cooling unit, and may not require a locally available power supply. Liquid nitrogen is also nonflammable and nontoxic, and it can be easily transported in tanks. These attributes make freezing with liquid nitrogen advantageous for emergency stabilization at remote sites. The liquid gas is pumped directly into copper freeze pipes installed in (or in emergencies, driven into) the ground, which immediately freezes adjacent surrounding ground as the liquid nitrogen vaporizes (Figures 13.6 and 13.7). The vaporized cold nitrogen (i.e., exhaust gas) further extracts heat as it flows back out of the ground (www.lindeus.com).

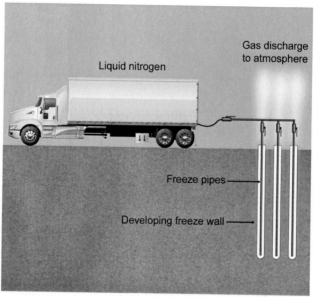

Figure 13.6 Schematic example of freezing by injecting liquid nitrogen. *Courtesy of Moretrench.*

Figure 13.7 Application of freezing by injecting liquid nitrogen. *Courtesy of SoilFreeze.*

This process may be practical for small, short-term projects and/or for emergency stabilization.

To reliably use and/or design freezing technology requires an understanding of the thermal, mechanical, and hydraulic properties of the frozen soil, the equipment used to freeze the ground, and the assessment of the construction process. Inherent in this approach is a fundamental understanding of the basic physics of thermal conductivity, heat capacity, and energy requirements to move between solid, liquid, and gaseous phases of the materials being treated. In reality, energy losses in practical applications result in a somewhat higher consumption than would be predicted by the basic physics alone. A number of variables that will affect the effectiveness of a freezing application include soil type (mineralogy and thermal properties), water content, velocity and inherent temperature of the ground water, and rate of freezing.

The design of a frozen earth barrier (or support) is governed by the heat capacity characteristics of soils, water, ice, and other earth materials, as described in Section 13.2. Formation of the frozen earth mass surrounding each pipe will depend on the thermal and hydraulic properties of each stratum being treated. Obtainable strengths will depend on soil type, moisture content, and temperature. For instance, when soft, relatively weak clay is cooled to below freezing, some portion of its pore water begins to freeze, and the clay begins to stiffen and strengthen. When the temperature is reduced further, more of the pore water freezes, and the soil strength can be dramatically increased. In contrast, a sandy soil with relatively higher initial strengths may be adequately frozen with substantial strength gains with less temperature differentials required. For example, a temperature of $-5\ ^\circ\mathrm{C}$ may be adequate for freezing granular soils, while temperatures as low as $-25\ ^\circ\mathrm{C}$ may be required for some fine-grained soils. The rate of freezing is also dependent on thermal and hydraulic properties of the soils materials involved. Fractured rock and coarse-grained soils will typically freeze significantly faster than finer-grained silts and clays with lower hydraulic conductivities. Freeze pipe spacing, freezing radius around pipes, and freezing times can be estimated by relatively straightforward computations, such as those provided by Harlan and Nixon (1978). But today most ground freezing projects are designed and analyzed by computer programs such as TEMP/W (www.geo-slope.com) because they can also provide detailed 2-D analyses (Figure 13.8).

Due to the number of specialty geotechnical contractors who have gained sufficient experience in thermal ground treatment (particularly artificial ground freezing), economical solutions are becoming much more common for a wide variety of applications. The increased number of contractors engaging in ground freezing has also generated some healthy competition for these services.

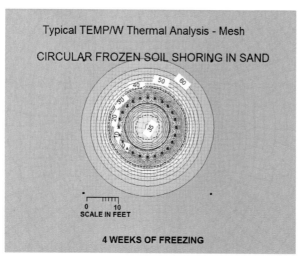

Figure 13.8 Typical ground freezing thermal analysis (TEMP/W). *Courtesy of SoilFreeze.*

A number of potential difficulties may be encountered in a ground-freezing project. Once the ground is frozen and earth excavated, the exposed frozen soil surface may be subject to thaw. In order to accommodate this, polyurethane foam or plastic sheeting may be applied to insulate the frozen soil and keep the earth frozen at the excavation site.

As previously mentioned, one consideration is to be aware and protect against possible deformations and ground movement. The potential for this comes both during the freezing, when there is potential for heave due to the expansion of the water in the ground as it freezes, and during the thawing process, when there may be settlement or slumping as the frozen ground contracts. Careful planning, along with detailed analyses and monitoring with contingency plans in place, is critical to ensure a successful project without disruption to existing infrastructure. Ground movements must be closely monitored throughout the duration of the project, and even for a period of time after completion.

Another attribute that must be closely monitored is continuity of a frozen mass when it is designed as a single structural component and/or if it is designed as a water cutoff. Detailed analyses using the as-built, installed freeze installation must be carefully compared to temperature monitoring to ensure that there are no "holes" or "windows" in the frozen barrier. If measurements indicate any deficiencies, they can be addressed prior to beginning construction by adding additional freeze pipes, providing extra time for freeze fronts to develop, or by taking other remedial measures.

13.4.3 Example Case Studies of Ground Freezing

Ground freezing has been successfully used as cost-effective solutions for thousands of projects worldwide, including: the Grand Coulee Dam in Washington State, The Chunnel connecting England and France, and on a number of components of Boston's "Big Dig." Ground freezing used for tunnel construction as part of reconstruction of Boston's Central Artery (a.k.a. the Big Dig) was at the time the largest, frozen earth-retaining application in history.

Newer technologies have now made ground freezing competitively priced while offering distinct advantages over other soil improvement options. Originally developed for use in mining, ground freezing continues to be successfully implemented in very large mining operations. Access shafts have been constructed to more than 650 m (2000 ft) deep, far below water table levels, using only freezing to both stabilize the excavation as well as provide a water cutoff (www.soilfreeze.com). In cases where underground construction in congested urban environments is planned, ground freezing has provided a workable solution when all other methods have been ruled out.

Because rapid stabilization can be attained by freezing using liquid nitrogen, this technique has been used for a number of "emergency" stabilization projects. The accurate control available from freezing has also made it a good choice when stabilizing soil masses adjacent to or below sensitive historical structures (Figure 13.9).

In another class of applications, ground freezing has been used for "rescue" of unmanned tunnel boring machines (TBMs) where equipment has either broken down or become "stuck" at great depths underground, sometimes requiring work in close proximity to existing infrastructure or residential neighborhoods.

A number of examples are described here to demonstrate the successful use of ground-freezing applications for a variety of interesting and sizeable projects where other alternatives were unfeasible.

13.4.3.1 Shafts and Tunnels

Ground freezing provides a strong, watertight shoring system ideal for construction of shafts and tunnels in soft or loose soils, particularly below the water table. It can also work in irregular ground that contains boulders or rock interfaces, where conventional sheet piling or soil mixing would not be accommodated.

Figure 13.9 Freezing for excavation adjacent to sensitive historic structure. *Courtesy of SoilFreeze.*

Design includes the spacing of freeze pipes and the thickness of frozen columns (there may be horizontal freeze pipes for tunneling support). As excavations are made deep below the water table, hydrostatic pressures on the walls and base may increase to critical levels. A critical design element then becomes how deep to install the freeze pipes (beyond the excavation depth) to create a frozen "plug" below the base of the excavation. This must be properly designed to ensure excavation stability. Here are some examples.

- South Bay Ocean Outfall, San Diego, CA

A good example of an exceptional frozen soil, deep-shaft construction project is the ocean outfall drop shaft in San Diego. Ground freezing was determined to be the only feasible method to create a stable, watertight, 14 m (46 ft) diameter shaft, over 60 m (200 ft) deep, excavated into sandy, unstable soils (www.layne. com) (Figure 13.10). In order to provide stability of the bottom (as well as the sides) of the shaft to very high earth and hydrostatic forces, a 3.7 m (12 ft) thick, frozen–earth wall was constructed to a depth of over 90 m (300 ft). Specialized drilling methods were required to ensure a high degree of vertical alignment through large zones of boulders and irregular soil conditions.

- Transit Tunnel, Boston, MA

Figure 13.10 60 m (200 ft) deep, 18 m (60 ft) diameter shaft enabled by 90 m (300 ft) deep freezing, San Diego. *Courtesy of Layne Christensen.*

An example of using freezing for tunneling in a "sensitive" environment involved the temporary excavation support and groundwater cutoff for the construction of a dual transit tunnel beneath historic buildings supported on timber piles. The tunnel was constructed through hundreds of the existing piles without any disturbance to the fully occupied buildings above (www.laynegeo.com).

• Northern Boulevard Crossing, New York City (Queens), NY

In this case, a 40 m (130 ft) section of tunnel was to be constructed beneath heavily traveled Northern Boulevard, an active subway structure. The site involved the presence of vehicular and rail traffic as well as thousands of pedestrians daily. To complicate the matter, the tunnel was to cut through existing pile foundations supporting the elevated structure above. All existing transit components were to remain active during tunnel construction.

Figure 13.11 Schematic design and photograph of horizontal freeze pipe array for support of tunnel section in a congested urban setting. *Courtesy of Moretrench.*

The solution was to use ground freezing with horizontal freeze pipes to support the excavation (Figures 13.11 and 13.12). Due to the sensitivity of the surrounding infrastructure, special care had to be taken to accurately monitor temperatures and deformations both during freezing and thawing after completion of the tunnel installation. A discussion of this project and the array of construction details and contingencies is contained in Schmall and Sopko (2014).

13.4.3.2 Mining
As mentioned previously in this chapter, ground freezing has been used in the mining industry for many years and continues to be utilized for both large and small mining applications today. A few examples of large-scale mining applications are presented here.
• Verglas Crown Pillar, Quebec, Canada

Figure 13.12 View of the complex conditions for tunneling through frozen ground beneath urban infrastructure. *Courtesy of Moretrench.*

To date, the largest structural excavation enabled by ground freezing was for access to an existing zinc and nickel mine in Quebec, Canada. The shaft was 72 m (220 ft) in diameter, and 41 m (125 ft) in depth through a high water content and very low strength (sensitive) clay. The excavation support consisted of a 12 m (35 ft) thick frozen "wall" made by using three rows of freeze pipes (www.laynegeo.com) (Figure 13.13).

• Aquarius Gold Mine, Timmins, Ontario

In order to provide a barrier around the perimeter of the Aquarius Mine to allow for dewatering, 4 km (2.5 mi) of barrier wall was constructed to depths of 38-145 m (125-480 ft). This project involved ~2500 freeze pipes and would use over 4500 metric tons (5000 tons) of refrigeration (www.laynegeo.com) (Figure 13.14). As the price of gold fell and ownership of the mine changed a few times, the actual freezing was deferred (www. blogs.wsj.com).

Figure 13.13 Shaft excavation in sensitive clay at Verglas mining project, Quebec, Canada. *Courtesy of Layne Christensen.*

13.4.3.3 TBM Rescues

As the use of TBMs and micro-tunnel boring machines (MTBM) increases with advancements in tunneling technology, contractors are more and more willing to tunnel in difficult ground conditions. As a result, there have been a

Figure 13.14 Ground freezing for (2.5 mi) perimeter barrier around the Aquarius Gold Mine, Timmins, Ontario, Canada. *Courtesy of Layne Christensen.*

number of cases of tunneling machines becoming stuck or breaking down. In order to facilitate the repair and/or "rescue" of the tunneling machines and get the projects completed, ground freezing has been utilized as a method to access often very difficult locations at significant depths and/or beneath existing development and infrastructure. A few case studies of TBM rescues are reported here.

- TBM Rescue, Renton, WA

A 3-foot diameter TBM became inoperable after tunneling under the Cedar River at a depth of ~25 ft in close proximity to the Renton Airport runway, and needed to be recovered and repaired to continue tunneling. Severe restrictions and limitations challenging this operation included equipment heights and operating times, among others. Ground freezing was successfully used to construct a stable, dry, rescue shaft to retrieve the TBM. The shoring wall was rapidly completed in two weeks. A 14-foot diameter precast manhole was then placed within the frozen shaft for future access by the repaired TBM (www.SoilFreeze.com). Figure 13.15 shows the freezing plan for this operation.

- Brightwater Conveyance System Tunnel 3, Lake Forest Park, King County, WA

In another TBM rescue operation, a machine had become disabled at a depth of about 100 m (300 ft) underground with more than 5 bars of hydrostatic pressure. Given the depth and pressure, and the fact that the stuck

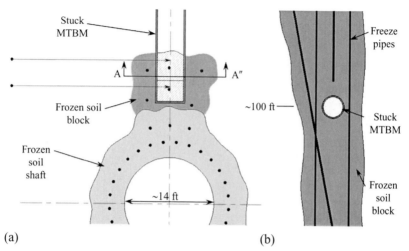

Figure 13.15 Ground freezing for tunnel boring machine (TBM) rescue. (a) Plan view. (b) Elevation view section A-A'. *Courtesy of SoilFreeze.*

TBM was located beneath a residential neighborhood, it was not feasible to access it from the ground surface. The successful solution was to complete the tunnel from the opposite direction with a second machine. At the connection point, the ground was frozen around the disabled machine by installing freeze pipes from the ground surface, supplemented by freeze pipes drilled from within the tunnel (Figure 13.16). This allowed workers to access and disassemble the stuck TBM, finally completing the tunnel (Gwildis et al., 2012).

13.4.3.4 Containment of Hazardous Contaminants

Ground freezing has also been proposed for containment and control of environmental contaminants. The Department of Energy (DOE) named ground-freezing technology among its top choices for containment of radioactive wastes. DOE considered using soil freezing at the 100-N Springs site at Hanford, WA, and several other locations, including the Savannah River, and nuclear sites in South Carolina and Idaho. Freezing was tested at several contaminated site including Oak Ridge, TN, and Hanford, but was eventually abandoned at Hanford for technical reasons (www.blogs. wsj.com).

• Fukashima Nuclear Power Plant, Japan

With continued leaking of radioactive contaminated groundwater into the Pacific, in August 2012 the Japanese government announced a plan to use

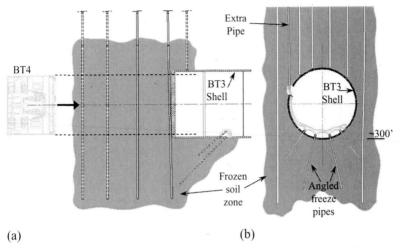

(a)   (b)

Figure 13.16 Ground freezing plan for deep TBM "rescue". (a) Side view of frozen soil. (b) End view of frozen soil block. *Courtesy of SoilFreeze.*

ground freezing to create a cutoff wall to a depth of 100 ft to prevent groundwater from entering the facility and becoming contaminated (www.geoprac.net). The anticipated cost of the wall will exceed $300 million (www.pbs.org). As of late 2013, this proposal was still being considered (www.blogs.wsj.com).

REFERENCES

Abu-Zreig, M., Rudra, R.P., Whiteley, H.R., 2001. Validation of a vegetated filter strip model (VFSMOD), Hydrological Processes. John Wiley & Sons, 15 (5), pp. 729–742.

Andersland, O.B., Landanyi, B., 2004. Frozen Ground Engineering, second ed. John Wiley & Sons, Hoboken, NJ.

Gwildis, U.G., Cochran, J., Clare, J.B., Mageau, D., Hauser, G.M., 2012. Use of ground freezing for connecting two tunnel boring machine-driven tunnels 300 feet underground. In: Proceedings, North American Tunneling 2012 Conference. Society for Mining, Metallurgy, and Exploration, Littleton, CO, pp. 753–762.

Harlan, R., Nixon, J., 1978. Ground thermal regime. In: Andersland, O.B., Andersen, D.M. (Eds.), Geotechnical Engineering for Cold Regions. McGraw-Hill, New York.

Hausmann, M.R., 1990. Engineering Principles of Ground Modification. McGraw-Hill, New York, 632 pp.

Mageau, D.W., Nixon, D.F., 2004. Active freezing systems. In: Esch, D.C. (Ed.), Thermal Analysis, Construction, and Monitoring Methods for Frozen Ground. ASCE Technical Council on Cold Regions Engineering Monograph. ASCE, Reston, VA, pp. 193–237, Chapter 6.

Schmall, P.C., Sopko, J.A., 2014. Ground freezing in a congested urban area. Geo-Strata. May/June, ASCE.

Schaefer, V.R., Abramson, L.W., Drumheller, J.C., Sharp, K.D. (Eds.), 1997. Ground Improvement, Ground Reinforcement and Ground Treatment: Developments 1987–1997. ASCE, New York, NY, Geotechnical Special Publication 69.

Turner, A.K., Schuster, R.L., 1996. Landslides: investigation and mitigation. In: Transportation Research Board Special Report 247. National Academy Press, Washington, DC, 673 pp.

Terashi, M., Juran, I., 2000. Ground improvement: state of the art. In: Proceedings of GeoEng 2000, Melbourne, Australia, 18 pp.

http://blogs.wsj.com/japanrealtime/2013/08/29/fukushima (accessed 25.03.14).

http://courses.washington.edu/cm420/Lecture12.pdf (accessed 02.12.13).

http://www.djc.com/special/design98/10047020.html (accessed 02.12.13).

http://www.foam-tech.com/case_studies/big_dig.htm (accessed 12.12.13).

http://www.geo-slope.com (accessed 15/12/13).

http://groundfreezing.net/ground-freezing-faq/ (accessed 12.2.13).

http://www.geoprac.net/geonews-mainmenu-63/38-failures/1416-ground-freezing-proposed-to-stop-radioactive-groundwater-at-fukushima-nuclear-plant (accessed 25.03.14).

http://www.lindeus.com/en/processes/water_and_soil_treatment/ground_freezing/index.html (accessed 2.12.13).

http://news.nationalgeographic.com/news/energy/2013/08/130819-japan-ice-wall-for-fukushima-radioactive-leaks/ (accessed 9.12.13).

http://www.soilfreeze.com/ (accessed 2.12.13).

http://www.technologyreview.com/news/518801/how-the-fukushima-ice-barrier-will-block-radioactive-groundwater/ (accessed 9.12.13).

http://www.pbs.org/wgbh/nova/next/tech/artificial-ground-freezing/ (accessed 2.12.13).

Modification with Inclusions and Confinement

(Photo courtesy of The Collin Group)

CHAPTER 14

Geosynthetic Reinforced Soil

This chapter provides an overview of earthwork construction where reinforcing materials (predominantly geosynthetics) are used to provide added strength and capacity to engineered fill, stability for embankments over soft ground, stability for steepened slopes, and for resistance to erosion and other deterioration. This type of soil reinforcement has become very popular and has continued to grow in use for a wide variety of applications. For many years, retaining structures were typically made of reinforced concrete and designed as gravity or cantilevered walls, which are rigid and cannot accommodate significant deformations. Earth walls stabilized with geosynthetic inclusions provide a cost-effective and technically sound alternative. Reinforced soil slopes offer a solution to myriad slope stability issues that have historically caused tremendous losses and/or expensive "fixes" and/ or redesigns. Other types of soil reinforcement include use of structural inclusions placed in situ and soil confinement mechanisms. These other methodologies are described in later chapters.

14.1 HISTORY, FUNDAMENTALS, AND MATERIALS FOR SOIL REINFORCEMENT

14.1.1 History of Soil Reinforcement

In ancient times soil reinforcement consisted of mixing straw with mud, reinforcing with woven reeds, and using branches and other plant material to improve strength and capacity to support greater loads. Modern soil reinforcement uses stronger and more durable materials, but employs many of the same fundamental mechanisms that provided strength in these early applications.

Early versions of "modern" soil reinforcement were developed in the early 1960s with Henri Vidal's patented Reinforced Earth® for construction of self-supporting retaining walls. These walls were constructed using galvanized steel strips with "ribs" to provide lateral resistance against earth pressures (Figures 14.1 and 14.2). These types of wall (and similarly slope) structures are generically referred to as *mechanically stabilized earth* (MSE). Construction of earth walls with geosynthetic reinforcing materials was

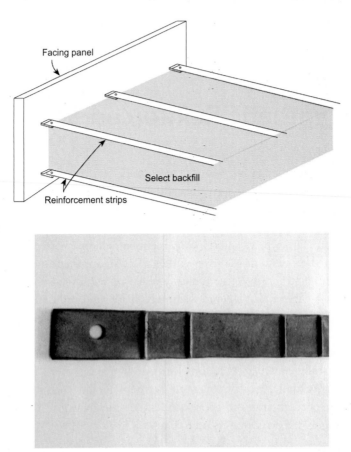

Figure 14.1 Representation of steel strip reinforced wall and detail of ribbed galvanized strip.

introduced in the 1980s (Federal Highway Administration, 2011). Since that time there has been an explosion of the use of geosynthetic reinforcement for soil structures as well as for many other geotechnical applications.

14.1.2 Soil Reinforcement Materials

A number of different geosynthetic materials have and continue to be used for soil reinforcement. As described in a previous section, early versions of MSE wall used steel strips as reinforcement. Many walls were constructed in this way and are still in service today with a generally good track record. Some issues and concern with the use of metallic reinforcement have arisen due to corrosion of these elements. As a result, use of metallic reinforcing members has been replaced by polymeric, geosynthetic materials for some applications. Corrosion of metallic inclusions is dependent on a number of factors, including salt and oxygen content in the ground, degree of

Figure 14.2 Construction of a metallic strip reinforced wall. *Courtesy of The Reinforced Earth Company.*

saturation, acidity, and sulfate content, among others. Corrosion rates may be predicted with some accuracy and some corrosion allowance is usually part of design. In addition, modern measurement techniques allow for in situ testing of metal reinforcement "condition," allowing "health" monitoring of structures built with these types of inclusions. The use of steel reinforcement is still widely practiced due to the high strength of these reinforcing members. A variety of steel reinforcement types include the discrete steel strips previously described, and welded wire bar and mat arrangements (Figure 14.3).

Geotextiles have long been recognized for their ability to reinforce engineered fill constructed as walls or slopes. They are also used to distribute loads beneath embankments and roadways over soft subgrade soils to reduce settlements and lateral deformations. Geotextiles have the additional advantage of providing a separation function keeping dissimilar material from mixing, such as when aggregate or base coarse is placed over a fine-grained subgrade. They may also function as filters depending on the application, as discussed in Chapter 8. Geotextiles used for reinforcement are usually woven, and are available in a wide range of weights, thicknesses, modulus, and strengths. For applications where small deformations (strains) are of

Figure 14.3 Welded wire bars and mats used for MSE wall reinforcement. *Courtesy of The Reinforced Earth Company.*

concern or should be monitored, some fabrics, such as Tencate's GeoDetect®, incorporate fiber optic inclusions that can measure strains as low as 0.02% (www.tencate.com).

For applications where higher reinforcement strength is required, polymeric *geogrids* may be utilized. The primary function of geogrids is clearly reinforcement. The open *apertures* of geogrids (relatively large openings between ribs) allows interconnectivity of the soil above and below, and therefore provides additional passive resistance along the sides of the transverse ribs. A full explanation of the reinforcing mechanisms of geogrids will be provided in the next section. Geogrids come in a variety of types described below. There are three fundamentally distinct categories based on the manufactured geometry of the geogrid:

- *Uniaxial geogrids* (Figure 14.4), are typically manufactured from a sheet of high-density polyethylene (HDPE) that has been punched and drawn in one direction. This unidirectional draw provides high tensile strength with minimum elongation in one direction. Uniaxial geogrids are ideal for applications where the stresses (loads) are primarily oriented in one direction, for example walls and embankment slopes.

- *Biaxial geogrids* (Figure 14.5) are commonly punched HDPE sheets drawn in two directions and so provide good reinforcement in orthogonal (or random) directions. While they may have a somewhat lower ultimate

Figure 14.4 Uniaxial geogrid and stabilized wall construction. *Courtesy of Tensar International Corp.*

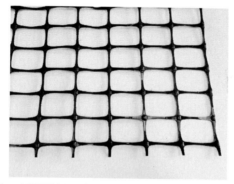

Figure 14.5 Traditional HDPE biaxial geogrid.

tensile strength than uniaxial grids, depending on design they may have nearly equal strength in the transverse direction as in the longitudinal direction. This makes them more suitable for resisting two-dimensional stresses. Many geogrids are manufactured by bonding two sets of orthogonal ribs together to form a grid matrix. A version of biaxial geogrids manufactured with fiberglass or polyester is used primarily in roadway applications and will be described later in this chapter. Other high-strength geogrids are used for foundation reinforcement or within reinforced soil masses.

- *Triaxial geogrids* (Figure 14.6) are relatively new on the market and provide a multidirectional reinforcement. With triangular apertures, increased rib thickness, and better junction efficiency, they provide a higher-strength alternative to biaxial geogrids, with the improved aggregate interlock and confinement of a reinforced soil mass. Research has shown that the use of triaxial geogrid beneath a roadway base coarse

Figure 14.6 Triaxial geogrid. *Courtesy of Tensar International Corp.*

has allowed for reduced base thickness on the order of 25-50% (www.tensar.com).

While the majority of geogrids have traditionally been made of HDPE, there are now many other materials and designs with a wide range of strengths, geometries, and attributes for an equally wide variety of application conditions (Figure 14.7). Tensile strengths of up to 1300 kN/m (92,500 lb/ft) are now readily available in grids constructed of tensioned multifilament polyester cores, coextruded and encased with polyethylene (HDPE) protection to maintain geometric stability (www.maccaferri-usa.com; Koerner, 2005) (Figure 14.8). These grids provide high strength reinforcement with

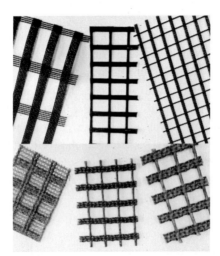

Figure 14.7 Other (non-HDPE) bonded and "green" woven geogrids.

Figure 14.8 Very high strength polyester/polyethylene composite geogrid from Maccaferri.

minimal deformation for high load and stress applications, such as basal reinforcement of embankments over soft ground.

Strengths of geogrids are commonly measured by single rib strength and wide-width tensile strength (ASTM D6637), as well as junction strength (where longitudinal and transverse ribs intersect). Anchorage (pullout) strength is computed as a combination of interface shear strength for both longitudinal and transverse ribs, plus the passive resistance provided by the bearing strength against the sides of transverse ribs. Allowable strengths used for design take into consideration a number of other potential issues including endurance properties of installation and creep, as well as possible degradation due to chemical and biological attack.

14.1.3 Soil Reinforcement Fundamentals

Soil is inherently weak in tension and stronger in compression and shear. The shear resistance of reinforcing materials placed within soil can be described as a combination of the interface friction between materials, adhesion between materials, and in some cases passive resistance of reinforcement inclusions. Since the development and implementation of patented soil reinforced walls in the 1960s, when metallic strips were used as reinforcing elements, there has been continued interest and growth in geosynthetically reinforced slopes and walls.

The general mechanics of geosynthetic soil reinforcement is based on a number of criteria usually involving specified test parameters such as material

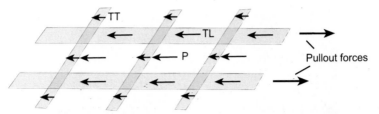

Figure 14.9 Forces acting on a geogrid or mat to resist pullout. TL, interface shear strength on the (top and bottom) surfaces of longitudinal ribs; TT, interface shear strength on the (top and bottom) surfaces of transverse ribs; P, passive bearing force of the leading edge of transverse ribs.

interface resistance, tensile strength, tear strength, elongation, and so forth. The friction and/or adhesion resistance between a geosynthetic material and a particular soil is commonly measured in a manner similar to a direct shear test, and is referred to as the *interface friction* or adhesion. The resistance is, therefore, the multiple of the unit interface resistance and the area of the material in contact with the soil. When a geogrid is used, there is typically a strikethrough and interlock of the soil material placed above and beneath through the open apertures of the grid. There is also an added passive resistance against the leading edge of the transverse member of a geogrid (or welded mat, or transverse ribbed surface of other geosynthetic material) Figure 14.9.

14.2 MSE WALLS AND SLOPES

14.2.1 Geosynthetic Reinforced Wall and Slope Basics

MSE refers to the use of reinforcement constructed between compacted soil layers to build earth structures such as retaining walls, bridge abutments, embankments, and steep, yet stable slopes. Various reinforcing materials have been used including steel strips, welded wire mats, geotextiles, and geogrids. Use of geotextiles for reinforcement began in the 1970s, while geogrids have been used since the early 1980s. As described earlier, versions of these MSE walls were developed in the early 1960s. Since that time, many tens of thousands of MSE walls have been and continue to be constructed due to a number of desirable attributes that they possess. It has been estimated that more than 850,000 m^2 (9 million ft^2) of MSE wall is constructed each year in the United States and has been used in every state (Federal Highway Administration, 2010). MSE structures of this kind are used not only for retaining walls, but also for bridge abutments, approach ramps, cut-and-cover tunnels, and noise walls.

MSE walls have several advantages over conventional gravity or reinforced structural walls:

- Relatively lightweight wall facing provides much lower bearing loads
- System has high flexibility, providing the ability to undergo small to moderate deformations
- Fundamentally simple construction
- Usually a very economical alternative to other earth-retaining structures
- Typically, significantly reduced construction time compared to structural walls

MSE walls may be faced in a number of ways and with a variety of different materials. These may be precast segmental panels (with or without) artistic designs; cast-in-place panels; rock-filled gabion cages; and welded wire mesh, timber, or integrated modular blocks (Figure 14.10). The majority of MSE walls in the United States are designed as permanent structures constructed with segmental precast facings connected to galvanized steel strips with heights up to 46 m (150 ft) (Federal Highway Administration, 2010). Details with respect to connections between facing units and reinforcement will differ depending on facing and reinforcement type, and loading conditions (Figure 14.11). Facing may also be constructed by a "wraparound" of geosynthetic material used as the primary or secondary reinforcement of the structure (Figure 14.12). The use of geosynthetic reinforcement in engineered earth slopes and embankments adds significant stability and strength, providing the ability to construct steep slopes that require a smaller footprint to achieve the same height.

14.2.2 Failure Design Modes

While the design of earth-retaining structures is based largely on lateral earth pressure theory, and slope designs are generally based upon slope stability analyses, designs for retaining walls and slopes reinforced with geosynthetic inclusions begin to have much greater similarities. Both applications depend on internal soil-reinforcement interaction (pullout resistance of the reinforcing members), and tensile rupture (tear strength) of the geosynthetic material. Designs must include calculated resistance to *internal stability*, as well as *external stability*, or "global" stability of a MSE mass. Designs for the reinforcement strength and placement within a stabilized soil mass will be described in the next section. Global stability must

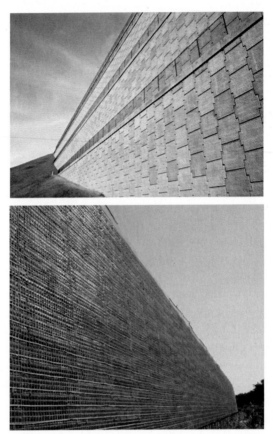

Figure 14.10 Example of some wall facings: top—precast panels, bottom—wire-wrapped. *Courtesy of The Reinforced Earth Company.*

adequately satisfy the requirements of overturning (usually taken as rotation about the toe), sliding (translation along the base of the reinforced mass over foundation soils), slope failure (encompassing the entire reinforced mass), and bearing (capacity of the underlying foundation soil to support the load of the MSE system). This last global stability mode may be enhanced or solved by reinforcing the subgrade/foundation soils, as will be addressed in Section 14.3.

A significant design detail that should not be overlooked, especially for geotextiles, is the survivability of the materials, both during installation and for working loads. This is generally related to strength and stiffness of materials as defined by grab tensile strength tests (ASTM D1682), but may also consider puncture or tear strength.

Figure 14.11 Typical wall connections for precast facing panels. *Courtesy of The Reinforced Earth Company.*

Figure 14.12 Geosynthetic wrapped-face wall. *Courtesy of Tencate-Mirafi.*

14.2.3 Reinforcement Design for MSE Walls

The design of MSE walls will depend on size load requirements and permanency of the structures. For temporary applications, such as limited-time access roads, construction staging areas, or surcharge fills, less expensive (and less capacity, durability, and strength) geotextiles may be employed. In many of these shorter-term applications, the wall facing is simply a wrapped tail of reinforcement held in place by the confining stress of material placed in above layers (wrapped geotextile wall) (Figure 14.13). For short-term protection, these types of walls may be covered with a thin shotcrete or other nonstructural facing material. For longer-term and permanent applications, MSE walls can be constructed with a design life of 75–100

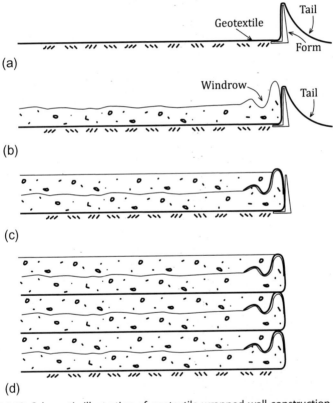

Figure 14.13 Schematic illustration of geotextile-wrapped wall construction. (a) Place geotextile on prepared subgrade/foundation surface with 1 m "tail" over facing form, (b) Place about 1/2 lift of backfill on geotextile and form windrow behind form to full lift height, (c) Place geotextile tail into windrow and complete lift to full height, (d) Completed geotextile reinforced wall.

years, and will typically have a semirigid facing of concrete panels, cribbing, or confined rockfill (e.g., gabion construction; see Chapter 16) (www. reinforcedearth.com).

The basic premise for internal stability design of MSE walls stems from lateral earth pressure theory. For simple design of soil reinforcement, a Rankine active earth pressure is assumed. Any additional loads (e.g., static surcharge loads or "live" vehicle loads) must be included, and added to the lateral earth pressures. Static surcharge loads are assumed to be "at-rest" conditions. Live loads are distributed to the soil mass using Boussinesq elastic theory as described in classic soil mechanics texts or design manuals (i.e., NAVFAC DM 7.2, Dept. of the Navy, 1982).

Design for internal stability involves calculating vertical spacing of reinforcement layers (Sv), embedment length to resist pullout from behind the active zone (Le), and for geotextile-wrapped walls, the overlap lengths needed to ensure integrity of the wall face. Figure 14.14 shows the basic layout for a MSE wall. Design details can be found in dedicated geosynthetic texts (i.e., Holtz et al., 1997; Koerner, 2005), or from guidelines and specifications provided by suppliers. Maximum vertical spacing is inversely proportional to lateral stress, so in general, layer spacing must be closer lower in a wall. Embedment length is a function of the lateral stresses, vertical spacing, and interface shear strength between backfill soil and geosynthetic, but it should always penetrate at least 1 m beyond the theoretical active slip surface (Koerner, 2005).

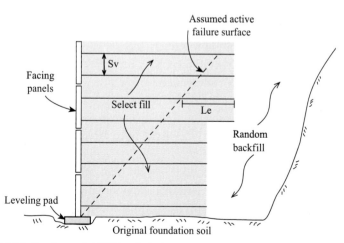

Figure 14.14 Fundamental MSE wall design components.

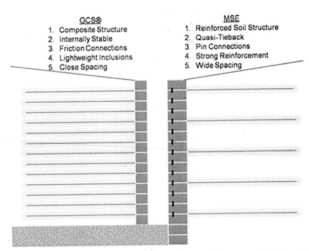

Figure 14.15 Geosynthetically confined soil (GCS) wall versus MSE wall. *Courtesy of GeoStabilization International.*

The Federal Highway Administration (FHWA) has recently embraced a newer version of the MSE wall referred to as geosynthetic reinforced soil (GRS) (or geosynthetically confined soil, GCS, by others) (Figure 14.15). These walls employ techniques similar to those for construction of MSE walls, but use much closer vertical spacing (typically 20 cm = 8 in.); lighter-weight reinforcement (typically a woven polypropylene geotextile); well-compacted, select granular fill; and smaller facing elements primarily secured to the reinforced soil mass only by friction. The close spacing of the reinforcement provides a similar type of pullout resistance as provided by traditional MSE walls, but in addition, GRS provides a significant component of confinement resulting in greatly increased capacity and stiffness. While the main thrust of using GRS walls by the FHWA has been for rapid construction of abutments for integrated bridge systems, other contractors have used this technology for stand-alone retaining walls, support of utility pipes, and rehabilitation of unstable slopes. Research reported by FHWA (2011) indicates that GRS (GCS) can have up to 5 times greater capacity than traditional MSE walls.

Design is nearly the same as for MSE walls with other types of geosynthetic reinforcement. Again, both external and internal stability must be adequately addressed. Internal stability is still a function of vertical spacing and embedment length (and connection strength when used with structural facing). The most significant difference is the strength, modulus, and pullout

Figure 14.16 Mechanically stabilized earth wall using uniaxial geogrid. *Courtesy GSE Environmental, LLC.*

resistance of the reinforcing material. Using geogrids for MSE wall reinforcement is generally a less-expensive alternative to steel-reinforced
MSE walls, but it usually provides greater pullout resistance than geotextile
reinforcement for the same configuration. Figure 14.16 shows a geogrid
MSE wall under construction.

14.2.4 Reinforcement Design for Reinforced Soil Slopes

MSE walls are often constructed with some angle of batter. At some point,
the angle of a wall face decreases such that the wall transitions into a steep
embankment slope. With this transition comes a change in design methodology from lateral earth pressures to one of slope stability. An angle of 70
degrees has been suggested as a transition point (Koerner, 2005). For steep
slopes, a planar potential failure surface may still be adequate, but as the slope
angle decreases, the assumed surface is generally taken to be curved. For simple limit equilibrium analyses, the assumed surface is taken as circular. Geosynthetic reinforcement is placed at specified locations (and vertical spacing),
but the spacing will not be dependent on lateral earth pressures; rather, it will
usually be designed to provide resistance to rotational slope failure
(Figure 14.17). In fact, reinforcement may not need to be placed evenly
or throughout a soil slope to gain sufficient stability. The increase in stability
is the added moment provided $\Sigma(P_i^* y_i)$. Secondary reinforcement may be
needed to resist shallow sliding (sloughing) near the surface of the slope
(Figure 14.18). Here, the pullout resisting moments of reinforcing elements
are added to the slope stabilizing moment of shear strength along the curved

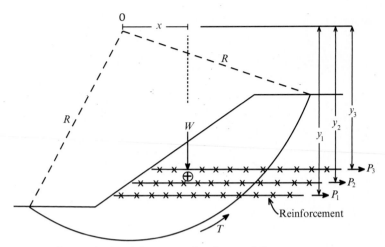

Figure 14.17 Geosynthetic reinforcement for slope stability.

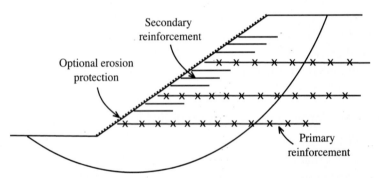

Figure 14.18 Secondary reinforcement (and erosion protection) for shallow slope stability.

surface (T) multiplied by the radius (R). The simplistic forces added by the pullout resistance offered by the reinforcement, is easily incorporated into solutions used by many stability programs. In fact, these calculations may be somewhat conservative because the added slope stability derived in this manner does not even account for the added shear strength afforded by the tensile "tearing" resistance of the geosynthetic material across the shear plane. Reinforced slopes have been constructed up to 74 m (242 ft) in the United States (Figure 14.19).

There are a number of advantages to constructing a *reinforced soil slope* (RSS). These come about from a combination of the smaller footprint required, resulting in a reduced right-of-way, and a savings in the total volume of material required to construct a slope or embankment to a particular

Figure 14.19 Steep, high reinforced soil slope for runway extension. *Courtesy of Tecate-Mirafi.*

height. Use of reinforcement may also allow construction of slopes with lower-quality fill material. Where space is available, a RSS can also serve as an economical alternative to MSE walls. It has been estimated that in some cases, reinforced slopes can be constructed at about one-half the cost of a MSE wall (Federal Highway Administration, 2010). Generally, RSS structures can be easily adapted to vegetated facings or even synthetic grass for an aesthetic advantage over the precast concrete facing typically used for MSE walls.

14.3 OTHER GEOSYNTHETIC REINFORCEMENT APPLICATIONS

In addition to engineered "earth structures" (primarily embankments, slopes, and walls), geosynthetics have been used for a wide variety of other reinforcement applications. Some of these are discussed in the following section.

14.3.1 Reinforced Foundation Soils

Sometimes it is desirable to build embankments or other structures over soft and/or weak ground. Traditional options for building upon these sites have included installing expensive deep foundations or load-bearing columns, excavating and replacing the poor soils with more suitable select material, stabilizing the soil with additives as described in Chapters 11 and 12, or preconsolidating the site. Each of these alternatives may be applicable, but may also be expensive, time-consuming, or both.

Reinforcement of the foundations soils with a high-strength geosynthetic placed between the weak subgrade and overlying engineered structure can often provide an economical solution to providing stability (capacity) to the foundation as well as a reduction in otherwise anticipated settlements (Figure 14.20). The reinforcement can be designed to spread the load to provide necessary bearing capacity, resist lateral spreading of the overlying embankment, and prevent deep rotational failures through the weak foundation soil. In some cases, some of the aforementioned alternatives (such as soil-mixed or aggregate columns, or installation of prefabricated vertical drains) may be used in conjunction with foundation reinforcement as a design solution.

14.3.2 Support of Load-Bearing Foundations

High-strength geosynthetics (usually geogrids or high-strength, high-modulus geotextiles) have been used to reinforce engineered fill beneath structural loads and foundations. In these cases, select granular (typically

Figure 14.20 Application of high strength polyester/polyethylene geogrid for a new highway interchange. *Courtesy of Maccaferri-USA.*

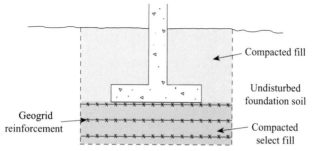

Figure 14.21 Geogrids for structural foundation bearing support.

aggregate) fill is compacted in relatively thin lifts with alternating layers of geogrids prior to placement of structural loads or spread footings (Figure 14.21). While the precise mechanisms providing the added support are not clearly defined for this application, both model tests and full-scale field applications have shown significant improvements in bearing capacity and reduced settlement deformations. In another version of this type of application, geogrids are placed within layers of compacted select fill to distribute (near) surface loads to concentrated support points offered by deep foundation schemes (Figure 14.22).

14.3.3 Roadway Applications

Geosynthetics for reinforcement (and separation) have been used for many years in both paved and unpaved roadway applications. For both situations

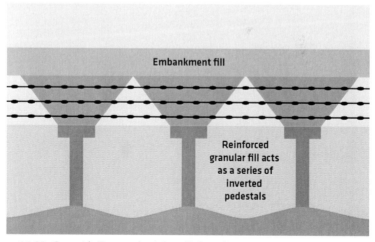

Figure 14.22 Geogrid "inverted pedestal" foundation support. *Courtesy of Tensar International Corp.*

they can provide a distribution of applied loads by transferring some of that load laterally as the geosynthetic materials are put into tension.

14.3.3.1 Unpaved Roads

Geotextiles have played an important role in providing a working solution for construction or improvement of secondary unpaved roads, haul roads, access roads, and roads in developing regions (Figure 14.23). For most cases, geotextiles were placed between soft, fine-grained subgrade soils, with sand or stone aggregate above. Acceptance of this particular application triggered a high-volume increase in the use of geotextiles beginning in the 1970s (Koerner, 2005). Based on field measurements of subgrade stability (e.g., from California Bearing Ratio, shear strength, or tests of resilient modulus, etc.), relatively simple calculations of the reduction in required aggregate thickness often show a net cost-savings by using the geotextile. In addition, the long-term performance is improved by the separation function of the geotextile keeping the aggregate from mixing into the subgrade below. For details of this type of analysis, refer to a reference on geosynthetic applications, such as Koerner (2005) or Holtz et al. (1997).

Geogrids have also been used to provide reinforcement for unpaved roads by increasing soil strength, spreading loads, and minimizing rutting by tensile membrane support (Koerner, 2005). Analytical methods described

Figure 14.23 Geotextile for unpaved roadway reinforcement. *Courtesy of Tencate-Mirafi.*

by Giroud et al. (1984) are still used as a basis for comparing the required thickness of a base aggregate with and without geogrids.

14.3.3.2 Paved Roadways

Geosynthetics have been widely used for reinforcement in the base layer of flexible pavement roadway applications for more than 30 years. Both geo-textiles and geogrids have been used, and have resulted in improved perfor-mance, reduced maintenance, and have allowed a reduction in base-layer thickness requirements. But there is some controversy regarding the actual reinforcing function and mechanism of a geotextile for this use, because there may not be enough deformation to mobilize its strength. Regardless, geotextiles still provide a benefit for paved roads through other functions, including separation and minimizing reflective cracking. Some geogrids (including coated fiberglass) are used directly beneath hot asphalt overlays (Figure 14.24). These materials are heat-resistant while maintaining high strength and are often manufactured with an adhesive on one side to help keep them in place during construction.

Reinforcement within a roadway base course takes advantage of the stri-kethrough offered by the apertures of a geogrid to provide confinement of the aggregate. This provides an increased modulus to the layer, which resists deformation from repeated traffic loads and serves the same function in rail-road ballast. Studies have shown that a significant load-spreading effect is

Figure 14.24 Biaxial geogrid for pavement support. *Courtesy of Tensar International Corp.*

Figure 14.25 Triaxial geogrid beneath base coarse for a paved roadway. *Courtesy of Tensar International Corp.*

realized as a function of increased stiffness and decreased long-term vertical and horizontal deformations, resulting in reduced cracking and cyclic fatigue behavior of asphalt pavement overlays (Koerner, 2005) (Figure 14.25).

14.3.4 Reinforcement for Erosion Control

Use of geosynthetic materials for erosion control crosses boundaries between "reinforcement" and "confinement." While confined soil materials (and rock-fill) are often used for combating scour and large-scale erosion due to flooding, storm surge, and so on (to be addressed in Chapter 16), the general topic of reinforcing surface soils in order to prevent erosion will be presented here.

There are hundreds of geosynthetic erosion control products on the market today with a wide range of materials, strengths, durability, and so forth. One of the first things to consider is the desired lifespan of the surface reinforcement. For permanent slope surface stabilization, composite geosynthetics with geogrid or coated twisted wire provide a strong and durable mat that can be secured to the ground with driven or drilled soil nails or pins (Figures 14.26 and 14.27). This is also a topic that crosses over into in situ reinforcement with inclusions. Soil nailing and anchoring will be addressed in Chapter 15. For temporary reinforcement, typically installed to retain surface soils until vegetation can be established, lighter-weight materials can be used. These are sometimes colored green for aesthetics, and often consist of loosely bound, loose poly fibers,

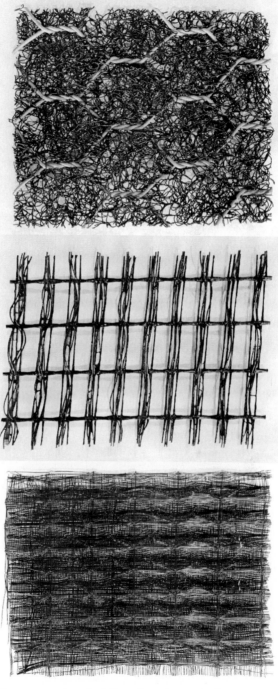

Figure 14.26 Some examples of reinforced erosion control mats.

Figure 14.27 Pinned reinforcement mat for stabilization (erosion control and sloughing) of shallow, surface soils of steep roadcut, Kailua, HI.

sometimes manufactured with a filter fabric backing to retain finer-grained material (Figure 14.28). Another category of erosion control mats is available for the environmentally conscious or for sensitive projects that utilize natural, biodegradable materials (Figure 14.29). Any of these erosion control mats may be hydroseeded or planted with plugs of reeds or similar vegetation (Figure 14.30). For immediate aesthetic enhancement, synthetic turf may also be used (Figure 14.31).

Figure 14.28 Lightly bound loose poly fibers. Composite on left has geotextile filter backing.

Figure 14.29 Natural and biodegradable erosions mats: (a) wood fiber (scrim), (b) straw, (c) shredded coconut, (d) woven coir.

Figure 14.30 Reinforced steep slope surface with vegetation "plugs."

Figure 14.31 Wall facing with synthetic turf. *Courtesy of Tencate.*

RELEVANT ASTM SPECIFICATIONS

D5261—10 Standard Test Method for Measuring Mass per Unit Area of Geotextiles, V4.13

D6637—11 Standard Test Method for Determining Tensile Properties of Geogrids by the Single or Multi-Rib Tensile Method, V4.13

D6638—11 Standard Test Method for Determining Connection Strength Between Geosynthetic Reinforcement and Segmental Concrete Units (Modular Concrete Blocks)

REFERENCES

Anderson, P.L., Gladstone, R.A., Sankey, J.E., 2012. State of Practice of MSE Wall Design for Highway Structures. ASCE, Reston, VA, Geotechnical Special Publication No. 226, pp. 443–463.

Carroll Jr., R.G., Walls, J.G., Hass, R., 1987. Granular base reinforcement of flexible pavements using geogrids. In: Proceedings of the Geosynthetics '87 Conference. IFAI, pp. 46–57.

Department of the Navy, 1982. Foundations and earth structures. Design Manual DM-7.2, Naval Facilities Engineering Command, Bureau of Yards and Docks, 244 pp.

Federal Highway Administration, 2010. Design and construction of mechanically stabilized earth walls and reinforced soil slopes. Publication no. FHWA-NHI-10-024.

Federal Highway Administration, 2011. Geosynthetic reinforced soil integrated bridge system synthesis report. Report no. FHWA-HRT-11-027.

Giroud, J.-P., Ah-Line, C., Bonaparte, R., 1984. Design of unpaved roads and trafficked areas with geogrids. In: Proceedings of the Symposium on Polymer Grid Reinforcement in Civil Engineering. Institution of Civil Engineers, pp. 116–127.

Hausmann, M.R., 1990. Engineering Principles of Ground Modification. McGraw-Hill, New York, 632 pp.

Holtz, R.D. (Ed.), 1988. Geosynthetics for Soil Improvement. ASCE, New York, NY, Geotechnical Special Publication No. 18, 213 pp.

Holtz, R.D., Christopher, B.R., Berg, R.R., 1997. Geosynthetic Engineering. BiTech Publishers Ltd, Canada, 451 pp.

Koerner, R.M., 2005. Designing with Geosynthetics, fifth ed. Pearson Education, New Jersey, 796 pp.

Wu, J.T.H., Lee, K.Z.Z., Helwany, S.B., Ketchart, K., 2006. Design and construction guidelines for geosynthetic-reinforced soil bridge abutments with a flexible facing. NCHRP report no. 556. Transportation Research Board, Washington, DC.

Zornberg, J.G., Gupta, R., 2010. Geosynthetics in pavements. In: Proceedings of the 9th International Conference on Geosynthetics, Guaruj, Brazil, vol. 3, pp. 379–400.

https://www.dot.state.oh.us/engineering/OTEC/2011%20Presentations/48C-NicoSutmoller.pdf (accessed 18.03.14).

http://www.geostabilization.com (accessed 16.03.14).

http://www.maccaferri-usa.com (accessed 16.03.14).

http://www.mdt.mt.gov/other/research/external/docs/research_proj/geo-reinforce.pdf (accessed 10.03.14).

http://www.reinforcedearth.com/applications/19 (accessed 16.03.14).

http://www.tencate.com/emea/geosynthetics (accessed 16.03.14).

http://www.ce.utexas.edu/prof/zornberg/pdfs/CP/Zornberg_Gupta_2010.pdf (accessed 16.03.14).

CHAPTER 15

In Situ Reinforcement

This chapter describes a variety of inclusions used to reinforce and/or support existing groundmasses in situ. This often includes high-strength steel rods placed into the ground for passive resistance, but can also include conventional piles or sheetpiling. Some in situ reinforcement approaches employing compacted gravel columns or rammed aggregate piers were covered in Chapter 6. In this chapter, an overview of the installation of structural members for reinforcement will be covered. For many construction applications, the earth pressure loads that are otherwise directly applied to conventional sheet piles or soldier piles and lagging, are transferred into the ground (or rock) beyond the potential failure surface by an anchoring system. In other situations, structural members may provide compressive (bearing) reinforcement. Common in situ reinforcement/anchoring schemes include ground anchors, soil nails, micropiles, helical piles, and bolts. Some smaller bearing reinforcement inclusions (not including deep pile foundations) are also described.

15.1 TYPES, INSTALLATIONS, APPLICATIONS

As suggested, a number of different types of in situ reinforcement systems may be utilized depending on the particular application, loads, permanency, and so forth. Some of these variations of reinforcement schemes are very similar in many respects, and sometimes the name refers as much to the function of the inclusion, such as "anchors" for pullout resistance, or piles/piers for bearing support. A number of in situ reinforcement types will be described here. Conventional deep foundation piles will not be covered.

15.1.1 Ground Anchors

Ground anchors are defined as structural units (typically grouted tendons) that transmit loads to stable soil or rock through tensile reinforcement. Grouted ground anchors are also sometimes called *tiebacks*, or *tiedowns* when subjected to uplift forces (Federal Highway Administration, 1999). These types of reinforcing systems are used to support temporary or permanent new wall construction, as well as for rehabilitation or reinforcement of critical and

Figure 15.1 Ground anchor supported soldier beam and lagging excavation wall. *Courtesy of Hayward Baker.*

potentially hazardous slopes, steep cuts, or fractured and/or weathered rock faces and tunnels. When used for slope stabilization or to remediate landslides, ground anchors are often used in conjunction with other structural inclusions (e.g., beams, blocks, stub walls).

The most commonly used anchored wall systems in the United States are soldier beam and lagging walls (Figure 15.1). These are nongravity cantilevered wall systems consisting of discrete vertical wall elements (usually driven wide-flange "H-piles" or double channel), spanned by timber lagging, reinforced shotcrete, or cast-in-place panels. In addition to providing support for lateral and downward forces as generated by excavations, cuts, and slopes, there are a number of applications where this type of tensile reinforcing system provides resistance to uplift forces. These may include hydraulic structures subjected to high internal water pressures, other structural slabs with high hydrostatic uplift pressure, buoyant underwater structures, foundations of tall, slender structures (such as transmission towers and wind turbines) subject to high overturning (i.e., wind) loads, support cables for utility poles, and so on.

Anchors are generally installed by inserting sleeved structural tendons into predrilled holes or trenched excavations, and grouted (or epoxied) into place over a length of the tip or deepest section of the tendon. Most commonly, anchors are placed at an inclined angle of between 15° and 30° below the horizontal (although some anchors have been installed between 0° and 45°). For most simple applications, the deeper portion of the hole is first grouted and the anchor tendon is then inserted into the uncured grout. The pullout resistance is a function of the shear resistance of the grout–ground interface, and so is dependent on soil conditions and grouting process. Once the grout has set, the tendons are tensioned at a bearing plate at the ground surface or excavation/ wall support. Typical grouted anchors are "fixed" over a bonded (grouted) or *anchored length* at depth, but are free to move axially along the shallower, unbounded length to allow tensioning (and sometimes retensioning). The bonded length should be fully located behind the critical failure surface. The unbonded length is either provided by a sheath or secondary weak grout (Figure 15.2). Once the tensioned anchor has been stress tested and "accepted" by meeting test specifications, the unbounded length may later be grouted to form a fully grouted anchor. Grout holes may be gravity fed or pressure-grouted, the latter more applicable to granular material and fissured rock. Some anchors may be postgrouted, often in stages, allowing the development of an enlarged grout bulb over the anchored length for increased pullout resistance in softer material. Some other anchoring systems may employ underreaming or installation of an anchoring plate at the anchor tip to provide additional support (Figure 15.3). Anchor tendons may be comprised of a single steel bar (available in various diameters from 26 to 64 mm) in lengths of up to 18 m (60 ft), or as multistrand tendons (typically groups of seven 15-mm diameter strands), which can be manufactured in any length (Federal Highway Administration, 1999). Another version of anchor, known as a "helical anchor," will be described in a later section of this chapter. Ground anchors may have typical capacities of up to 1000 kN (112 U.S. tons) in soil, and in excess of 10,000 kN (1120 U.S. tons) if grouted into rock (Hausmann, 1990).

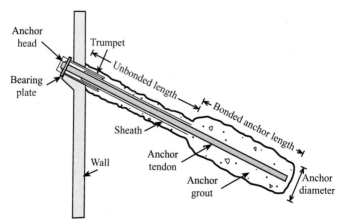

Figure 15.2 Components of a grouted soil anchor. *After FHWA (1999).*

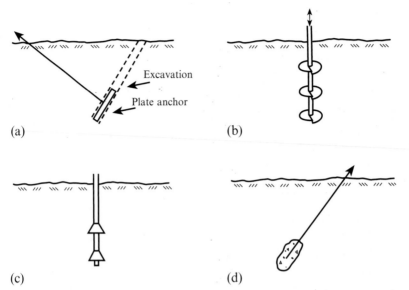

Figure 15.3 Various types of ground anchors: (a) plate anchor, (b) helix (screw) anchor, (c) underreamed anchor, (d) grouted anchor. *After Hausmann (1990).*

Permanent anchors have proven to be an integral part of an economical construction and permanent wall support system, where anchored walls are installed, tested, and finished in sequence with advancing excavation depths. Anchors are also used in conjunction with continuous walls such as sheet pile, secant pile, slurry, or soil-mixed walls. These types of structures are more often used for temporary excavation support (Federal Highway Administration, 1999).

15.1.2 Soil Nailing

Soil nailing consists of driving, screwing, drilling, or "shooting" a series of steel or fiberglass bars into the ground, most commonly for excavation support or stabilization of steep slopes or cuts. Nails have also been used for rock slope stabilization (see Section 15.1.3). The "nails" are typically fully grouted in place to secure the inclusions and provide additional pullout resistance, although some versions of nails have barbed ends (Figure 15.4) and are driven into the ground without grouting. There are a number of principle differences between soil nails and anchors. Soil nails are usually much shorter than tieback anchors, and generally have no structural (vertical) wall member (e.g., soldier pile or structural wall) for reaction. They utilize both passive

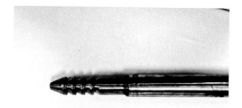

Figure 15.4 Detail of a launched soil nail with "barbed" end.

and active forces mobilized on either side of a slip surface, rather than the active anchored wall systems. They are generally not tensioned like an anchor system at a distance (depth) below/behind the ground surface. Instead, nails reinforce a groundmass through resistance along their entire length. Soil nail walls may be "finished" with reinforced shotcrete, precast panels, heavy steel mesh, or vegetated "cells." Nail tendons typically have lower tendon strength and must be spaced closer together due to their lower capacity.

With their closer spacing relative to anchors, they form a coherent, reinforced soil mass capable of providing support for excavations or slope stabilization. For shallower-depth slides, both shear and pullout resistance of the nails reinforces potential sliding mass. Soil nails have also been used for rehabilitation of historic stone retaining walls (Figures 15.5).

Design of soil nailing systems consists of proper dimensioning of nail spacing, length, and inclination to assure that the ground mass is stabilized. When used to secure a rotational-type slide, the nails must be secured well beyond any potential failure surface. Monitoring of any movements of the stabilized ground mass is vital to ensuring that any problems are detected early. In extremely corrosive environments, fiberglass nails may be used to resist deterioration.

Dynamically *"launched" soil nails* are placed into the ground by a compressed air cannon at speeds of up to 400 kph (250 mph), usually without a need for grouting. The use of launched soil nails for slope stabilization has many benefits over the construction of more conventional retaining systems. Costs may be on the order of one-tenth of a traditional system, and launched soil nails can be installed in a fraction of the time without nearly as much disruption to the environment or to ongoing serviceability of the structure under

Figure 15.5 Rehabilitation of historic rock retaining walls with soil nails. *Courtesy of GeoStabilization International.*

repair (www.geostabilization.com). In addition, as the nails are launched into the ground, they generate a shock wave that causes the earth materials to elastically deform without the full static resistance that might normally have been expected. After insertion, the earth materials collapse onto the bar in a relatively undisturbed state with increased normal stresses. In addition, the displacement of the ground surrounding the nail is densified. The combination of these effects results in pullout capacities up to 10 times that of driven piles (www.geostabilization.com) without the need for grouting. A version of the launched soil nail employs a perforated hollow rod that can be injected with grout after installation, resulting in further densification/stabilization of the surrounding soil, as well as providing increased pullout resistance. Another

Figure 15.6 Launched soil nails for emergency slope stabilization. *Courtesy of GeoStabilization International.*

modification employs insertion of a solid rod inside the grout-filled, perforated, hollow launched nail to provide added strength. These versions have been termed SuperNails® (www.geostabilization.com).

Some of the advantages to using launched soil nails is the ability to mobilize rapidly for emergency repairs, the rapid rate of installation, and reduced need for design planning (Figure 15.6). Traditional soil nailing includes a lengthy delay while cement grout hardens. Launched soil nails can provide effective stabilization almost immediately after installation. Several private and government-sponsored research projects have verified and supported the effectiveness and economy of utilizing soil nails rather than more conventional soldier piles and lagging for slope and retaining structure repairs. Case studies have shown that the use of launched soil nails in place of traditional temporary shoring or slope stabilization methods has realized cost savings of around 50-80% (in some cases, millions of dollars), with installations in days as opposed to several weeks (www.geostabilization.com). This type of solution can be critical, especially when disruption of service of roadways, rail lines, or utilities is a serious concern. For highly corrosive environments (e.g., coastal bluffs, highly acidic soil, etc.) fiberglass nails have been employed (Figure 15.7).

15.1.3 Rock Bolts

Rock bolts are a type of drilled soil nail or anchor used when the ground to be stabilized consists mostly of rock materials. These types of inclusions are typically grouted in place and posttensioned, similar to soil anchors

Figure 15.7 Stabilization of coastal bluff with fiberglass launched soil nails: Before and after stained shotcrete finish. *Courtesy of GeoStabilization International.*

(Figure 15.8). The bolts hold potentially unstable jointed or fractured rock masses together in compression with bearing faceplates providing passive resistance, forming a more stable structural entity. Rock bolts have been used for many years as temporary roof support in the mining industry and for tunneling, and are now used routinely for stabilizing roadcuts, rock cliffs, steep slopes, bridge abutments, and reinforced concrete dams. Two other versions of rock bolts sometimes used are those that have expansive shells (like a drywall anchor) rather than grout for resistance, and untensioned steel rods grouted into boreholes referred to as *dowels* (Hausmann, 1990).

15.1.4 Micropiles

Micropiles (a.k.a. minipiles, pin piles, root piles) are essentially small-diameter piles (often steel bars or pipes) grouted into predrilled holes to form short friction piles with high capacity and a generally lower amount of settlement

Figure 15.8 Rockfall mitigation with nails/rock bolts. *Top: Courtesy of Layne Christensen; Bottom: Courtesy of GeoStabilization International.*

compared to much more expensive driven piles (Figure 15.9). They can be installed in almost any type of soil, and even rock. Micropiles are most commonly used for structural foundation support, underpinning, wall support, and slope stabilization. One of the significant advantages of using micropiles is the lack of required overhead or lateral site constraints that would prohibit installations requiring much larger equipment. As opposed to the array of structural reinforcement methods described previously, micropiles can provide significant compressional capacity as well as tensile restraint. The industry reports that micropiles can have working capacities up to 2200 kN (250 tons) (www.rembco.com). Traditional micropiles are installed in concrete-filled predrilled holes. They are often used in groups to transfer bearing loads to subsurface soils in place of expensive deep foundations (Figure 15.10). For higher capacity, micropiles are pressure-grouted in

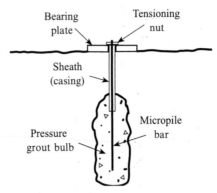

Figure 15.9 Components of a micropile.

place, which increases lateral pressures and densifies surrounding soil (if compressible), greatly increasing side resistance.

Launched micropiles, installed with the same type of equipment used for launched soil nails, can be rapidly installed and used for soil reinforcement, as is required for shallow excavation support, for support of retaining structures and embankments, or even for scour protection around bridge piers or culvert discharge channels (Figure 15.11).

Figure 15.10 Minipiles (vertical soil anchors) for roadway support. *Courtesy of GeoStabilization International.*

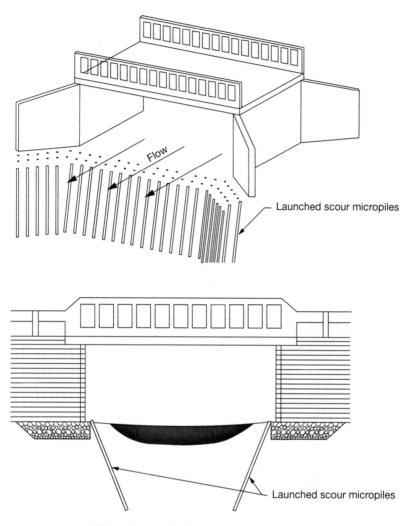

Bridge abutment/Culvert scour protection

Figure 15.11 Launched micropiles for scour protection. *Courtesy of GeoStabilization International.*

15.2 DESIGN BASICS

15.2.1 Capacity Estimates

For the most part, the final design of anchors, bolts, and nails is empirical and may rely on actual field testing of some selected, installed test anchors. Some empirical guidelines are available for initial feasibility estimates based upon soil/rock type and condition, identified by field borings and standard soil/

rock testing procedures such as SPT, CPT, pressuremeter test (PMT), and flat-plate dilatometer (DMT). Anchor pullout resistance for walls may be estimated as a function of interface frictional and effective stress, similar to evaluation of the shaft resistance of a deep foundation pile or shaft. For driven or "launched" inclusions, the stresses normal to the elements may be increased by displacement. When designing for slope stabilization, the required resisting forces may be significantly greater than for walls and should be estimated based on limit equilibrium analyses.

15.2.2 Spacing

Both vertical and horizontal spacing of in situ reinforcing inclusions must be designed to cover a load proportionate to the area attributable to each inclusion. In addition, closer spacing may be needed to attain a stiffer composite wall if smaller deformations are required. The Federal Highway Administration (FHWA) (1999) recommends that horizontal spacing should be no <1.2 m to avoid group effects between adjacent inclusions, while maximum spacing must consider the flexural capacity and tolerances of the wall. A minimum depth should also be observed for the uppermost grout application to prevent heave and provide sufficient confining stress to contain grout pressures (if used) and provide sufficient pullout resistance capacity.

15.2.3 Other Considerations

Designs should also consider other possible failure modes, including bond strength between the tendon and grout, tendon strength, stiffness/bending potential of the wall material(s), and depth of most critical ground failure (which could be significantly deeper than a Rankine active failure wedge). This last parameter is essentially addressed by evaluating maximum lateral earth pressures as well as deeper possible ground failure (limit equilibrium rotational stability) to ensure adequate embedment depth of anchors. Some general guidelines are provided by FHWA (1999). Where appropriate, seismic ground forces should be added to capacity requirements.

Protection of metallic elements from corrosion is also an important consideration for long-term durability and performance of permanent inclusions. This may be accomplished by providing one or more physical barriers, including corrosion-inhibiting compounds, sheaths, epoxy coatings, and grouts (FHWA, 1999).

Figure 15.12 Multistrand ground anchor testing. *Courtesy of Hayward Baker.*

15.2.4 Testing and Monitoring

As a measure of quality assurance, a percentage of installed anchors, bolts, or nails are tested to loads of ∼120-150% (typically 133% according to FHWA, 1999) of the design working load based upon the projected service life or level of risk associated with failure (Hausmann, 1990) (Figure 15.12). According to FHWA, all ground anchors installed as part of a complete structure is load tested to verify its capacity and response behavior before being put into use. This is also called acceptance testing, and may be part of the design specifications. Permanent anchor installations may require continual monitoring of loads and deformations, and so may be designed to include appropriate instrumentation. These types of monitoring tests can indicate any stress changes and/or creep phenomenon. If necessary, some anchors may be retensioned (if appropriate).

15.3 HELICAL ANCHORS AND PILES

Helical anchors and piles (*piers*) typically consist of square or tubular steel shafts with one or more helical screw plates (also called *flights*) securely welded at regularly spaced intervals (for multiple helix plates). Standard separation distance between two adjacent plates is three times the diameter of the lower plate (Atlas Systems, Inc., 2003; www.abchance.com). Figure 15.13 shows two typical helical anchors. Anchors are commonly available in single,

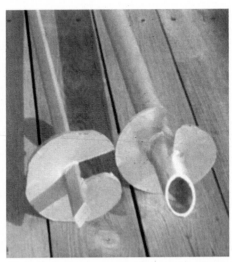

Figure 15.13 Typical square and tubular shaft helical soil anchors. *Courtesy of Magnum Piering.*

double, triple, or quad plate arrangements. When multiple helical plates are used, the anchors are typically designed with plates of varying diameters that increase in size from near the shaft tip up toward their connection point, where they are securely fastened. Helical plate diameters typically range from 15 to 40 cm (6-16 in.) (www.helicalanchorsinc.com).

Tubular rods provide superior compressive strength (for bearing loads), with greater torsional, buckling, and lateral capacity than square shafts, making them preferable for support beneath larger column loads (www.magnumpiering.com). They will have several times the capacity of a square shaft rod. Tubular rods are typically manufactured to meet ASTM A513. Some rods may be manufactured with perforations in the tube at selected location(s), which allows injection of grout through the tube annulus. Square bar shafts are more commonly used for tieback anchors or helical nails. They will provide good tensile resistance at a significantly lower cost. In some cases, a combination of square and tubular shafts may be used.

The helical plates are rotated as they are pushed into the ground, much in the way a screw is installed (Figures 15.14 and 15.15). Installation is fast with no disturbing vibrations or drilling, easily installed in limited access areas, creates minimum site disturbance, and can be immediately tested and/or loaded. The mechanism of support against the plates provides passive resistance in both compressional bearing or tension for pullout resistance, which lends them favorably to a wide array of applications (Figure 15.16). Typical

Figure 15.14 Schematic of helical pile for foundation support. *Courtesy of Hayward Baker.*

Figure 15.15 Installation of a helical pier. *Courtesy of Hayward Baker.*

applications include foundation support for new and remedial construction; underpinning of floors, walls, and columns; tieback anchors for walls or other earth support; uplift resistance for supporting guy wires and

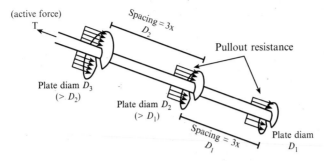

Figure 15.16 Forces acting on a helical anchor.

transmission towers; marine buoyancy applications; and soil nail applications. Helical nails differ from anchors in that the flights are all the same diameter and spaced at equal intervals along the full length of the nail shaft. As for traditional soil nails, they are spaced closer together than anchors and are, therefore, more evenly distributed on the wall. This helps to reduce stress concentrations on wall and/or excavation facings and throughout the reinforced soil mass. Helical nails are used for essentially any application where traditional soil nails may be considered.

15.4 OTHER IN SITU REINFORCEMENT

A few other approaches or applications have been used for in situ ground reinforcement. Koerner (2005) described use of geosynthetic webbing put into tension by driven soil nails to reinforce a slope and provide surface erosion protection. The mechanics of this type of reinforcement actually cross the line between structural inclusions and soil confinement, to be described in Chapter 16.

Stabilization has also been accomplished using driven piles, drilled shafts, soil-mixed columns, and reticulated micropiles, placed through the critical slip surface of active, incipient, and potentially unstable slopes.

RELEVANT ASTM STANDARDS

A513/A513M—12 Standard Specification for Electric-Resistance-Welded Carbon and Alloy Steel Mechanical Tubing, V1.01
A981/A981M—11 Standard Test Method for Evaluating Bond Strength for 0.600-in. [15.24-mm] Diameter Steel Prestressing Strand, Grade 270 [1860], Uncoated, Used in Prestressed Ground Anchors, V1.04

D4435—13e1 Standard Test Method for Rock Bolt Anchor Pull Test, V4.08

D7401—08 Standard Test Methods for Laboratory Determination of Rock Anchor Capacities by Pull and Drop Tests, V4.09

REFERENCES

Abramson, L.W., Lee, T.H., Sharma, S., Boyce, G.M., 2002. Slope Stability and Stabilization Methods, second ed. John Wiley & Sons, New York, 717 pp.

Atlas Systems, Inc., 2003. Technical Manual—Rapid Foundation Support Products. Atlas Systems, Independence, MO.

Federal Highway Administration, 1999. Ground anchors and anchored systems. Geotechnical Circular No. 4. Publication No. FHWA-IF-99-015, Office of Bridge Technology, 287 pp.

Hausmann, M.R., 1990. Engineering Principles of Ground Modification. McGraw-Hill Inc., 632 pp.

Koerner, R.M., 2005. Designing With Geosynthetics, fifth ed. Pearson Education Inc., 796 pp.

United States Department of Agriculture—Forest Service, 1994. Application guide for launched soil nails. Engineering Management Series Report No. EM 7170-12A, USDA-Forest Service, Washington, DC, 63 pp.

CHAPTER 16

Soil Confinement

The practice of confining soil to create high-capacity, load-bearing structures, and to provide erosion control, temporary flood protection, and lateral earth retaining functions, is described in this chapter. This general method has existed and been used for many years with very simple designs. The advent of geosynthetics and ingenuity in construction methods has expanded the use of soil confinement to many other areas and geotechnical applications. Newer materials have allowed rapid and relatively easy construction of temporary and permanent roadway structures, slope stabilization and/or rehabilitation systems, and retaining structures. They have also provided a means to contain grout to desired locations and have even provided a method to drain (consolidate) saturated materials such as dredged material or mine spoils.

16.1 CONCEPTS AND HISTORY

It is well understood from fundamental shear strength theory that the strength and loading capacity of granular (cohesionless) soil is a function of confining stress, and will increase (roughly) proportionately with increased confinement. In fact, virtually all soil types will have greater load-bearing capacity and shear resistance if mechanically (or otherwise) confined. The use of confinement for constructing various types of earth structures and retaining systems has been demonstrated for many years with the use of timber cribs filled with rocks (rockfill) for support of bridge spans and railroad trestles. Rock-filled, wire mesh "cages" called *gabions* have been widely utilized to buttress slopes and provide slope erosion protection while allowing good drainage of groundwater or rainfall. Gabions have also been used as gravity retaining walls with aesthetically pleasing faces, while again providing important drainage capacity.

Confined soil in the form of conventional sandbags has been utilized for many years for flood control (Figure 16.1), emergency repair of water conveyance structures, or as gravity "bunker" walls. Sandbags have even been used for economical home construction in low-income regions, such as parts of South Africa (Figure 16.2).

Soil Improvement and Ground Modification Methods

Figure 16.1 Use of sandbags as a temporary "earthen" flood control levee. *Courtesy of FEMA.*

Figure 16.2 Low-cost, sandbag home construction in South Africa (www. architecturalist.com).

16.2 SOLDIER PILES AND LAGGING

While soldier piles may be considered a form of structural inclusion, systems of soldier piles integrated with lateral lagging is a methodology that provides excavation support through a combination of lateral earth pressure resistance and confinement of retained soil. Piles are typically driven H-piles with wood, steel, or concrete panels inserted between piles to complete the retaining structures (Figure 16.3). Soldier pile and lagging retention

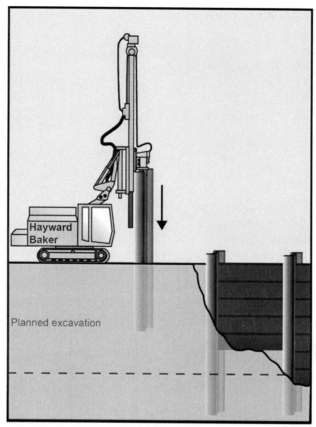

Figure 16.3 Soldier piles and lagging installation schematic. *Courtesy of Hayward Baker.*

structures are most often used for temporary excavation support, and may be further enhanced with anchors or internal bracing, especially for larger wall heights (Figure 16.4).

16.3 CRIBS, GABIONS AND MATTRESSES

Cribs have been used, particularly by the railroad industry, for hundreds of years (Figure 16.5). Traditionally constructed by stacked timbers filled with large stone, today's cribs are often made of concrete or (recycled) plastic elements, providing confinement of the stone for construction of structural piers or retaining walls (Figure 16.6). Another form of historical, timber-retaining structure utilized driven timber piles to confine the rockfill behind it. While many of the wooden crib structures lasted for many years, under less

Figure 16.4 Photo of a soldier piles and lagging excavation support. *Courtesy of Hayward Baker.*

Figure 16.5 Historic use of cribs for railroads. *Image by Bill Bradley.*

than ideal conditions deterioration of the wooden confining structures has been an issue. Timber crib walls continue to be constructed worldwide, although more resilient and durable (albeit heavier) concrete crib walls up to 30 m (100 ft) high have become popular in many regions (Figure 16.7).

Gabions are stone-filled rectangular baskets, typically made of (usually PVC-coated) twisted wire mesh. Gabions are commonly used as gravity retaining walls for earth retention or as buttresses for slope support (Figures 16.8 and 16.9). They have also been used for scour and/or erosion

Figure 16.6 Crib wall under construction. *Courtesy of Maccaferri, Inc.*

Figure 16.7 High concrete crib wall. *Courtesy of Maccaferri, Inc.*

protection along channel linings. As confined stone they have very high strength, load capacity, and high durability. Gabions have a number of advantages over conventional retaining structures in that they are very flexible; they can conform to irregular topography or geometries and can easily tolerate differential settlements without distress. Often, they can be used for erosion control for stream or riverbank applications. Gabions are typically

Figure 16.8 Gabion (buttress) retaining wall construction schematic. *Courtesy of Hayward Baker.*

Figure 16.9 Gabion buttress walls for channel bank stabilization/protection. *Top: Courtesy of Hayward Baker; Bottom: Courtesy of Maccaferri, Inc.*

Figure 16.10 "Green" gabion wall. *Courtesy of Maccaferri, Inc.*

free-draining and so usually will require no additional drainage construction. To ensure their drainage ability, a filter may be used between them and the soil retained behind them. Gabions can be filled with stone of various colors or textures to provide a choice of aesthetics. The surface layers and/or facings may be lined with a natural fiber (i.e., coconut or coir) and baskets filled with a combination of stone and topsoil materials so that they may be vegetated (Figure 16.10).

Gabion *mattresses* are constructed in a similar manner, but in relatively thin, large-footprint rectangles. These mattresses are convenient for use as channel linings, shoreline protection, and other high-energy erosion protection environments (Figure 16.11). As they can be easily placed underwater, they have also been used for offshore and other submerged applications, including foundations for breakwaters, jetties, and groins; pipeline protection; scour protection; and shoreline revetments. Where corrosion is a concern or in other harsh environments with salt or acid, gabions, mattresses, and confining structures are now often constructed using copolymer polypropylene geogrids (www.maccaferri-usa.com) (Figure 16.12).

Sack gabions are a version of gabions constructed by pouring aggregate into prepared wire (or geosynthetic) "sacks" as a rapid and low-effort means of gabion construction. These sacks are mostly used for temporary (and sometimes permanent) erosion/scour protection in high-energy hydraulic environments, and can be placed directly in moving water (Figure 16.13).

Figure 16.11 Gabion mattress for channel lining. *Top: Courtesy of Maccaferri, Inc.; Bottom: Courtesy of Tensar International Corporation.*

A number of ASTM standards have been devised for the wire mesh, rockfill, and placement of gabion structures. These standards are listed at the end of this chapter.

16.4 GEOCELLS

Geocells are manufactured as 3-D sheets of HDPE membranes (or geogrids) that are shipped as compact, collapsed bundled units. When stretched out (typically to 6.6 m (20 ft) lengths), the "sheets" form a series of individual cells into which soil is placed and compacted (Figure 16.14). Presto Products Co., together with the U.S. Army Corps of Engineers, developed this technology in the late 1970s and early 1980s. The infilled soil is confined by the cells (cellular confinement) such that the combined system can provide

(a)

(b)

Figure 16.12 Geogrid used for confinement of rockfill: (a) gabion mattresses and (b) shoreline confinement structures (Courtesy of Tensar International Corporation)

significant load-bearing, lateral load resistance, and erosion resistance (Figure 16.15).

A number of advantages have led to the use of geocells in a wide variety of applications. Similar to other types of confinement systems such as gabions or mattresses, geocell systems are flexible, easily transported, able to be vegetated, and uncomplicated to install. They can also be manufactured in a variety of colors to meet aesthetic requirements, and may be textured or perforated to provide additional frictional resistance. If infilled with clean

Figure 16.13 "Sack gabion" being filled and placed in water. *Courtesy of Maccaferri, Inc.*

granular material and perforated, they will also be free-draining and have very high load capacity (Figure 16.16). Geocells have the additional advantage of providing adequate support for many applications using local on-site soils, rather than select fill material that would otherwise need to be imported. Furthermore, the walls constructed with geocells generally provide a significant, relative cost savings compared to other retaining wall types (Figure 16.17).

Geocells are often used for retaining wall, free-standing earth berm, or steep slope (embankment) construction. They are capable of handling significant bearing loads, by stacking filled horizontal layers of the expanded sheets (Figure 16.18). This technique will typically result in a lighter

Figure 16.14 Cellular confinement for rapid construction of high capacity unpaved roads. *Courtesy of Presto Geosystems.*

structure than other conventional walls, and would apply lower loads to soft, weak and/or compressible foundation soils. Geocell wall/embankment construction has been successfully used in locations with poor soil and/or site conditions, including marshlands and rice fields with highly organic soils. When placed directly over soft soils, a geotextile is often placed beneath the geocells to provide separation (and some load distribution). In this type of construction the wall and/or slope faces may be very steep, allowing for construction where rights-of-way may be an issue, but are typically always battered to some degree. For added stability, walls and slopes may be constructed with intermittent layers of geogrid reinforcement (Figure 16.19).

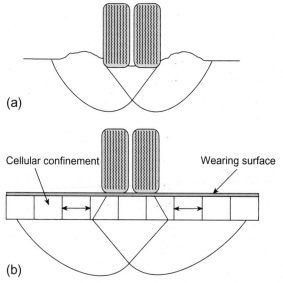

Figure 16.15 Comparison of distribution and lateral transfer of tire loads with cellular confinement: (a) unconfined; (b) confined.

Figure 16.16 Very high load capacity gravel-filled geocells. *Courtesy of Presto Geosystems.*

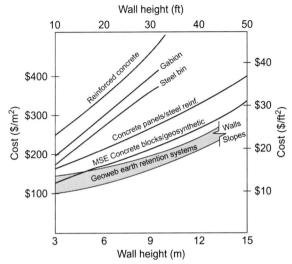

Figure 16.17 Relative costs for various earth retention walls. *Courtesy of Presto Geosystems.*

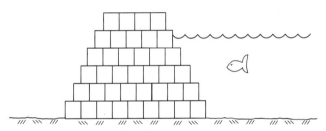

Figure 16.18 Free-standing geocell wall.

Another advantage is that the outermost cells may be vegetated by filling with topsoil and/or seeding to provide a natural "green" appearance (Figure 16.20).

Geocell confinement systems have been used for rapidly installed load support for emergency and temporary roadways in loose sandy sites, such as desert and beach environments, as well as for permanent support over weak foundations (Figure 16.21). These systems were employed by the military during Desert Storm and in Afghanistan operations to create expedient roadways and other transportation facilities (www.prestogeo. com; www.prs-med.com) (Figure 16.22). Several other load-support applications include base stabilization for paved roads, surface stabilization

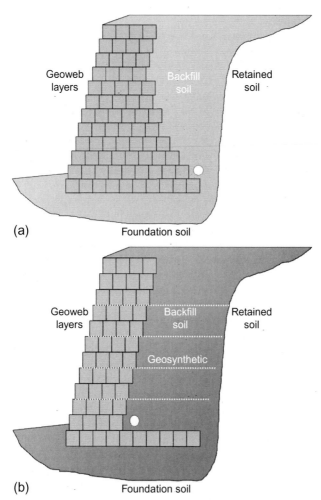

Figure 16.19 Geocell walls: (a) gravity wall; (b) geosynthetic reinforced wall. *Courtesy of Presto Geosystems.*

for unpaved roads, support of railroad ballast, and foundation support for embankments constructed over soft soils. The 3-D confinement creates a relatively stiff slab that greatly reduces the rutting and "washboarding" of unpaved roads, and allows for much thinner base layers beneath paved roads, while retaining integrity and reducing necessary maintenance.

Geocells have also been used very effectively in single sheets as erosion control for protection of slopes and channels, and for protection of geomembrane liners. The cells, typically staked down, hold soil securely in place on

Figure 16.20 Composite geocell wall before and after vegetation. *Courtesy of Presto Geosystems.*

Figure 16.21 Geocell reinforcement over weak foundation soils. *Courtesy of Presto Geosystems.*

Figure 16.22 Geocells used in rapid road construction for military mobilization in desert environments. *Courtesy of PRS Mediterranean Ltd.*

slopes, allowing for the establishment of vegetation (Figure 16.23). When used with coarse granular fill, cellular confinement can eliminate the need for riprap or "hard" armor in canals, drainage ditches, storm water swales, and culvert outflows. The cells may also be filled with concrete to create flexible, highly resistant concrete mats.

16.5 GEOSYNTHETICALLY CONFINED SOIL/GEOSYNTHETIC REINFORCED SOIL

Geosynthetically confined soil (GCS®, www.geostabilization.com; geosynthetic reinforced soil (GRS), FHWA) was introduced in Chapter 14 during the discussion of mechanically stabilized earth (MSE) walls. GCS/GRS is a version of a traditional MSE wall, but acting more as a composite structure employing close spacing (200 mm or 8 in.) of lighter reinforcement. The

Figure 16.23 Cellular confinement (geocells) for stabilization of surface soils on steep slopes. *Courtesy of Presto Geosystems.*

close spacing induces a confining effect in the soil within ∼100 mm (4 in.) of each reinforcement layer forming a continuously confined soil mass. Typical lightweight facing blocks are each held in place only by friction between them and the reinforcement layers (Figure 16.24). A schematic illustrating the differences between the two configurations is shown in Figure 16.25. Figure 16.26 shows a GCS wall supporting a roadway. This type of structure has offered a low-cost alternative for new or rehabilitated bridge abutments as well as other earth structures.

The Federal Highway Administration includes GRS walls as an integral component of an accelerated integrated bridge system (Wu et al., 2006). While they may at first appear very similar to MSE construction, there are a number of distinct differences between these two types of retaining walls. The stability of MSE walls relies on the pullout resistance of relatively widely spaced, high-strength reinforcement and to the added shear

Figure 16.24 Geosynthetically reinforced wall with light concrete facing blocks. *Courtesy of Federal Highway Administration.*

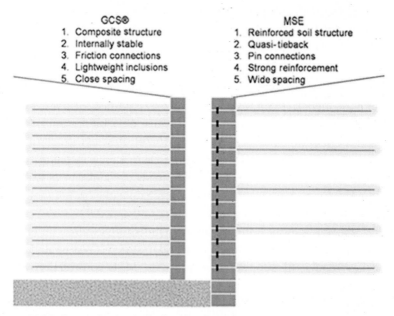

GCS®
1. Composite structure
2. Internally stable
3. Friction connections
4. Lightweight inclusions
5. Close spacing

MSE
1. Reinforced soil structure
2. Quasi-tieback
3. Pin connections
4. Strong reinforcement
5. Wide spacing

Figure 16.25 Geosynthetic confned soil (GCS) versus mechanically stabilized earth (MSE) wall. *Courtesy of GeoStabilization International.*

Figure 16.26 Geosynthetically confined soil (GSC/GRS) supporting a roadway. *Courtesy of GeoStabilization International.*

resistance afforded by the tensile (tear) strength of the reinforcing members. Reinforcement in MSE walls is typically physically attached to facing elements, which themselves may be secured to each other. GCS walls use close spacing (typically 200 mm or 8 in.) of relatively lightweight geotextiles within a compacted select granular fill. The mechanism of support comes from confinement of the fill within ~100 mm (4 in) of the reinforcement, which provides an internally stabilized soil mass. Research has indicated that GCS structures have bearing capacities of up to 20 times those of traditional MSE walls (www.geostabilization.com).

16.6 FABRIC FORMWORK AND GEOTUBES

Another form of confinement involves the use of geotextile "tubes" for containment of grout materials placed around and beneath existing foundations distressed by scour, erosion, or material deterioration. This method of controlled confinement is sometimes called *fabric formwork*, as the geotextile creates a confined form for the placed material. This has been shown to be particularly successful when construction or grouting is performed in flowing water environments. The flexible geotextile provides a form, which can take irregular shapes, fill voids, or follow undulating topography.

Geotubes® introduced earlier in Chapter 8 as a means to dewater saturated dredged or spoil materials, have also been used successfully for cost-effective shoreline protection; beach restoration; containment berms; wave barriers (breakwaters); jetties; the creation of wetlands; and for the construction of artificial islands, reclaimed land, or other marine structures (Figure 16.27) (www.infralt.com; www.tencate.com). The tubes are constructed of a high-strength, durable (but flexible) woven fabric. If the fabric

Figure 16.27 Construction of sand filled geotubes for a hurricane protection (storm surge) barrier. *Courtesy of Infrastructure Alternatives, Inc.*

is expected to be exposed for extended periods it may be coated or covered with a UV protective layer.

In one case study, contaminated dredged materials were reused in place of 450,000 m³ (15.9 million ft³) of expensive imported fill for construction of the largest private containment port in South America that services over two million containers per year (www.tencate.com). It is estimated that reusing the dredged material saved tens of millions of dollars as well as greatly reducing the carbon footprint for that project.

16.7 EROSION CONTROL

In addition to the confinement methods described earlier, there are a wide variety of erosion control mats designed to confine or hold surface soils in place and resist the forces of surface water flows and wind. These mats range from lightweight temporary meshes that are staked to the ground, intended only to stabilize the surface soils until vegetation can be established, and often consisting of "green" biodegradable natural materials, to heavy duty reinforced mats securely anchored to the ground for long-term resilience (Figures 16.28 and 16.29). Confinement of surface soils on slopes and high-energy surfaces provides a means to retain soils subjected to harsh erosional forces. Similar confinement schemes have also been used on weathered and/or fractured rock faces (Figure 16.30).

Presented at the end of Chapter 15 is a slope stabilization method introduced by Koerner (2005) for relatively shallow, potential slide masses. The method works by securing a geosynthetic netting over the slope surface with soil anchors or nails that extend beyond an assumed, potential failure surface,

Figure 16.28 High-strength erosion control mat installation. *Courtesy of Maccaferri, Inc.*

Figure 16.29 Confinement with nailed high capacity steel mesh. *Courtesy of GeoStabilization International.*

and then post-tensioning the netting and anchor/nail. The soil nails act to secure the potential slide mass with the same mechanisms described in Chapter 15, with the addition of confinement of the soil now put in compression by the tensioned netting and nails (Figure 16.31). As described previously, the confined soil mass will have improved strength and stability characteristics.

Figure 16.30 Confinement of weathered rock face with steel mesh and soil nails. *Courtesy of GeoStabilization International.*

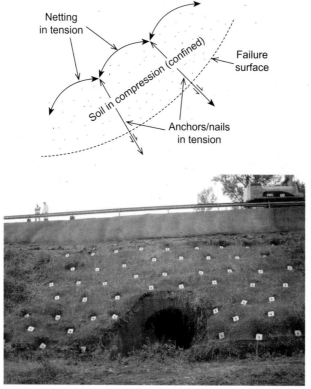

Figure 16.31 Slope stabilization incorporating anchored and tensioned netting to provide confinement. *Top: After Koerner, 2005; Bottom: Courtesy of GeoStabilization International.*

RELEVANT ASTM STANDARDS

A974—97(2011) Standard Specification for Welded Wire Fabric Gabions and Gabion Mattresses (Metallic Coated or Polyvinyl Chloride (PVC) Coated), V1.06

A975—11 Standard Specification for Double-Twisted Hexagonal Mesh Gabions and Revet Mattresses (Metallic-Coated Steel Wire or Metallic-Coated Steel Wire With Poly(Vinyl Chloride) (PVC) Coating), V1.06

D6711—01(2008) Standard Practice for Specifying Rock to Fill Gabions, Revet Mattresses, and Gabion Mattresses, V4.09

D7014—10 Standard Practice for Assembly and Placement of Double-Twisted Wire Mesh Gabions and Revet Mattresses, V4.09

REFERENCES

Federal Highways Administration, 2011. Geosynthetic reinforced soil integrated bridge system synthesis report. Report no. FHWA-HRT-11-027, Washington, DC.

Hausmann, M.R., 1990. Engineering Principles of Ground Modification. McGraw-Hill, New York, 632 pp.

Koerner, R.M., 2005. Designing with Geosynthetics, fifth ed. Pearson Education, New Jersey, 796 pp.

Koerner, R.M., 2012. Designing with Geosynthetics, sixth ed. Xlibris Corp, Bloomington, ID, 914 pp.

Wu, J.T.H., Lee, K.Z.Z., Helwany, S.B., and Ketchart, K. 2006. Design and Construction Guidelines for Geosynthetic-Reinforced Soil Bridge Abutments with a Flexible Facing. NCHRP Report No. 556, Transportation Research Board, Washington, DC.

http://www.architecturelist.com/2008/04/25/sand-bag-house-in-cape-town/ (accessed 01.04.14).

http://www.geostabilization.com/SNL/Design_Build/tools_gsi.html#4 (accessed 08.01.14).

http://www.haywardbaker.com (accessed 18.01.14).

http://www.infralt.com/content/114/Shoreline-Protection.html (accessed 07.01.14).

http://www.maccaferri-usa.com/products (accessed 18.01.14).

http://www.moretrench.com (accessed 18.01.14).

http://www.prestogeo.com (accessed 25.01.14).

http://www.prs-med.com/road-construction/military-roads-applications (accessed 02.04.14).

http://www.tenaxus.com (accessed 15.01.14).

http://www.tencate.com/amer/geosynthetics/solutions/marine-structures/default.aspx (accessed 07.01.14).

http://www.tensarcorp.com (accessed 20.01.14).

CHAPTER 17

Lightweight Fill Materials

Using lightweight fill materials has long been recognized as a means of reducing mass in order to reduce the gravitational loads, which in turn reduce bearing loads, settlement, and slope driving forces. A number of lightweight materials have been used in embankment and fill construction including chipped bark, sawdust, dried peat, fly ash, slag, cinders, cellular (foamed) concrete, shredded tires, natural lightweight aggregate (i.e., pumice), and expanded polystyrene (EPS or *geofoam*) (Holtz and Schuster, 1996). ASTM D4439 defines geofoam as "Block or planar rigid cellular foamed polymeric material used in geotechnical engineering applications." EPS geofoam is by far the lightest of all the aforementioned lightweight fill materials, typically 50 to 100 times lower than soil, and so has the advantage of requiring much less substitute material to achieve a desired reduction in load. Expanded shale, clay and slate, "ash-rock" aggregate generated from 100% coal ash, and tire-derived aggregate have also become more popular due to their generally low cost and an interest in using recycled materials. Several small as well as large, high-profile projects have used lightweight inclusions, including I-15 in Salt Lake City, Utah; Boston's "Big Dig"; and the Woodrow Wilson Bridge in Virginia. At this time, all states have evaluated the use of EPS geofoam as a lightweight fill alternative.

17.1 TYPES OF LIGHTWEIGHT FILLS

As outlined above, there are actually a number of lightweight materials used as "fill" for various geotechnical applications. Some of these materials are organic and susceptible to deterioration, while others are more stable with varying levels of uniformity, densities, and overall installation costs. The most-used lightweight fill materials for geotechnical applications are expanded polystyrene (EPS) and extruded polystyrene (XPS) (Horvath, 1999). Some other lightweight materials, such as the polyurethane foam grouts described in Chapter 12, are also used for filling voids. But for cost-effective construction applications involving large volumes of fill, and for the lightest and stiffest material, EPS geofoam is by far the most common. With this in mind, the duration of this chapter will concentrate on the attributes and applications of EPS geofoam.

17.2 PROPERTIES OF EPS GEOFOAM

Geofoam is EPS manufactured into large blocks (Figure 17.1) or sheets that typically weigh only 16-32 kg/m³ (1-2 pcf) (www.fhwa.dot.gov), although some versions may be slightly heavier to provide superior strength and stiffness. For the lighter versions, that is about 100 times lighter than most soil and at least 20 to 30 times lighter than other lightweight fills, making it an attractive alternative and solution to construction over soft or loose soils and other applications where a load reduction (vertical and/or horizontal) is desired. Once installed, it is covered either by soil and vegetation to appear like a normal earthen slope and/or embankment, or finished to look like a wall. It may be produced with various, but uniform, stiffness, density, strength, and drainage capabilities. Geofoam is manufactured to meet ASTM D6817 specifications and will not deteriorate or leach with time. It has a number of advantages in addition to its very light weight:

- Much more uniform than any natural or recycled fill material
- Low compressibility and high stiffness
- Rapid construction possible, which can dramatically reduce project schedules

Figure 17.1 Geofoam blocks. *Courtesy of Geolabs Hawaii.*

- Installation requires minimal labor and is insensitive to weather conditions
- Minimal transmission of applied loads to lateral pressures
- Moisture resistant
- Freeze-thaw resistant
- Offers insulating properties
- Inert for long-term applications with no leachates
- 100% recyclable

In addition, geofoam is essentially maintenance free and environmentally friendly. The stable foam contains no CFCs, HCFCs, HFCs, or formaldehyde, and may even be reused years after initial installation. In one reported case, geofoam blocks used for load-bearing and settlement mitigation beneath a commercial structure were removed and reused several years after their first installation and burial. Testing of the excavated EPS material showed that it had retained most all of its initial engineering properties and still retained specified values. Geofoam blocks excavated from the first known EPS embankment in Norway showed no signs of deterioration after 24 years of service (www.insulfoam.com).

EPS geofoam is produced in various types with a range of uniform stiffness, density, and drainage capabilities. Geofoam is manufactured to meet ASTM D6817 specifications and will not deteriorate, decompose, decay, or produce undesirable gasses or leach with time. To control long-term deflection or creep from sustained loads, stiffness of EPS should be high enough to resist 1% strain. As a result of this, compressive strengths are usually reported for 1% deformation.

Originally designed for insulation purposes, EPS foam has a consistently high thermal-insulative value (R-value) measured according to ASTM C578. These insulation properties have been utilized in regions that experience seasonal freezing, permafrost, and frost heaves, by creating an insulated barrier to keep the subgrade soil beneath from freezing and thawing. For these situations, care must be taken to account for possible freezing of surface layers above the geofoam insulation.

For most geofoam products, fluids do not readily flow through them and, in fact, EPS foam is sometimes shaped to retain or channel water. This (together with the insulating properties) is the premise behind Styrofoam® coffee cups. However, drainage properties may be increased. EPS foam products have been manufactured where the individual expanded beads are coated, such as with an asphalt, so that there is an open matrix of interconnected voids through which fluids may flow (Figure 17.2). This has been used to provide a "free-draining" geofoam material.

Figure 17.2 Example of "free-draining" EPS.

An additional attribute of EPS geofoam is vibration and noise dampening. Due to the high ratio of stiffness to density, this material is relatively efficient at dampening small-amplitude vibrations and noise typical from vehicular or train traffic (Horvath, 1999).

17.3 GEOFOAM APPLICATIONS

Due to its extreme light weight and stiffness, geofoam applies minimal vertical and lateral stresses. As a result it can significantly reduce settlements, spread concentrated loads, minimize lateral loads on retaining walls, provide minimal slope loads, and "fill" large volumes (i.e., embankments, grade fills) without adding any "real" stresses. Geofoam has also been used to protect underground utility conduits, pipes, and drainage culverts. Figure 17.3 shows the schematics of a variety of applications.

Geofoam has been used mainly for transportation projects, including roadway embankment widening, new alignments and new embankments, bridge abutments and approaches, retaining walls, airport taxiways, and so

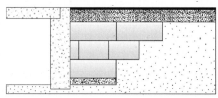

Bridge abutment

Figure 17.3 Various Geofoam applications. *Courtesy of AFM Corporation.*

Figure 17.4 Example of Geofoam for fill of transportation projects. *Courtesy of AFM Corporation.*

forth (Figure 17.4). Anywhere that soft, loose foundation soils may pose a problem due to insufficient bearing capacity or settlement issues that would be imposed by traditional soil fill loads, the use of geofoam may be an effective solution. It avoids the need for staged construction or preconstruction.

Having now been used for roadway projects in more than 20 states, applications have now spread to reducing loads on buried features such as culverts and utility pipes (Figure 17.5). Geofoam has also been used in a

Figure 17.5 Commuter rail embankment over a box culvert, Draper, UT. *Courtesy of ACH Foam Technologies.*

number of hillside and slope rehabilitation projects to reduce or minimize the driving weight of potential slide masses. In some cases, use of Geofoam has been used in excavated ground to compensate for new building loads.

This material is still relatively new to the marketplace and new and emerging uses continue to be developed for it. One of these is as a compressible inclusion to provide controlled deformation between structural elements and soil or rock. This can be applied between a rigid concrete slab or wall and expansive soil. or simply soil that will tend to deform. Foam inclusions are also resistant to dynamic (seismic) loads, and effective at muting noise and vibrations. Another growing application is to use EPS for concrete formwork and as facing for MSE walls.

17.3.1 Construction with Geofoam

Efficiency and cost savings come from a number of application attributes, including very rapid construction schedules, no need for heavy construction equipment, minimal labor force (Figure 17.6), and no need to compact, monitor, or inspect layered engineered fill. This does not even account for material transportation (which includes a significant carbon footprint). It has been estimated that it would take 12 dump trucks of conventional soil fill for every (light) flatbed truckload of geofoam (Figure 17.7).

While sometimes used in conjunction with geogrids and geotextiles for added support and/or load distribution and separation, the foam blocks are

Figure 17.6 Minimal labor force needed for Geofoam installation. *Courtesy Atlas EPS.*

Figure 17.7 Bulk Geofoam site delivery. *Top courtesy of FHWA; bottom Courtesy of Geolabs Hawaii.*

often simply placed directly on a smoothly graded subgrade surface. The use of geosynthetics only really adds to stability if there is a tendency for the foam blocks to spread laterally or deform due to very unstable subsurface conditions. Completed geofoam structures are typically covered with soil overfill, plastic membrane, or concrete, then finished to look like a traditional embankment, slope, wall, or roadway. The use of a hydrocarbon-resistant membrane is common for almost all roadway applications, for protection against susceptibility to chemical or solvent attack, or accidental fuel spills.

Geofoam is easily cut and shaped with chainsaws or hot wire cutting equipment (faster and cleaner) (Figure 17.8), enabling working around and "fitting" against utilities, drainage components, corners, and maintaining correct grades (Figure 17.9).

Figure 17.8 Cutting and shaping Geofoam blocks with a chainsaw (above) and hot wire saw (below). *Courtesy of Atlas EPS.*

Figure 17.9 Cut and shaped geofoam fit around a drainage conduit. *Courtesy of Atlas EPS.*

Standard specifications are available for geofoam installations from a number of suppliers like Harbor Foam Inc. (www.harborfoam.com).

17.3.2 Other Construction Considerations

In its pure form, EPS foam is inherently flammable. While this is not usually an issue once the materials have been buried, a few cases have been reported of losses due to fires occurring during construction with geofoam blocks. In the United States, specifications require that all geofoam be manufactured in a flame-retardant form (Horvath, 1999).

Another concern for certain project environments is possible infestation by insects (termites or other animals or organisms). Even though the foam material is not an edible source of food for these pests, some manufacturers provide an option to manufacture geofoam with an additive to resist insect infestation.

To prevent separation between blocks and aid in the integrity of a foam block monolithic mass, three different methods have been employed. First, experience has shown that, at a minimum, two layers of blocks should always be placed, and that all vertical planes should be offset (Figure 17.10). Second, galvanized metallic "grips" (Figure 17.11) or polyurethane adhesive may be used to securely hold blocks together. The use of grippers has also enhanced worker safety by providing added friction between layers of blocks (Tobin, 2014). Finally, shear keys can be constructed between layers of geofoam blocks to prevent horizontal movement between layers. Lateral sliding resistance may be a factor due to seismic loading (NCHRP, 2013).

Due to the very light weight and high buoyancy of geofoam, designs should also consider possible flooding conditions that could "float" an EPS embankment or fill. This can be prevented by covering the foam with enough ballast weight to offset any potential buoyant forces, while still providing a structure that has a greatly reduced weight as compared to traditional fill.

17.3.3 History and Case Studies

The first major use of EPS foam for a geotechnical fill application was reportedly for a highway bridge project near Oslo, Norway, in 1972 (www.achfoam.com), although Horvath (1999) states that other geotechnical geofoam applications date back to the 1960s. The next big market for geofoam applications was in Japan, where more than 1.7 million m^3 of EPS was used for airport taxiways from 1985-1997.

Figure 17.10 Staggering of blocks for stability. *Top courtesy of AFM Corporation; bottom Courtesy of FHWA.*

Figure 17.11 "Gripper" plate used to secure Geofaom blocks together. *Courtesy of AFM Corporation.*

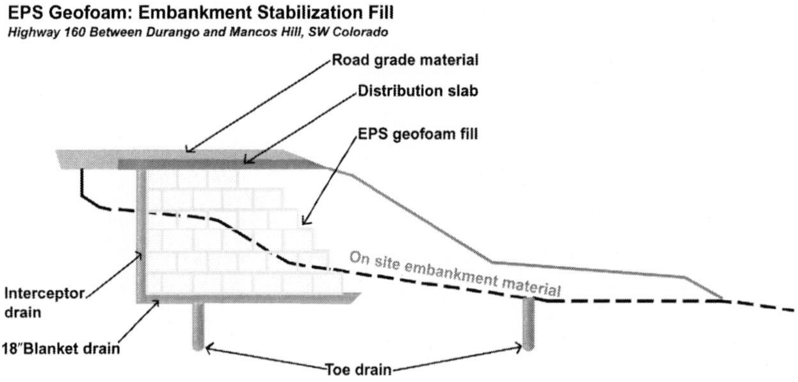

EPS Geofoam: Embankment Stabilization Fill
Highway 160 Between Durango and Mancos Hill, SW Colorado

Figure 17.12 First U.S. application of geofoam as lightweight fill. *Courtesy of ACH Foam Technologies.*

The first use of geofoam in the United States was for repair of a failed highway embankment slope on UD Highway 160 between Mesa Verde National Park (Mancos Hill) and Durango, Colorado (Figure 17.12). Repair costs using geofoam were approximately $160,000 rather than an anticipated $1,000,000 for traditional slope repair alternatives (Yeh and Gilmore, 1989). Since that time there has been an explosion of projects and applications throughout the United States and worldwide. The Federal Highway Administration has fully embraced its use to the point of promoting it to all state DOTs.

17.3.4 Case Studies

As mentioned previously, geofoam has been used for some major, high-profile transportation projects in recent years. A few of these are described here:

• Woodrow Wilson Bridge, Alexandra, Virginia

When it was decided to expand and upgrade the capacity of I-495 (Capitol Beltway) between Virginia and Maryland, geofoam was selected as part of the solution to founding the Woodrow Wilson bridge approach and interchange over highly compressible, low-strength soils on the Virginia side of the passage (Figure 17.13). Use of geofoam was also instrumental in allowing a required, tight project schedule.

• "Big Dig," Boston, Massachusetts

Reconstruction and new construction of the freeways, interchanges, and tunnels in Boston involved numerous innovative and advanced soil improvement technologies. 5300 m^3 (3.5 million ft^3) of EPS geofoam

Figure 17.13 Construction of the new Woodrow Wilson Bridge, approach, and interchanges. *Courtesy of FHWA.*

was used for some highway ramps and abutments over soft clay deposits, allowing expedient construction without the need for compacted fill and deep foundation support (www.geofoam.com) (Figure 17.14).

• Interstate I-15, Salt Lake City, Utah

In preparation for the 2002 Winter Olympics in Salt Lake City, major improvements were made to the transportation facilities, including reconstruction of over 27 km (17 mi) of the Interstate I-15 freeway, which provides the main N-S corridor through the Salt Lake valley. Due to the soft, deep lake sediments, designs had to consider settlement and stability issues

Figure 17.14 Use of Geofoam for a highway ramp as part of reconstruction of Boston's Central Artery ("Big Dig") project. *Courtesy of FHWA.*

Figure 17.15 Aerial view of a portion of the I-15 project under construction, Salt Lake City, UT. *Courtesy of AFM Corporation.*

that would be exerted from new and widened highway embankments, bridge abutments, and approach ramps (Figures 17.15 and 17.16). To complicate the problem, a plethora of sensitive utilities were buried directly under much of the construction alignment (www.insulfoam.com) that would not tolerate the expected 0.5–1.0 m of settlement expected from conventional fill. With geofoam, the settlements were only a few centimeters (www.achfoam.com). In addition, some of the soil deposits underlying

Figure 17.16 During and after construction of bridge abutment for I-15 project, Salt Lake City, UT. *Courtesy of AFM Corporation.*

the valley are also known to be thixotropic and may be subjected to significant seismic loading stemming from the close proximity of the Wasatch Fault. This provided an added concern for the stability of facilities constructed upon them. With a tight project schedule, the use of geofoam was an obvious choice over traditional fill materials. According to the contractor, the project was completed 6 months ahead of schedule and at a significant cost savings (Tobin, 2014).

Geofoam was also constructed against concrete abutment and approach walls, applying only minimal lateral pressure against these components. This allowed lighter and less robust wall designs as well as rapid construction, stability, and minimal net vertical loads. In all, over 100,000 m^3 (3.5 million ft^3) of EPS geofoam was installed for this project (www.achfoam.com), making it the largest EPS installation ever undertaken at that time.

• TRAX Lightrail Expansions, Salt Lake City, Utah

With the great success of its application for the I-15 reconstruction, geofoam has since been used for a number of other transportation projects in Salt Lake City. Of particular note are projects to expand the TRAX light rail system. For one project, extending the line to West Valley (2008-2009) included using 60,000 m^3 (2.1 million ft^3) of geofoam, primarily for construction of embankments up to 12 m (40 ft) high (Figure 17.17) (www.achfoam.com). For a second project, 53,000 m^3 (1.9 million ft^3) of geofoam was installed in 2010-2011 for an extension to the Salt Lake City International Airport.

• New Orleans Airport, New Orleans, Lousiana

Rehabilitation and enlargement of taxiways at the Louis Armstrong International Airport in New Orleans, Lousiana, involved installation of more than 19,000 m^3 (680,000 ft^3) of EPS geofoam over highly compressible and variable peat soils (Figure 17.18). Concerns over possible insect infestation were addressed by using a treatment for the EPS material.

• Kaneohe Interchange, Oahu, Hawaii

A 21 m (70 ft) high embankment was constructed as part of the H-3 Freeway construction project in 1994 (Figure 17.19). Analyses of site conditions showed an anticipated settlement of over 3 m (10 ft) and insufficient bearing capacity of the soft tropical soils to support a traditional earthfill embankment of that size. Originally planned deep wick drains would penetrate artesian formations, complicating an already difficult situation. The redesign used approximately 17,000 m^3 (600,000 ft^3) of geofoam to provide a solution that resulted in minimizing settlements without any stability problems.

RELEVANT ASTM STANDARDS

C165—07(2012) Standard Test Method for Measuring Compressive Properties of Thermal Insulations, V4.06

C203—05a(2012) Standard Test Methods for Breaking Load and Flexural Properties of Block-Type Thermal Insulation, V4.06

C578—14 Standard Specification for Rigid, Cellular Polystyrene Thermal Insulation, V4.06

D1621—10 Standard Test Method for Compressive Properties of Rigid Cellular Plastics, V8.01

Figure 17.17 Geofoam embankment constructions for TRAX lightrail system, Salt Lake City, UT. *Courtesy AFM Corporation.*

Figure 17.18 Geofoam application for expanded taxiway at Louis Armstrong International Airport, New Orleans, LA. *Courtesy of AFM Corporation.*

D1623—09 Standard Test Method for Tensile and Tensile Adhesion Properties of Rigid Cellular Plastics, V8.01

D4439—14 Standard Terminology for Geosynthetics, V4.13

D5321/D5321M—14 Standard Test Method for Determining the Shear Strength of Soil-Geosynthetic and Geosynthetic-Geosynthetic Interfaces by Direct Shear, V4.13

D6817/D6817M—13a Standard Specification for Rigid Cellular Polystyrene Geofoam, V4.13

D7180/D7180M—05(2013)e1 Standard Guide for Use of Expanded Polystyrene (EPS) Geofoam in Geotechnical Projects, V4.13

D7557/D7557M—09(2013)e1 Standard Practice for Sampling of Expanded Polystyrene Geofoam Specimens, V4.13

Reference: ASTM Book of Standards, ASTM International, West Conshohocken, PA, www.astm.org.

Figure 17.19 Highway embankment construction, Kaneohe, Hawaii. *Courtesy of Geolabs Hawaii.*

REFERENCES

EPS Industry Alliance, 2012. Expanded Polystyrene (EPS) Geofoam Applications & Technical Data. EPS Industry Alliance, Crofton, MD, 36 pp.

Holtz, R.D., Schuster, R.L. 1996. Stabilization of soil slope. In: Turner, A.K., Schuster R.L. (Eds), Landslides: Investigation and Mitigation. Transportation Research Board, Special Report 247, National Academy Press, Washington DC, pp. 439–473.

Horvath, J.S., 1999. Lessons learned from failures involving geofoam in roads and embankments. Manhattan College research report no. CE/GE-99-1, 28 pp.

Koerner, R.M., 2005. Designing With Geosynthetics, fifth ed. Pearson Education, New Jersey, 796 pp.

NCHRP, 2013. Guidelines for geofoam applications in slope stability projects. Research Results Digest 380, National Cooperative Highway Research Program, Transportation Research Board, 26 pp.

Tobin, M., 2014. Personal communications.

Yeh, S.-T., Gilmore, J.B., 1989. Application of EPS for slide correction. In: Stability and Performance of Slopes and Embankments II. ASCE, New York, NY, pp. 1444–1456, Geotechnical Special Publication 31.

http://www.achfoam.com/Geofoam-for-transportation.aspx (accessed 02.02.14).

http://www.afmtechnologies.com/EPS/geofoam.asp (accessed 11.03.14).

http://atlaseps.com (accessed 05.02.14).

http://benchmarkfoam.com/wp-content/uploads/2009/07/geofoam-brochure1.pdf (accessed 12.03.14).

http://www.civil.utah.edu/~bartlett/Geofoam/Presentation%20-%20EPS-Civil%20Applications.pdf (accessed 02.02.14).

https://www.dot.state.oh.us/engineering/OTEC/2011%20Presentations/48C-NicoSutmoller.pdf (accessed 14.03.14).

http://www.fhwa.dot.gov/research/deployment/geofoam.cfm (accessed 01.02.14).

http://www.geofoam.com (accessed 25.01.14).

http://harborfoaminc.com/pdf/Harbor-Foam-Geofoam-Specification.pdf (accessed 14.03.14).

http://insulfoam.com/images/stories/cases/cs_I-15_Corridor-V2-Mar4.pdf (accessed 14.03.14).

CHAPTER 18

Emerging Technologies, Trends, and Materials

As technology advances, equipment improvements are made, environmental concerns become mainstream, and sustainability becomes ingrained throughout engineering practice, new advancements continue to be made in soil and ground improvement. This final chapter addresses some of the approaches being developed in looking forward to implementing new methods, ideas, and materials into engineering practice. The desire for LEED geotechnical construction has also provided an impetus to contractors to reuse and reduce the carbon footprint from that of more traditional methods.

18.1 WHAT'S NEW—WHAT'S AHEAD?

Throughout this text, along with discussion of the various methods of soil improvement, admixtures, inclusions, and so forth, there have been references to new and emerging technologies, materials, equipment, and practices. While this has been noted, it seems appropriate to include one additional chapter devoted to addressing some these subjects.

In virtually all ground improvement methods, there continue to be advancements that increase efficiency, lower costs, and address environmental concerns by making use of recycled materials. Take, for example, explosive replacement, which is a recent advancement for deep densification. This technique uses explosives to create voids, which are then filled with crushed stone. This method was successfully applied to improve strength and settlement characteristics of foundation soils to support highway embankments in China (Shuwang et al., 2009). The constantly growing use of geosynthetics and new types of geosynthetic materials make these areas of improvements ever changing.

Because these emerging technologies are ongoing, the content of this chapter will undoubtedly be out-of-date by the time of publication. With that said, only a few areas of particular note will be addressed here.

18.2 UTILIZATION OF WASTES

As environmental concerns continue to grow worldwide and limitations on available disposal sites become more apparent, there have been tremendous advances in recycling or utilization of waste materials. While not all waste is suitable for geotechnical applications, much of the waste stream has been shown to be useful, often providing cost savings as well as environmental benefits. These waste materials range from surplus soil and waste slurry from construction projects, to industrial waste and by-products, to municipal waste. A number of federal and state initiatives in the United States have promoted and provided incentives for the use of recycled materials, particularly for transportation projects. Use of recycled materials is even required for some federally funded projects. Advancements and improvements in material processing and field construction techniques have also improved the reliability and cost-effectiveness of these materials for general civil engineering construction (Aydilek and Wartman, 2004).

The waste stream can be divided into three major categories: (1) waste material that can be utilized "as is" without treatment and poses minimal environmental concerns, (2) waste that can be stabilized or treated so that the resulting material will be stable and nonhazardous, and (3) materials such as waste sludge, waste oil, waste plastics, and so on, the treatment of which are very difficult for various technical and economic reasons (Kamon et al., 2000).

Large amounts of *fly ash* are generated annually from the burning of coal as fuel for electricity production. Fly ash has been known for a long time to be a useful admixture for cement and, with the right composition, has also been demonstrated to improve a number of properties, especially for certain "poor" soil types. The attributes of fly ash were described in Chapter 11, but its use continues to increase and, therefore, merits a mention here.

Municipal solid waste (MSW) ash is generated by combustion of the municipal waste stream. This process has been gaining popularity because it has the advantage of reducing the volume of waste that otherwise would be placed in limited landfills, as well as providing a complementary power supply fuel source. The residual ash has some of the same qualities as other ash by-products, but may also have irregular levels of hazardous components due to the variability of the source materials. This variability will be mostly dependent on regional location, but may be fairly uniform locally. Studies have shown a number of soil improvement/use attributes for MSW ash, including landfill covers, fill material, and as a soil-stabilizing admixture. Its future use will depend on monitoring of hazardous contaminant levels.

Waste paper sludge, or *fiber-clay*, as it is sometimes called when used as a recycled material, has been shown to serve as a cover material for sanitary landfills due to its high residual clay content (Simpson and Zimmie, 2005). It has also been used as a component of secondary roadway materials and as kitty litter.

Recycled concrete is now becoming regularly used in new construction, either as aggregate fill, for rammed aggregate piers, as roadway base, as riprap, or as aggregate for new concrete (www.geopier.com; www.en. wikipedia.org). This offers environmentally safe and sustainable LEED point enhancement to those using it by reducing the "carbon footprint" associated with cement production. At the time of this writing, it is estimated that over 140 million tons of concrete are recycled annually in the United States alone (www.cdrecycling.org).

Ground granulated blast furnace slag (GGBFS) is a by-product from the blast furnaces used to make iron. It is used as a "substitute" or filler for cement that allows for water reduction of 3-5% in concrete without any loss in workability. In the same manner that fly ash is added to cement, GGBFS may also be a suitable additive when cement is used as a soil improvement admixture. It may aid in the ability to mix cement with certain soil types and may add to the ultimate strength gains for treated soils.

Steel slag is a by-product of smelting and refining steel. Steel slag fines are produced from the crushing and screening process, where the larger sizes of steel slag are used as aggregate for transportation construction or structural fill as specified in ASTM D5106. Steel slag fines were demonstrated to be a useful additive to stabilize and treat dredged material for use in highway embankments, while immobilizing arsenic (Grubb, 2011; Grubb et al., 2013) and copper (Ruiz et al., 2013) in contaminated sediments. It was further shown that steel slag was potentially beneficial at immobilizing copper-contaminated sediments by capping in place. Recent research has also shown that steel slag can successfully immobilize elevated concentrations of phosphate, which may cause algal blooms and pfisteria in aquatic environments (Ruiz et al., 2013).

Utilization of *crushed glass*, either by itself as an aggregate or as an additive, has been studied, but has not yet seen significant application in geotechnical construction. Crushed glass, however, has been used as a substitute for sand and fine gravel in asphalt pavements for nearly 20 years. Recycled glass has potential applications as fill material and drainage material in road works, although coarse sizes have been found to be unsuitable for most geotechnical applications (Disfani et al., 2011). Preliminary test studies indicate that recycled glass is most suitable when mixed with other materials, such as natural aggregate or waste rock for base material with up to 30% recycled glass. Some concerns have been expressed regarding environmental risk, including handling.

18.3 BIOREMEDIATION

Bioremediation includes emerging trends, practices, and research using living plants and organisms to stabilize and make improvements for geotechnical and geoenvironmental applications.

18.3.1 Biostabilization Applications

One application of bioremediation is through the use of *vegetation* for stabilization of slopes, particularly shallow surface materials. Added vegetation can have some significant added benefits as well as some adverse effects. The benefits include (Abramson et al., 2002):

- Interception of rainfall by foliage (including evaporative losses)
- Reduction of soil moisture and increase of soil suction by uptake from plant root systems and transpiration
- Physical soil reinforcement by root systems
- Reduction of sloughing and loss of loose surface materials by "catchment" by shrubs and trees
- Stabilization by buttressing and arching between adjacent trees

But at the same time, a number of adverse conditions are generated that should be addressed and/or accounted for. Because of this, some experts advocate for minimal "heavy" vegetation on slopes and embankments (especially earthen dams and levees). These provide:

- Increased potential for water infiltration into a slope
- Increased seepage paths, especially when root systems biodegrade
- Surcharging slopes with added weight of heavy vegetation

In addition, vegetating slopes provide an aesthetically pleasing environmental attribute and can quickly beautify a constructed, cut, or reworked slope.

18.3.2 Contaminant Remediation

Other types of bioremediation being studied utilize microorganisms such as algae, bacteria, fungi, and other microorganisms to break down organic matter (including hydrocarbons) in efforts to "clean up" environmental contaminants (ei.cornell.edu; oilandgas.ohiodnr.gov). This may be done by enhancing the growth of pollution-eating microbes to speed up the natural biodegradation processes, or by introducing specialized microbes to degrade the contaminants.

18.3.3 Inorganic Precipitation

DeJong et al. (2010) describe "biomediated" treatments to improve soil properties without addition of synthetic materials, by harnessing natural biological processes. The premise is that inorganic calcite precipitation facilitated by biological activity can significantly improve stiffness and strength, while decreasing compressibility and permeability of in situ soil formations.

Research at the University of Western Australia has analyzed the use of natural and synthesized calcium (calcite) precipitates. These studies have also shown promise for strengthening calcareous soils for use with offshore structure foundations in tropical and subtropical regions where coral and calcareous sands and gravels are present (Kucharski et al., 1997).

REFERENCES

Abramson, L.W., Lee, T.H., Sharma, S., Boyce, G.M., 2002. Slope Stability and Stabilization Methods, 2nd ed., John Wiley & Sons, Inc., 717p.

Aydilek, A.H., Wartman, J. (Eds.), 2004. Recycled Materials in Geotechnics. ASCE, Reston, Virginia, Geotechnical Special Publication No. 127, 229 pp.

DeJong, J.T., Mortensen, B.M., Martinez, B.C., Nelson, D.C., 2010. Bio-mediated soil improvement. Ecol. Eng. 36 (2), 197–210.

Disfani, M.M., Arulrajah, A., Bo, M.W., Hankour, R., 2011. Recycled crushed glass in road work applications. J. Waste Manage. 31 (11), 2341–2351, Elsevier.

Grubb, D., 2011. Recycling on the waterfront. Geo-Strata March/April, 24–29, ASCE.

Grubb, D., Wazne, M., Jagupilla, S., Malasavage, N., Bradfield, W., 2013. Aging effects in field-compacted dredged material: steel slag fines blends. J. Hazard. Toxic Radioact. Waste 17 (2), 107–119.

Holtz, R.D., Schuster, R.L., 1996. Stabilization of slopes in landslide investigation and mitigation. Transportation Research Board special report 247, National Academy Press, Washington, DC, pp. 429–473.

Kamon, M., Hartlén, J., Katsumi, T., 2000. Reuse of waste and its environmental impact. In: Proceedings of Geo Eng 2000, Melbourne, Australia, 28 pp.

Kucharski, E., Price, G., Li, H., Joer, H.A., 1997. In: 30th International Geological Congress, Beijing, China. Engineering Properties of CIPS Cemented Calcareous Sand, vol. 23. International Science, pp. 449–460.

Ruiz, C.E., Grubb, D.G., Acevedo-Acevedo, D., 2013. Recycling on the waterfront II. Geo-Strata July/August, 40–44, ASCE.

Shuwang, Y., Wei, D., Jin, C., 2009. In: Use of Explosion in Improving Highway Foundations. ASCE Publications, Reston, VA, pp. 290–297, Geotechnical Special Publication 188.

Simpson, P.T., Zimmie, T.F., 2005. In: Waste Paper Sludge—An Update on Current Technology and Use. ASCE, Reston, VA, pp. 75–90, Geotechnical Special Publication No. 127.

http://www.cdrecycling.org/concrete-recycling (accessed 3.4.14).

http://ei.cornell.edu/biodeg/bioremed/ (accessed 3.4.14).
http://en.wikipedia.org/wiki/Concrete_recycling (accessed 3.4.14).
http://www.geopier.com/Geopier-Systems/Rammed-Aggregate-Pier-System   (accessed 3.4.14).
http://oilandgas.ohiodnr.gov/portals/oilgas/pdf/Bioremediation.pdf (accessed 3.4.14).

STANDARD SIEVE SIZES

Sieve opening (1 μm = 0.001 mm)	Sieve opening (0.001 in.)	US standard mesh no. (ASTM E11) (AASHTO M92)
20 μm	0.8	635
25 μm	1.0	500
32 μm	1.3	450
38 μm	1.5	400
45 μm	1.8	325
53 μm	2.1	270
63 μm	2.5	230
75 μm	3.0	200
90 μm	3.5	170
106 μm	4.2	140
125 μm	4.9	120
150 μm	5.9	100
180 μm	7.1	80
212 μm	8.3	70
250 μm	9.8	60
300 μm	11.8	50
355 μm	14.0	45
425 μm	16.7	40
500 μm	19.7	35
600 μm	23.6	30
710 μm	28.0	25
850 μm	33.5	20
1000 μm (1.00 mm)	39.4	18
1180 μm (1.18 mm)	46.5	16
1400 μm (1.40 mm)	55.1	14
1700 μm (1.70 mm)	66.9	12
2000 μm (2.00 mm)	78.7	10
2360 μm (2.36 mm)	92.9	8
2800 μm (2.80 mm)	110.2	7
3350 μm (3.35 mm)	131.9	6
4000 μm (4.00 mm)	157.5	5
4750 μm (4.75 mm)	187.0	4
5600 μm (5.60 mm)	220.5	3-1/2

APPROXIMATE CONVERSIONS TO SI UNITS

Symbol	When You Know	Multiply By	To Find	Symbol
Length				
in	inches	25.4	millimeters	mm
ft	feet	0.305	meters	m
yd	yards	0.914	meters	m
mi	miles	1.61	kilometers	km
Area				
in^2	square inches	645.2	square millimeters	mm^2
ft^2	square feet	0.093	square meters	m^2
yd^2	square yards	0.836	square meters	m^2
ac	acres	0.405	hectares	ha
mi^2	square miles	2.59	square kilometers	km^2
Volume				
fl oz	fluid ounces	29.57	milliliters	mL
gal	gallons (U.S.)	3.785	liters	L
ft^3	cubic feet	0.028	cubic meters	m^3
yd^3	cubic yards	0.765	cubic meters	m^3
ac–ft	acre feet	1233	cubic meters	m^3

NOTE: volumes greater than 1000 L shall be shown in m^3

Symbol	When You Know	Multiply By	To Find	Symbol
Mass				
oz	ounces	28.35	grams	g
lb	pounds	0.454	kilograms	kg
T	short (U.S.) tons (2000 lb)	0.907	megagrams (or "metric ton")	Mg (or "t")
Temperature (exact degrees)				
°F	Fahrenheit	(F–32)/1.8	Celsius	°C

Continued

Symbol	When You Know	Multiply By	To Find	Symbol
Force				
lbf (lb)	poundforce (pounds)	4.448	newtons	N
T	short (U.S.) tons (2000 lb)	8.896	kilonewtons	kN
Pressure or stress				
lbf/in^2 (lb/in^2)	poundforce per square inch	6.89	kilonewtons per square meter (kilopascals)	kN/m^2 (kPa)
lbf/ft^2 (lb/ft^2)	poundforce per square foot	0.0479	kilonewtons per square meter (kilopascals)	kN/m^2 (kPa)
Unit weight (density)				
lbf/ft^3	poundforce per cubic foot (pcf)	0.1572	kilonewtons per cubic meter	kN/m^3
lbf/in^3	pounds per cubic inch	271.43	kilonewtons per cubic meter	kN/m^3
Length				
mm	millimeters	0.039	inches	in
m	meters	3.28	feet	ft
m	meters	1.09	yards	yd
km	kilometers	0.621	miles	mi
Area				
mm^2	square millimeters	0.0016	square inches	in^2
m^2	square meters	10.764	square feet	ft^2
m^2	square meters	1.195	square yards	yd^2
ha	hectares	2.47	acres	ac
km^2	square kilometers	0.386	square miles	mi^2

Symbol	When You Know	Multiply By	To Find	Symbol
Volume				
mL	milliliters	0.034	fluid ounces	fl oz
L	liters	0.264	gallons	gal
m^3	cubic meters	35.314	cubic feet	ft^3
m^3	cubic meters	1.307	cubic yards	yd^3
Mass				
g	grams	0.035	ounces	oz
kg	kilograms	2.202	pounds	lb
Mg (or "t")	megagrams (or "metric ton")	1.103	short (U.S.) tons (2000 lb)	T
Temperature (exact degrees)				
°C	Celsius	1.8C + 32	Fahrenheit	°F
Force				
N	newtons	02.225	poundforce	lbf
kN	kilonewtons	0.1124	short (U.S.) tons (2000 lb)	T
Pressure or stress				
kN/m^2 (kPa)	kilonewtons per square meter (kilopascals)	0.145	poundforce per square inch	lbf/in^2 (lb/in^2)
kN/m^2 (kPa)	kilonewtons per square meter (kilopascals)	20.885×10^{-3}	poundforce per square foot	lbf/ft^2 (lb/ft^2)
Unit weight				
kN/m^3	kilonewtons per cubic meter	6.361	pounds per cubic foot	lbf/ft^3
kN/m^3	kilonewtons per cubic meter	0.00368	pounds per cubic inch	lbf/in^3

After FHWA (http://www.fhwa.dot.gov/publications/convtabl.cfm).

Note: Page numbers followed by *f* indicate figures and *t* indicate tables.

Edwards Brothers Malloy
Thorofare, NJ USA
January 14, 2015